Aspects of Signal Processing

Part 2

NATO ADVANCED STUDY INSTITUTES SERIES

*Proceedings of the Advanced Study Institute Programme, which aims
at the dissemination of advanced knowledge and
the formation of contacts among scientists from different countries*

The series is published by an international board of publishers in conjunction
with NATO Scientific Affairs Division

A	Life Sciences	Plenum Publishing Corporation
B	Physics	London and New York
C	Mathematical and Physical Sciences	D. Reidel Publishing Company Dordrecht and Boston
D	Behavioral and Social Sciences	Sijthoff International Publishing Company Leiden
E	Applied Sciences	Noordhoff International Publishing Leiden

Series C – Mathematical and Physical Sciences

*Volume 33 – Aspects of Signal Processing
Part 2*

Aspects of Signal Processing

With Emphasis on Underwater Acoustics

Part 2

Proceedings of the NATO Advanced Study Institute
held at Portovenere, La Spezia, Italy
30 August–11 September 1976

edited by

G. TACCONI,
University of Genoa, Italy

D. Reidel Publishing Company

Dordrecht-Holland / Boston-U.S.A.

Published in cooperation with NATO Scientific Affairs Division

Library of Congress Cataloging in Publication Data

Nato Advanced Study Institute on Signal Processing with
    Emphasis on Underwater Acoustics, Portovenere, Italy, 1976.
    Aspects of signal processing with emphasis on
underwater acoustics.

    (NATO advanced study institutes series : Series C,
Mathematical and physical sciences ; v. 33, pts. 1-2)
    Sponsored by NATO Division of Scientific Affairs and
Selenia s. p. a., Rome.
    Includes bibliographical references and index.
    1. Signal processing--Addresses, essays, lectures.
2. Sonar--Addresses, essays, lectures. 3. Underwater
acoustics--Addresses, essays, lectures. I. Tacconi, G.,
1925-   II. North Atlantic Treaty Organization.
Division of Scientific Affairs. III. Selenia s. p. a.
IV. Title. V. Series.
TK5102.5.N36   1976   621.38'043   77-3238
ISBN 90-277-0798-7   set of two parts
ISBN 90-277-0799-5   part 1   ISBN 90-277-0800-2   part 2

Published by D. Reidel Publishing Company
P.O. Box 17, Dordrecht, Holland

Sold and distributed in the U.S.A., Canada, and Mexico
by D. Reidel Publishing Company, Inc.
Lincoln Building, 160 Old Derby Street, Hingham, Mass. 02043, U.S.A.

TABLE OF CONTENTS

SUBJECT 6 - MODERN PROCESSOR ARCHITECTURE AND TECHNIQUES

[+] Author unable to attend; paper not presented at the Institute but included in these proceedings.

CONTENTS OF PART 1

OPTIMUM ANTENNA PROCESSING: A MODULAR APPROACH

C. Giraudon

CIT - Alcatel, Arcueil, France

1. INTRODUCTION

1.1 - Problem

We have N+1 sensors with which we want to detect or to estimate a plane wave signal imbedded in noises. We assume that
sensors have been correctly delayed such that the signal
components are identical on all sensors. Noises on sensors
are assumed to be locally stationary and uncorrelated with
the signal.

1.2 - Solution

At about the same time several people found solutions to this
problem. Here we use Dr. MERMOZ formalism which appears to be
very pratical and physical (ref. {2}). The answer is to filter the N+1 sensors by the column filter matrix

$$\{p(F)\} = \{\gamma_n(F)\}^{-1} \cdot \{N+1\} \cdot W(F) \qquad \{1\}$$

where $\{p(F)\}$: column matrix of the N+1 sensor filters
    $\{\gamma_n(F)\}$: noise cross-spectral square matrix

    $\{N+1\}$: column matrix containing only 1's.
    $W(F)$: scalar filter

This filtering can be implemented in 2 parts put in cascade.
1)st part : $\{\gamma_n(F)\}^{-1} \cdot \{N+1\}$
    is a filter which only depends on noises on the sensors.
    It is a spatial filtering only.

G. Tacconi (ed.), Aspects of Signal Processing, Part 2, 401-410. All Rights Reserved.
Copyright © 1977 by D. Reidel Publishing Company, Dordrecht-Holland.

$2)^{nd}$ part : W(F) is a scalar filter which depends on :
 - the criterion used for optimality
 - the signal nature.

For the remaining of the paper we will focus our attention on detection of a known signal, we thus put : $W(F) = S^{\times}(F)$.
The {1} matrix equation can be pictured as follows

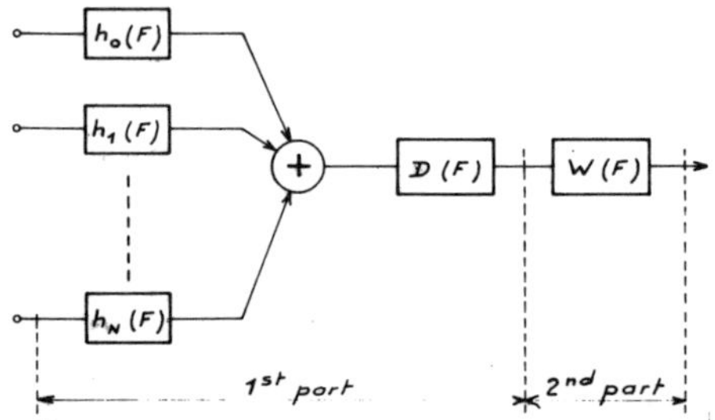

- $h_i(F)$: an elementary branch filter is given by $h_i = \sum_j M_{ji}(F)$

  where $M_{ij}(F)$ is a cofactor (submatrix determinant) of $\{\gamma_n(F)\}$
- $D(F)$: a filter common to all branches is given by $D(F) = \dfrac{1}{\Delta(F)}$

  where $\Delta(F)$ est $\{\gamma_n(F)\}$ matrix determinant
- $W(F) = S^{\times}(F)$ is the signal matched filter.

1.3 - <u>Implementation in a real situation</u>

So far we have assumed that the noise was stationary, which
allows to use predetermined filters for the $h_i(F)$ and $D(F)$
filters. When the noise is not stationary these filters must
be recomputed or refreshed with noise sources supposed to
vary slowly in time. One difficulty in evaluating the spatial
filters on the above sketch is that no one knows if the signal
is present or absent, which leads to a bias in the estimation
of the D(F) filter. Subsequently it is necessary to invert a
matrix which appears to be a difficult and cumbersome operation.

2. <u>DERIVATION OF AN EQUIVALENT SCHEME</u>

Proceeding through linear filters, it is possible to transform
the preceding scheme into an equivalent one which follows.
However, in taking the inverse matrix we did not divide by the
determinant, thus avoiding any signal corrupted estimate, this
will be commented hereafter. This implementation scheme reads
from right to left, due to the formalism of column matrices as
input vectors.

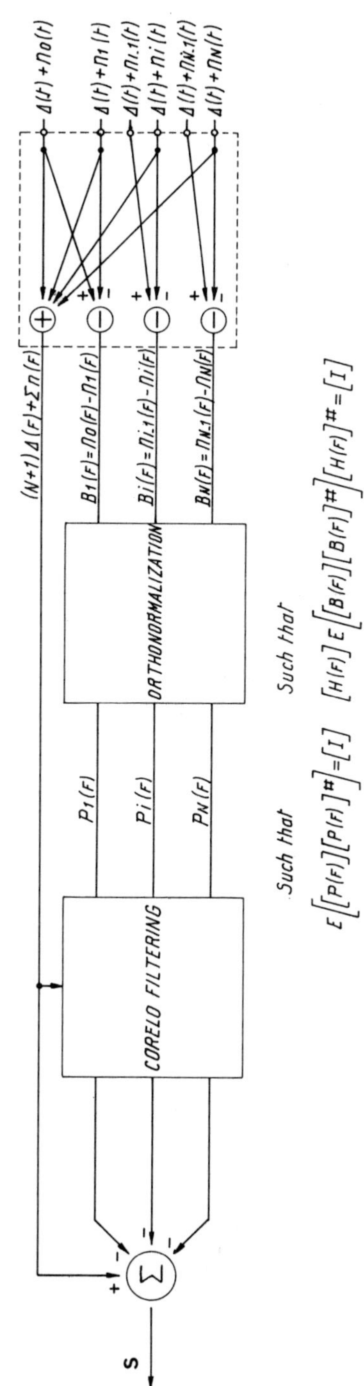
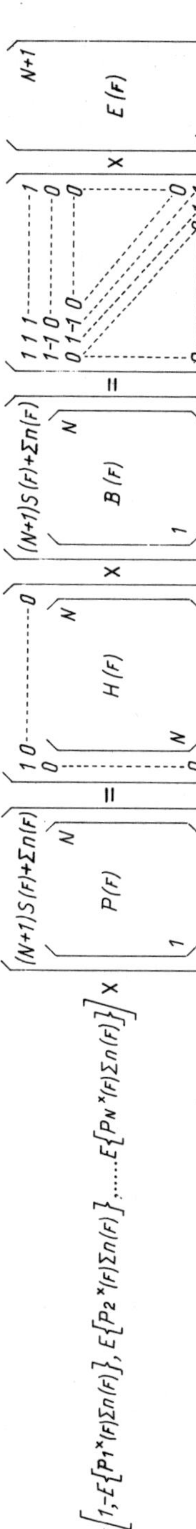

Figure: 1

3. ORTHONORMALIZING AND FILTERING

3.1 - Orthonormalization using a Gram-Schmidt matrix

We can use triangular $\{H(F)\}$ matrices of the Gram-Schmidt type.

For example, for 3 sensors we have :

$$\{\gamma_B(F)\} = \begin{pmatrix} \gamma_{11} & \gamma_{12} \\ \gamma_{12}^{\times} & \gamma_{22} \end{pmatrix}$$

$$\{H(F)\} = \begin{pmatrix} \dfrac{1}{\sqrt{\gamma_{11}}} & 0 \\ \dfrac{-\gamma_{12}^{\times}}{\gamma_{11}\sqrt{\gamma_{22} - \dfrac{|\gamma_{12}|^2}{\gamma_{11}}}} & \dfrac{1}{\sqrt{\gamma_{22} - \dfrac{|\gamma_{12}|^2}{\gamma_{11}}}} \end{pmatrix}$$

A possible scheme for two noises is then :

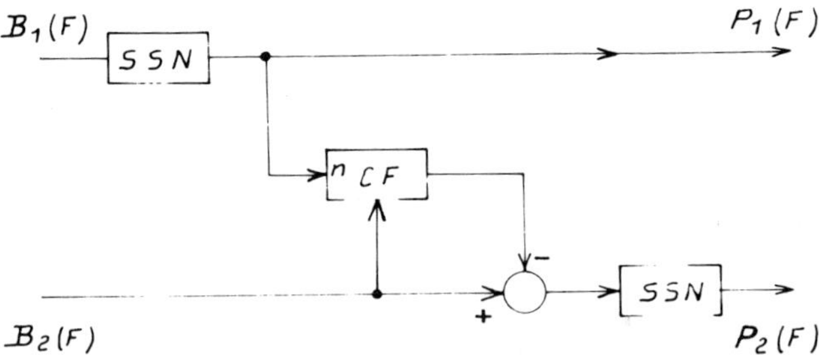

- the self spectrum normalizer (SSN) has the following
  functions :

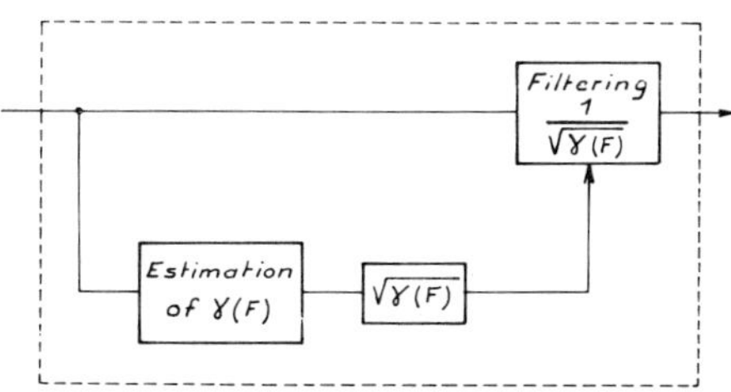

- the correlo-filtering device works as follows :

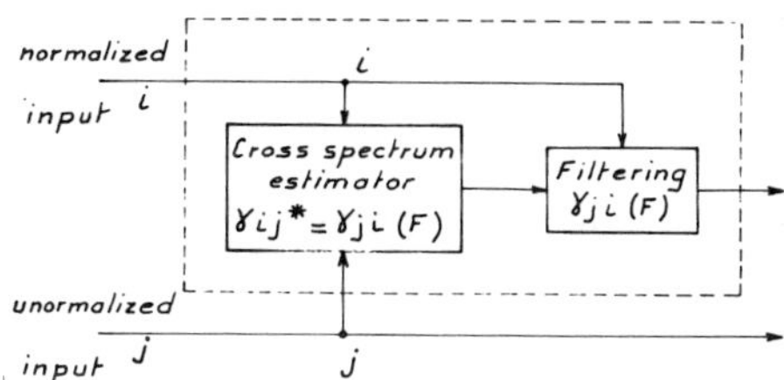

Combining these two modules, we can operate any kind of
orthonormalization.

3.2 - <u>Ultimate filtering via the "correlo-filter"</u>

The ultimate filtering by the line matrix

$$\left[1, \; - E\{P_1^{\mathbf{x}}(F)\Sigma n(F)\}, \; - E\{P_2^{\mathbf{x}}(F)\Sigma n(F)\}, \; \ldots, \; E\{P_n^{\mathbf{x}}(F)\Sigma n(F)\} \right]$$

can be implemented with the CF box as follows :

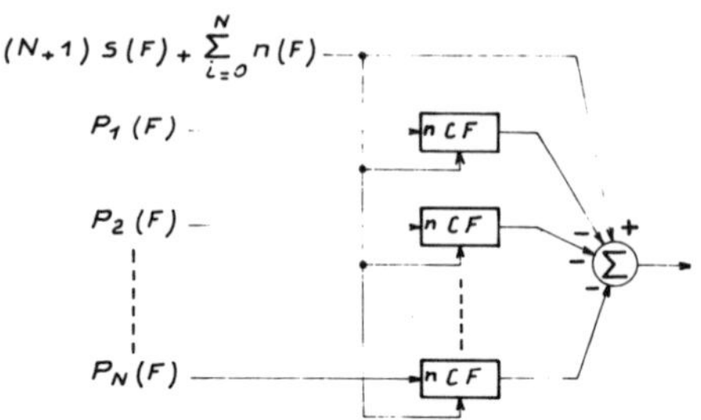

4. ADAPTIVITY AND OPTIMALITY

4.1 - It has been shown (ref.{3}) that complete optimality and complete adaptivity cannot exist together at the same time. In our preceding description of the process, complete optimality has been left out at chapter (2) when we inverted the noise cross spectral matrix.

In this inversion process, there is no problem in computing the cofactor terms which only involve terms of the

$E\{P_i^{\times}(F)\Sigma n(F)\}$ form.

i.e. Since all $P_i(F)$ are uncorrelated with the S(F) signal, these terms are obtained by correlating the $P_i(F)$ branch with the signal branch.

However the determinant computation which involves an :

$E\{\Sigma n(F)\Sigma n^{\times}(F)\}$ term has been omitted since this estimation entails the assumption that no signal is present in the upper branch.

Since this assumption is not fulfilled, no one can estimate the determinant.

We then transgress complete optimality for complete adaptivity. One must stress that although the used processing is suboptimal, it is the best one that we can imagine since it gives at any frequency the optimum ratio between power spectral densities of signal and noise.

4.2 - The above described processing is completely optimal and adaptive in certain special cases of noise field distribution.

1st case : The noise field consists only of N discrete noise sources (Njammers), in which case the noise spectral matrix

is degenerate and the determinant does not need to be evaluated.
It can be shown that the processing completely eliminates N
discrete noise sources.

2d case : The system works in a narrow bandwidth (active sonar case). In this case, the determinant beeing merely a complex number can be skipped off as a filter. The system is
then completely adaptive and optimal under a narrow bandwidth
condition. In this case the scheme of figure 1 simplifies :
the SSN turning out in an AGC and the CF in a complex correlator.

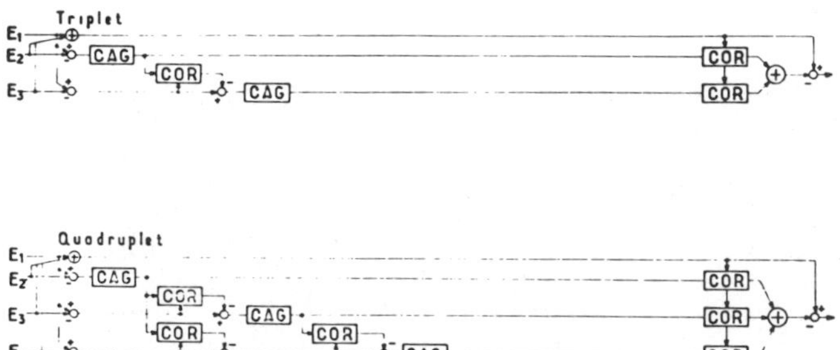

5. SOME FIELD TRIAL RESULTS

A great number of laboratory experiments has be conducted during
the last years. They all have been in perfect agreement with the
theory. Rather than displaying clean and clear cut images of
these laboratory experiments we choose to show some field trial
pictures. These trials took place in a lake and at sea.

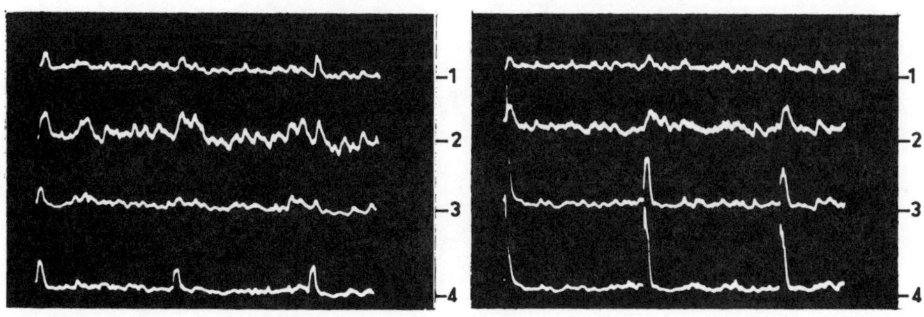

Fig. 2   Fig. 3

Experiments reported in figures 2 and 3 took place in a lake.
A 9 hydrophones array was used at 4900 Hz ± 220 Hz. The usual
beamformer, whose beamwidth is $2\theta_3 = 14°$, had an output which
is displayed on trace 1.
The signal was a plane wave pulse signal (θ = 50 ms) which pro-
pagates in the direction of the axis of the beamformer.
The jammer is a plane wave noise signal, in the band F_o ± 30 Hz

whose direction of propagation presents with the beamformer
axis an angle of 1,7° (fig. 2) and 8° (fig. 3).
The 9 hydrophones signals are processed through two layers of
(3 + 1) triplets whose outputs are shown on trace 3 (summed
outputs of 1st layer) and trace 4 (2nd layer output). Trace 2
is irrelevant.

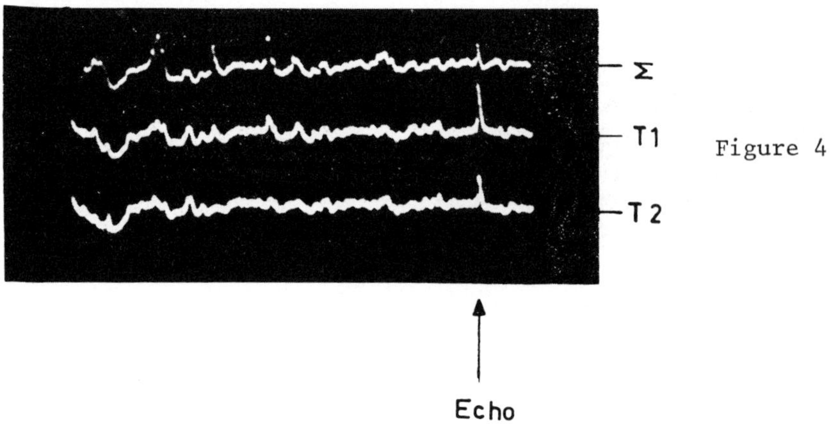

Figure 4

One figure 4, one finds a submarine echo (distance $\simeq$ 8000 yards)
which is processed through the same system as above. Trace Σ
shows the usual beamformer output. On T_1 and T_2 one finds the
outputs of respectively the 1st layer of optimum processing
(picked after summation) and the 2nd layer. One remarks the
decreasing of noise variance from Σ to T_1, T_2 and the echo
amplitude on T_1 as compared to T_2. The second layer benefit is
not always obvious.

Figure 5   Figure 6

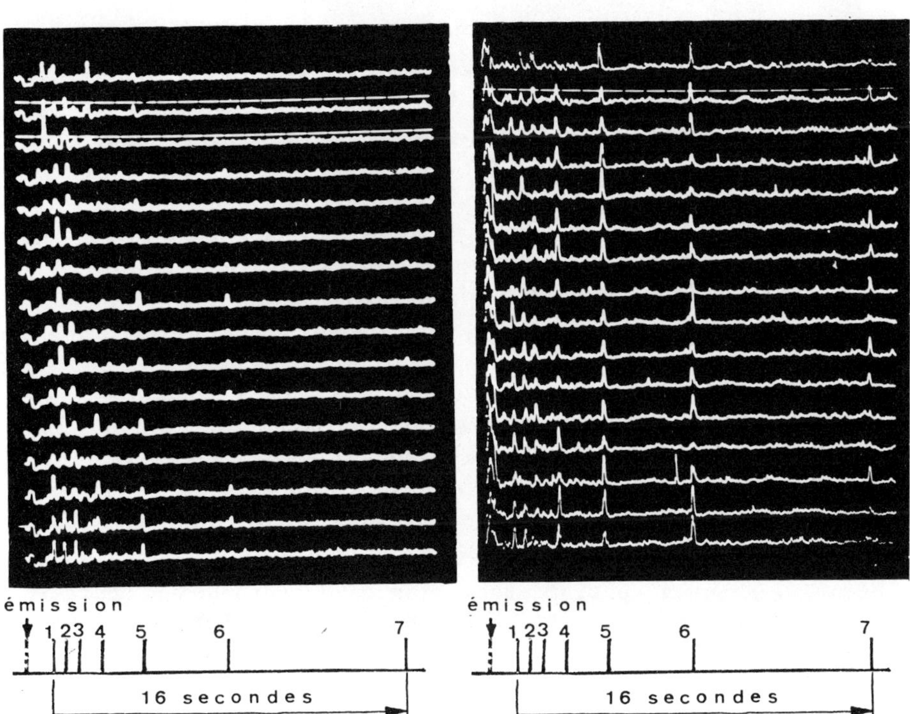

When using a 6 hydrophones linear vertical array for receiving
signals from a transponder, we get either by the beamformer
(fig.5) or by the optimum processing (fig.6) a series of echoes.
These echoes were recorded during shallow water operations.
Most of the transponder echoes were clustered at the beginning
of the sweep to observe the gain in reverberation.

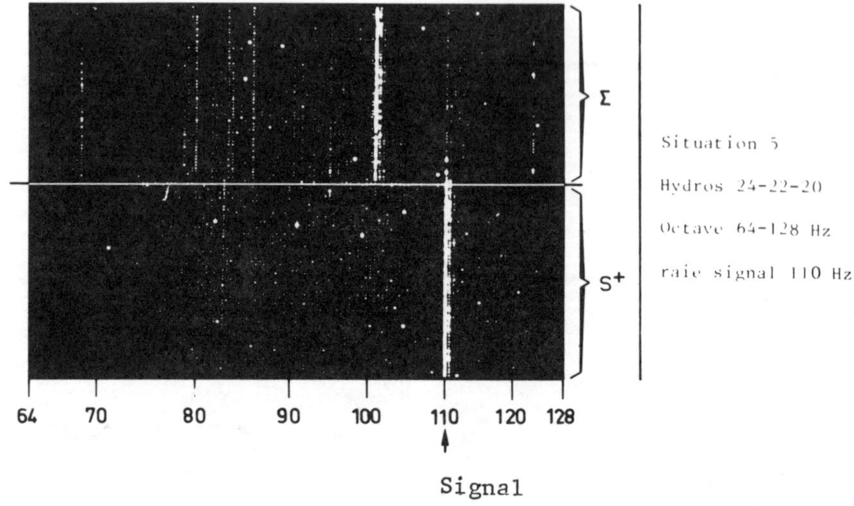

Figure 7

Figure 7 shows a spatial processing on 3 hydrophones operating
as a self noise canceller. Σ is the usual beamformer output
(frequency is in x and time in y coordinates), S^+ is the optimum processing output.

6. <u>CONCLUSION</u>

The idea of optimum antenna processing has gained a wider acceptance with the recent interest in this new technique from
radar people. Although this technique has not been so far
widely used, due to lacks of credibility and its cost, it
should finally reach its maturity. Much of the work reported
herein has been done under DRME contracts.

<u>REFERENCES</u>

{1} MERMOZ : "Essai de synthèse sur les antennes de détection
            optimales et adaptatives".
            Annales des Télécommunications 1970 n° 7-8

{2} MERMOZ : "Antennes de détection optimales et adaptatives
            Théorie et Applications" 1971 Collection du CNET.

THE FORGOTTEN ALGORITHM IN ADAPTIVE BEAMFORMING

E. B. Lunde

Norwegian Defence Research Establishment
Horten, Norway

1. INTRODUCTION

Many quite different algorithms can be used in adaptive beamforming, but the one mostly treated in the literature, is the
gradient descent (GD) algorithm. Its behaviour, stability and
performance are well known. In this paper, emphasis is put on
the Woodbury identity (WI) algorithm, whose name is motivated
by the algorithm's use of the Woodbury identity (matrix inversion
lemma). The GD algorithm will be used only as a reference.

Among the many possible criteria for optimum beamforming,
the constrained least mean square (LMS) criterion is the one
used in this paper. It requires a distortionless frequency
response in the beam steering direction while minimizing output
noise power.

The numerical stability of the WI algorithm, and methods
to assure this stability, will be discussed. The two algorithms
are then compared with regard to computer requirement, performance, and flexibility.

2. ARRAY PROCESSING

Consider the K-channel array processor shown in figure 2.1. The
inputs to the processor are assumed to be zero mean discrete
fourier transformed sensor signals, and they are conveniently
written as a data input vector :

$$X^T = [x_1, x_2, \ldots, x_K] \tag{2.1}$$

G. Tacconi (ed.), Aspects of Signal Processing, Part 2, 411-421. *All Rights Reserved.*
Copyright © 1977 by D. Reidel Publishing Company, Dordrecht-Holland.

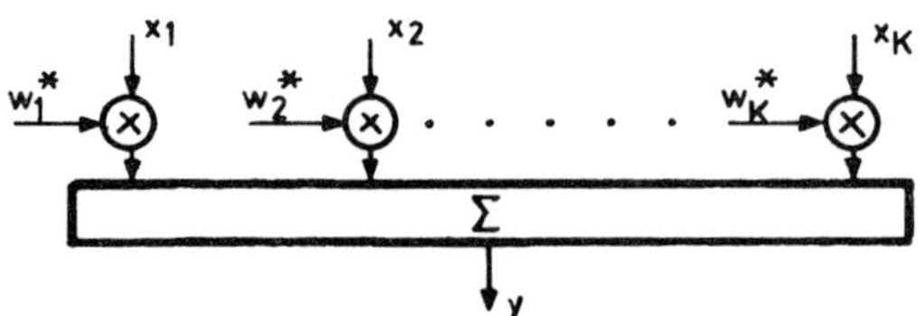

Fig. 2.1 Array processor

In the same way the weight vector is defined :

$$W^T = [w_1, w_2, \ldots, w_K]$$ (2.2)

The processor output is now given by :

$$y = W^\dagger X$$ (2.3)

the superscripts $*$, T, and $\dagger$ indicate complex conjugate, transpose, and complex conjugate transpose, respectively.

2.1 Optimum array processing

The expected cross-spectral density matrix of X is defined by :

$$R = E\{X \cdot X^\dagger\}$$ (2.4)

Under ideal assumptions, a point source Q in space will yield a contribution to R given by the dyad [1]:

$$R_Q = \sigma_Q^2 \, d \cdot d^\dagger$$ (2.5)

where σ_Q^2 is Q's power spectral density at the sensors, and d is the phase vector of Q with respect to the given array geometry.

The optimum weight vector according to the mentioned criterion, and focusing the sensor array on the point Q, is [1]:

$$W_{opt}^\dagger = m^\dagger R^{-1} / (m^\dagger R^{-1} m)$$ (2.6)

where the focusing vector m must be chosen equal to the phase vector of the point Q:

$$m = m(r, \phi) = d$$ (2.7)

Usually m is chosen to focus on infinite distance, $m = m(\phi)$, and is therefore called the steering vector. With L different steering vectors, the processor (beamformer) is said to have L beams.

2.2 Adaptive array processing

The mission of an adaptive algorithm is to give an updated estimate of W_{opt} when R is not exactly known. The gradient and the Woodbury algorithms solve this problem quite differently. The gradient algorithm updates the weight vector W directly [2].

$$y_\ell(n) = W_\ell^\dagger(n)\, X(n) \qquad\qquad \ell = 1.2,\ldots,L \qquad (2.8)$$

$$W_\ell(n+1)=(I-m_\ell m_\ell^\dagger/K)\cdot[W_\ell(n)-\mu_{GD}y_\ell^*(n)\cdot X(n)] + m_\ell/K$$
$$\ell = 1,2,\ldots,L \qquad (2.9)$$

where μ_{GD}, $0<\mu_{GD}<<1$, is a parameter determining the time constant of the adaptation.

The Woodbury algorithm does not update W directly. Instead it updates the matrix R^{-1} and L scalars defined by :

$$b = m^\dagger R^{-1} m \qquad\qquad (2.10)$$

The processing can be written [3] :

$$D(n) = R^{-1}(n)\cdot X(n) \qquad\qquad (2.11)$$

$$h(n) =[X^\dagger(n)\cdot D(n) + (1-\mu_{WI})/\mu_{WI}]^{-1} \qquad\qquad (2.12)$$

$$R^{-1}(n+1) = [R^{-1}(n)-h(n)\cdot D(n)\cdot D^\dagger(n)]/(1-\mu_{WI}) \qquad (2.13)$$

$$g_\ell(n) = m_\ell^\dagger D(n) \qquad\qquad \ell =1,2,\ldots,L \quad (2.14)$$

$$y_\ell(n) = g_\ell(n)/b_\ell(n) \qquad\qquad \ell =1,2,\ldots,L \quad (2.15)$$

$$b_\ell(n+1) = [b(n)-h(n)\cdot g_\ell^*(n)\cdot g(n)]/(1-\mu_{WI}) \quad \ell=1,2,\ldots,L \quad (2.16)$$

Woodbury's identity has been used to obtain (2.13), therefore the particular name of this algorithm. Notice that it is common for all beam steering directions. Before the two algorithms are compared, the stability of the WI algorithm is discussed.

3. NUMERICAL STABILITY OF THE WOODBURY IDENTITY ALGORITHM

It can be proved that under certain conditions, the updating of R^{-1} is numerically stable [4]. In a real situation these conditions are satisfied, provided that sufficient wordlength is used in the computation. Assume that R^{-1} is represented by an integer matrix P and an integer q, so that:

$$R^{-1} = 2^q P + \text{roundoff error matrix} \qquad\qquad (3.1)$$

Sufficient wordlength B_1 for the elements of P is a function of three parameters:

$$2^{B_1} \sim c/(g \sqrt{\bar{\mu}_{WI}}) \tag{3.2}$$

where μ_{WI} is the parameter determining the time constant of the updating. g is the maximum relative output error:

$$g = \max_{\ell}[\{|y_\ell(n)|^2 - |y_{\ell,opt}|^2\}/|y_{\ell,opt}|^2] \tag{3.3}$$

c is the conditional number of the matrix R (or R^{-1}):

$$c = \lambda_{max}/\lambda_{min} \tag{3.4}$$

where λ_i, i=1,2,...,K are the eigenvalues of the matrix. B_1 is also an implicite function of the matrix dimension K, since it can be shown:

$$c \leq const \cdot K - 1 \tag{3.5}$$

Both μ_{WI} and g are known (chosen) quantities. Therefore, the only quantity to control in order to enforce stability, is the conditional number c. If the data are such that $c>c_o$, where $c=c_o$ is used to calculate B_1, then the data must be modified so that $c_{mod} \leq c_o$. The simplest and most natural way to do this, is to add artificial white noise to the data:

$$X_{mod} = X + X_{art} \tag{3.6}$$

where the cross-spectral density matrix of the artificial noise is:

$$R_{art} = E\{X_{art}X^\dagger_{art}\} = \sigma_{art}^2 I \tag{3.7}$$

This will decrease the conditional number:

$$c_{mod} = (\lambda_{max}+\sigma_{art}^2)/(\lambda_{min}+\sigma_{art}^2) < c \text{ for } \lambda_{max}>\lambda_{min} \tag{3.8}$$

If λ_{max} and λ_{min} are known, the value to be used on σ_{art}^2 is found by setting $c_{mod}=c_o$, which yields the following rule:

$$\sigma_{art}^2 = (\lambda_{max}/c_o) - \lambda_{min} \quad \text{for } c>c_o$$
$$= 0 \qquad\qquad " \quad c \leq c_o \tag{3.9}$$

This method requires the computation of λ_{max} and λ_{min}. Another method requiring less computation, but also yielding poorer performance, is to use:

$$\sigma_{art}^2 = tr(R)/c_o \tag{3.10}$$

For a given conditional number c, this method yields c_{mod} between a maximum and a minimum value, as indicated in figure 3.1. c_{mod}

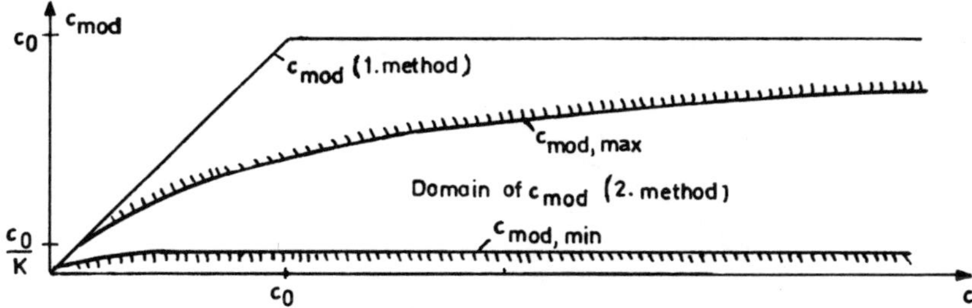

Fig. 3.1 Domain of c_{mod} as a function of c

resulting from the first method is also indicated, showing this
method's better performance.

In contrast to the updating of R^{-1}, the updating of $b=m^\dagger R^{-1}m$
is not numerically stable. Therefore, after a number of adap-
tations. M, a direct calculation of b must be carried out. M is
a function of two parameters [4]:

$$M \sim \Delta B_1/\mu_{WI} \qquad (3.11)$$

where ΔB_1 is the number of bits, by which the wordlength B_1 is
larger than necessary, in order to obtain a stable updating of
R^{-1}. Hence there is a tradeoff between the wordlength B_1 and
how often b has to be calculated directly. Sufficient wordlength
B_2 in the representation of b is determined by:

$$2^{B_2} \sim (e^{\mu_{WI}M}-1)/(\mu_{WI}g) \qquad (3.12)$$

4 COMPARISON OF THE TWO ALGORITHMS

The two algorithms will be compared with respect to computer requirement, performance, and flexibility.

4.1 Computer requirement

The computer requirement of the two algorithms will be judged
from two factors, computational load represented by the number of
real multiplications, and the size of the memory. Different
wordlengths are not taken into consideration. With this in mind,
figures 4.1 and 4.2 show the number of real multiplications
versus the number of sensors and beams, respectively. CON represents conventional beamforming (delay and sum). Figure 4.3 shows
the ratio of the number of real multiplications for the two
algorithms. As a rule of thumb, the cross-over curve between the
two algorithms ($\beta=1$) is defined by $K \approx L$, the number of sensors

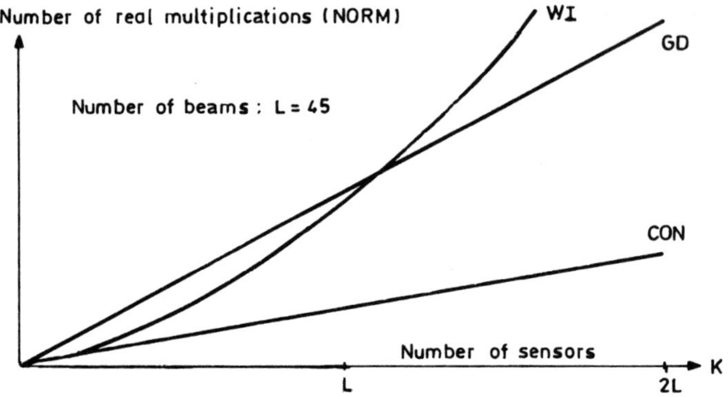

Fig 4.1   Number of real multiplications versus number of sensors

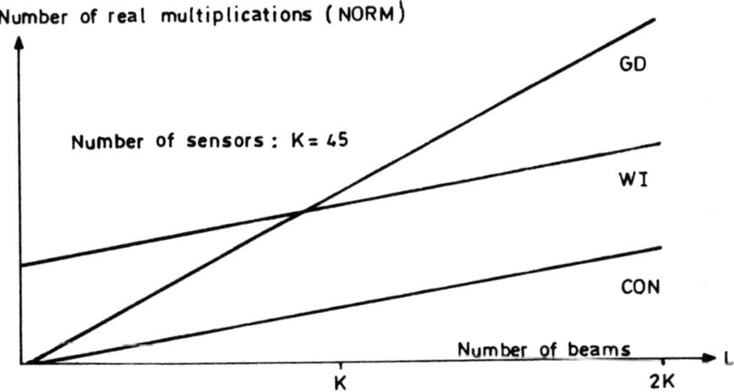

Fig. 4.2 Number of real multiplications versus number of beams

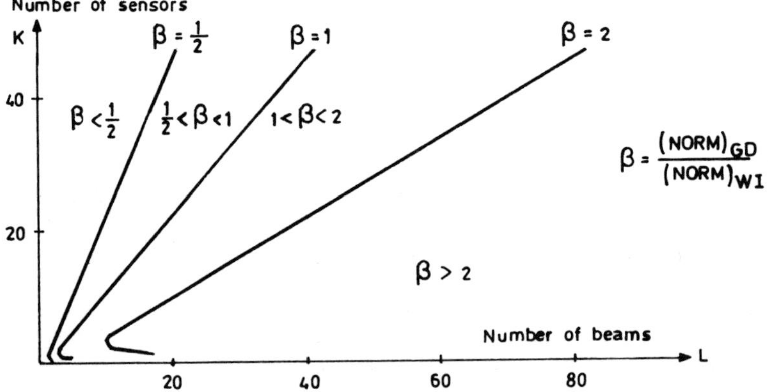

Fig. 4.3   Ratio of number of real multiplications versus number
           of sensors and number of beams

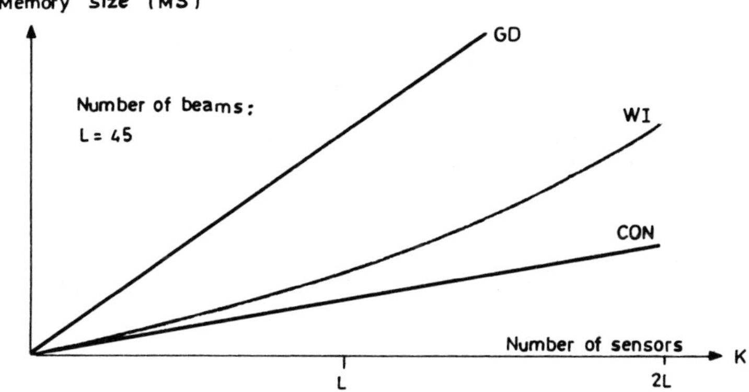

Fig. 4.4   Memory size versus number of sensors

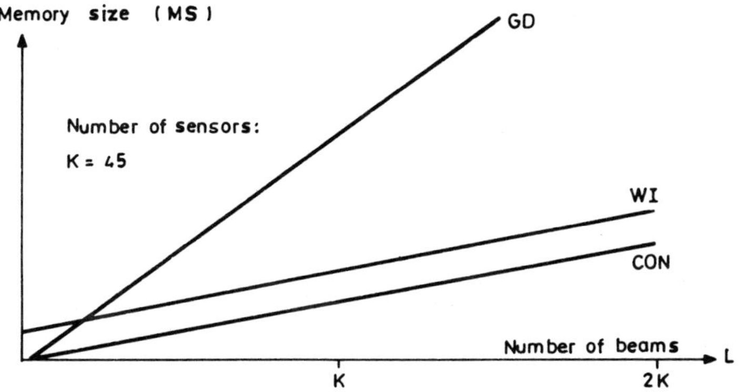

Fig. 4.5   Memory size versus number of beams

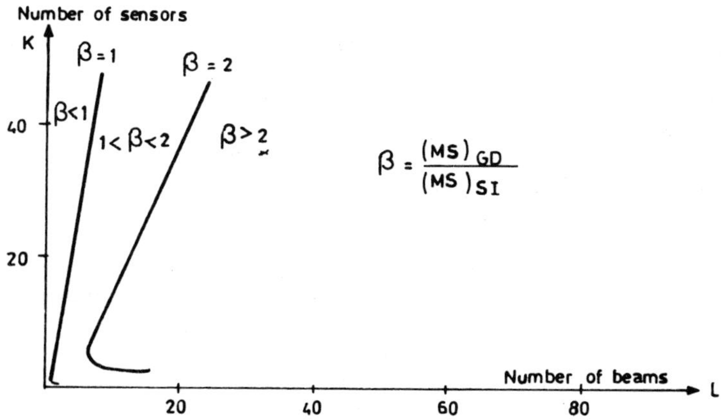

Fig 4.6   Memory size ratio versus number of sensors and
number of beams

approximately equal to the number of beams.

The WI algorithm is favourable for large L since the updated matrix R^{-1} is common to all beams.

Similarly, figures 4.4 and 4.5 show the memory requirement versus the number of sensors and beams, respectively. Figure 4.6 shows the memory requirement ratio.

All these figures and curves must be taken for what they are, coarse indications. But at least they show from a computational point of view that one algorithm is not always superior to the other.

4.2 Performance

Performance is judged by the ability of satisfying two conflicting desires, keeping the updating as noise-free as possible (long time constant), and at the same time adapt as fast as possible in nonstationary situations (short time constant). A fairly well behaved quantitative representation of performance is the so-called learning curve, which in this paper is defined by:

$$P(n)=\{ \sum_{\ell=1}^{L} [W_\ell^\dagger(n) \ R \ W_\ell(n)-W_{\ell,opt}^\dagger \ R \ W_{\ell,opt}]\}/L \tag{4.1}$$

where ℓ is the beam index. $W_\ell(n)$ is directly available for the gradient algorithm. For the Woodbury algorithm, it is calculated by:

$$W_\ell^\dagger(n) = m_\ell^\dagger R^{-1}(n)/[m_\ell^\dagger R^{-1}(n) \ m_\ell] \quad \ell=1,2,\ldots,L \tag{4.2}$$

One algorithm is said to have the potential of better performance than the other, if there exist reasonable parameters, μ_{GD} and μ_{WI}, so that the average learning curve of the first algorithm is below the learning curve of the second one both in stationary and nonstationary situations. One simple simulation is to keep the data stationary, then adapt from an initial condition through a transient state (nonstationary situation) into steady state (stationary situation). The weight vectors of the conventional beamformer may be used in the initial condition.

Such a simulation was carried out with a linear array consisting of 10 sensors with geometrically increasing spacing between sensors, the spacing ratio beeing 1.1. The ratio between the shortest spacing and the wavelength was 0.3. The noise field consisted of two point sources at 30° and 40° bearing, respectively. They had a transversal coherence width three times the array length. In addition there was spatially incoherent noise. The relative strength of the three sources at the sensors was 10 : 3 : 1.

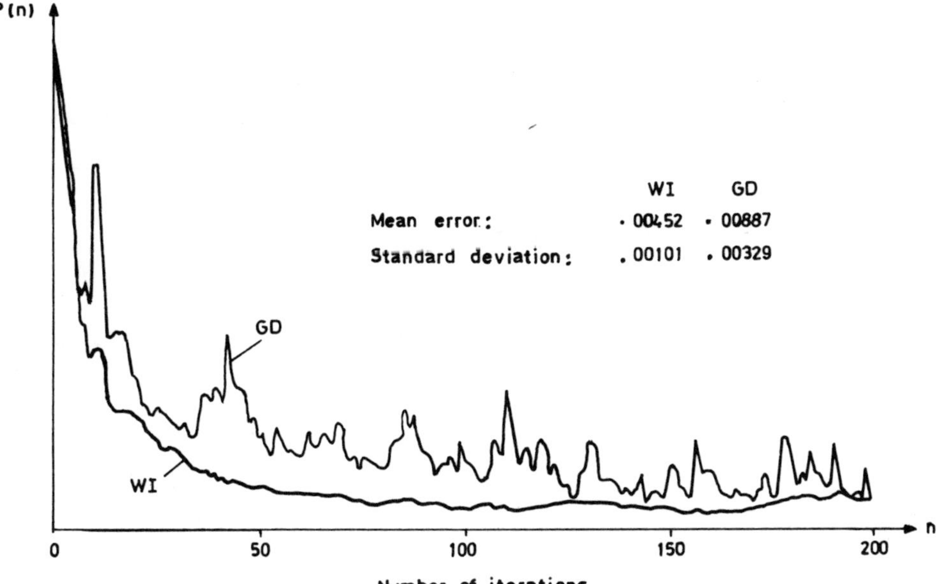

Fig. 4.7 Learning curves

In this particular example, the Woodbury algorithm was found to
perform better than the gradient one. Figure 4.7 shows the
learning curves, with indication of mean error and standard
deviation in the stationary state. Figure 4.8 shows the responses after 1, 10, 50, and 200 iterations, respectively. It
should, however, be warned against extrapolating conclusions
from this single simulation, especially to situations with a
much larger number of sensors.

4.3 Flexibility

A weight vector W is dependent on the focusing vector m, see
(2.9). Hence the updating with the GD algorithm has to go
through a transient state, if there is a desire to change m.
This is not the case with the WI algorithm, because R^{-1} is independent of m, and because the other quantity to be adapted
with this algorithm, $b = m^{\dagger} R^{-1} m$, has to be calculated directly
from time to time.

5 CONCLUSION

The Woodbury identity algorithm shows some interesting qualities,
both with regard to computer requirement, performance, and flexibility, when compared with the gradient descent algorithm. It may

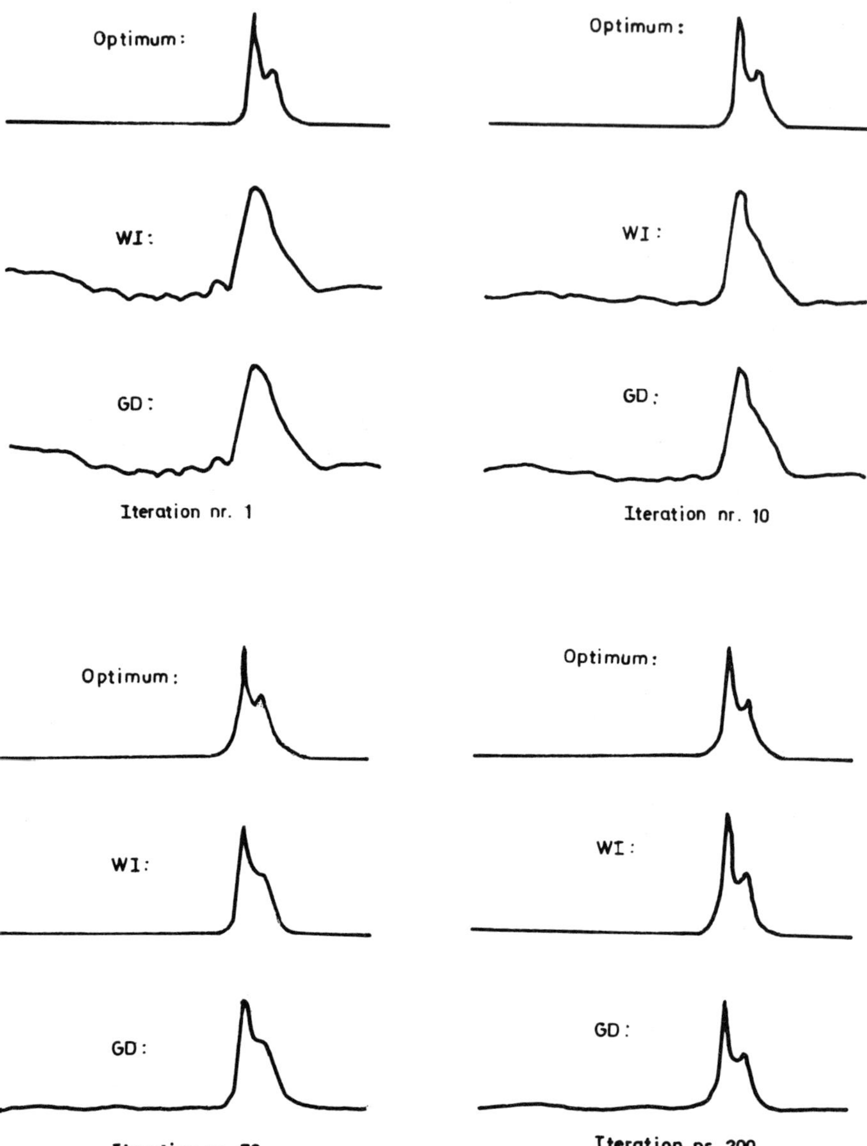

Fig. 4.8 Response development

deserve a higher degree of theoretical investigation than the open literature indicates. It does not deserve to be the forgotten algorithm.

REFERENCES

1.   H. Cox, Sensitivity Considerations in Adaptive Beamforming, in: Signal Processing, ed, by J. W. R. Griffiths, P. L. Stocklin, and C. van Schooneveld, Academic Press, London and New York, 1973

2.   N. L. Owsley, A Recent Trend in Adaptive Spatial Processing for Sensor Arrays: Constrained Adaptation, in: Signal Processing, see reference 1.

3.   E. B. Lunde, Comparison of Two Methods for Constrained Adaptive Beamforming Employing an Auxiliary Array, NUSC Technical Memorandum No. TC-6-75, 1975

4.   A. Sagehaug, Analysis of Some Numerical Problems in Adaptive Beamforming, NDRE Technical Memorandum, 1975.

THE EFFECTS OF SIGNAL AND WEIGHT COEFFICIENT QUANTISATION IN ADAPTIVE ARRAY PROCESSORS

J. E. Hudson

Dept. Electronic and Electrical Engineering,
Loughborough University, Leics., U.K.

SUMMARY

This paper discusses two areas in which quantisation can effect
the performance of an adaptive array processor. The first
concerns a passive sonar using all-digital processing in which
the input hydrophone data is digitised by A/D converters, which
introduce random quantisation errors into the signal channels.
The second quite separate problem arises when the weight
coefficients in an adaptive array processor are quantised, such
as would occur in an adaptive radio antenna which utilised digitally
controlled analogue weights, and which for practical reasons, might
be fairly coarsely quantised. In either case the effect of quantisation is to limit the degree of rejection of interference sources
which can be achieved in the adaptive null-steering system.

1.   INTRODUCTION

This paper is associated with the one presented earlier in the
proceedings by Prof.J.W.R.Griffiths|1|. The fundamental ideas
have been presented by Widrow|2|, Lacoss|3|, Frost|4| and others.
The application of these ideas in two areas have been picked out
for analysis in depth because they represent problems which might
arise using currently available hardware devices. These two
areas were described in the summary. In either case the problem
is to determine the smallest number of bits of quantisation
which give acceptable performance.

The measure of performance used is the ability of the processor to reduce its gain in the direction of a plane-wave interference which lies in the sidelobes of the conventional (i.e.

uniformly weighted) array pattern.

2.   SIGNAL QUANTISATION

A perfectly coherent wave illuminating the sensor array, if
it is sufficiently narrow-band, can be represented as the product
of a time-invariant spatial vector S and a scalar complex time
function s(t). Thus the sample of received signals at time t_i
is the vector $s(t_i).S$. The effect of the A/D converter is to
introduce a vector of quantisation errors $U(t_i)$, so that the processor input vector $X(t_i)$ is $s(t_i).S + U(t_i)$. If the number
of bits of quantisation is greater than about six, the vector
of quantisation errors can be regarded accurately as being uncorrelated with the signal and its elements as being mutually
incoherent $|5|$. Thus the covariance matrix of the processor
inputs is

$$R_{xx} = E(XX^T) = SS^T.E(|s(t_i)|^2) + E(U(t_i)U(t_i)^T) \qquad (2.1)$$

$$= P_s SS^T + \sigma_q^2 I \qquad (2.2)$$

where S^T is the complex conjugate transpose of S, P_s is the mean
signal power per channel, I is the identity matrix, and σ_q^2 is
the variance of the quantisation errors in each channel. The
vector S is normalised so that $S^T S = N$ where N is the number of
sensors. If δ is the input voltage change at the A/D converter
inputs which corresponds to one unit at the output, then
$\sigma_q^2 = 2\delta^2/12$.

It can easily be shown $|6|$ that the output power of a
linearly constrained least-mean-square processor is given by
$P_{out} = (C^T R_{xx}^{-1} C)^{-1}$ in which C is the constraint or steering
vector such that $W^T C = 1$ where W is the weight vector and R_{xx} is
the input covariance matrix. Using a matrix identity we can
write for R_{xx}^{-1}

$$R_{xx}^{-1} = \sigma_q^{-2} I - P_s SS^T/(\sigma_q^2(\sigma_q^2 + P_s S^T S)) \qquad (2.3)$$

$$\text{and } P_{out} = \left[\frac{C^T C}{\sigma_q^2} - \frac{P_s C^T SS^T C}{\sigma_q^2(\sigma_q^2 + P_s S^T S)} \right]^{-1} \qquad (2.4)$$

The conventional, uniform weight processor uses weight vector C/N
and its output power for the signal S is $E(|X^T C|^2)/N^2 = \rho^2 P_s$
say, where ρ^2 is unity if the array is steered toward the signal
by variation of C to make C = S. We are specifically interested
in the output power of the adaptive processor when ρ is reasonably
small, i.e. the signal is an interference lying in the sidelobes
of the conventional array directional pattern. Making an

assumption that $P_s >> \sigma_q^2$, which is reasonable if at least six bits of signal quantisation are used, we can write,

$$P_{out} \simeq \left[\frac{N}{\sigma_q^2} - \frac{\rho^2 N}{\sigma_q^2} \right]^{-1} = \frac{\sigma_q^2}{N(1-\rho^2)} \qquad (2.5)$$

This output power can be related to the quantisation noise output power of the conventional processor which would be σ_q^2/N. Since ρ^2 is small, the output power of the adaptive processor is not very much greater than this and is independent of the power of the interference. It is reasonable to conclude that the interference itself is precisely nulled but the processor cannot reduce the quantisation noise which the interference sets up during A/D conversion. The variation of the weight vector required to null the interference increases the white noise gain of the array. This explains the $1/(1-\rho^2)$ term in the output power and shows that nulling an interference in the main beam of the conventional pattern causes the output of high quantisation power.

An approximate estimate of the number of bits B of quantisation required can now be made. The overload voltage at the A/D converter input is set (arbitrarily) to $6^{\frac{1}{2}}$ times the rms.(real) signal level to allow only a minute probability of overflow. For two's compliment binary arithmetic this gives $\delta = 6^{\frac{1}{2}} 2^{-B+1} P_s^{\frac{1}{2}} 2^{-\frac{1}{2}}$ Substitution in Eq. (2.5), assuming that $\rho^2 << 1$ gives,

$$P_{out} = 6 P_s 2^{-2B+2}/12N = N^{-1} 2^{-2B+1} P_s \qquad (2.6)$$

The output power of the processor when directed at the interference itself is just P_s. Thus the depth of the null is $10\log_{10}(2^{2B-1}N)$. This varies from 67 dB for a ten element array with 10 bit quantisation to 40 dB for a five element array with 6 bits, and good interference cancellation is not difficult.

In simulation tests for interferences in the sidelobes output powers are usually one or two dBs higher than expected, mostly due to incomplete convergence of the processor in the presence of loop noise. Various approximations in this result breakdown when the interference enters the main beam.

3.   WEIGHT VECTOR QUANTISATION

We assume that a radio antenna is to have variable weighting coefficients in a weight/sum linear array processor. An electronically phased array already has suitable phase shifters, the adaptive array also requires variable attenuators. The required complex weights, for convenience in calculation, are assumed to be realised by splitting the output from each element

into two phase-quadrature channels and passing each through
variable stepped attenuators.

The amplitude gain of each attenuator is set by a binary
word so that linear steps of amplitude transfer are achieved. The
presence of a phase inverter is assumed which is set by the sign
bit of the binary word. We can regard each coefficient as a
complex variable whose real and imaginary parts are set in
linear steps from -1.0 to +1.0. Simulation tests of other
arrangements of coefficients show that the final result is not
very sensitive to the method adopted. For example one could
use a combination of a stepped phase shifter and a single
stepped attenuator.

Analysis of the effects of the weight quantisation poses
some difficulties since quantisation errors are fixed at system
equilibrium and are not treatable as random. Furthermore the
errors tend to be related to various correlations present within
the processor. To enable a simple approximate solution to be
obtained it is assumed that the errors in the coefficients are
mutually incoherent and uncorrelated with the signals.

The control of the coefficients is done as follows. The
adaptive processor maintains an internal complex vector V which
is made to converge to the desired vector of weighting coefficients.
Vector V is varied by a standard stochastic steepest descent
algorithm $|3|$ and after every iteration the actual weight vector
components are set individually to their closest approximation
to the elements of $V; W = Q(V) = V+E(V)$.

The error signal is derived from the real signal path so
that errors in W are part of the feedback loop. The algorithm
to minimise the output power is as follows:

(a)   Acquire signal vector at time t_k : X_k

(b) $V'_{k+1} = V_k - \beta X_k X_k^T W_k$ (steepest descent)

(c) $V_{k+1} = V'_{k+1} + C(1 - C^T V'_{k+1})/C^T C)$ (look direction constraint)

(d) $W_{k+1} = Q(V_{k+1})$ (quantise weights)   (3.1)

This technique causes the integration of errors in vector V until
they are sufficiently large to cause a jump in one of actual
weighting coefficients of W.

We now derive the expected depth of nulls. Let the space
vector corresponding to the interference be S and suppose the
internal vector V to be exactly set so as to null the interference

$S^TV = O$. The actual signal output is given by $y = S^TW = S^T(V+E)$ $= S^TE$. For the variance of the output signal y we have,

$$\langle yy^* \rangle \;=\; \langle S^TEE^TS \rangle \;=\; S^T \langle EE^T \rangle S \qquad\qquad (3.2)$$

where $\langle\,\rangle$ is an ensemble average over all possible weight errors given by Eq.3.1 (d). The assumption of mutually incoherent errors gives $\langle EE^T \rangle = \sigma_w^2 I$ where

$$\sigma_w^2 \;=\; (\Delta W_{real})^2/12 + (\Delta W_{imag})^2/12 = (\Delta W)^2/6 \qquad\qquad (3.3)$$

in which ΔW is the magnitude of each quantisation increment. Thus,

$$\langle yy^* \rangle \;=\; \sigma_w^2 S^TS \;=\; \sigma_w^2 N \;=\; N(\Delta W)^2/6 \qquad\qquad (3.4)$$

ΔW is calculated as follows. To minimise the errors, the largest element in V, V_{max}, is found. W is set to $Q(V)$ but is scaled so the largest element of W, corresponding to V_{max} is unity. For M levels of quantisation we thus have $M-1$ intervals extending from $-V_{max}$ to $+V_{max}$ and $\Delta W = 2V_{max}/(M-1)$.

The output power of the system, from Eq.3.4, is thus $\frac{2}{3}NV^2_{max}/(M-1)^2$. This output power corresponds to the depth of the null since the system gain is unity when it is directed at the interference.

V_{max} is estimated in the following way. It is assumed that the weight vector V is a linear combination of S and C such that $V^TC = 1$ and $V^TS = O$ (this can be proved for the case of weak incoherent background noise). We find that

$$V = (C - \rho S)/N(1 - |\rho|^2) \qquad\qquad (3.5)$$

(noting that $C^TS = \rho N$) satisfies this condition. Since the elements of both C and S have moduli of unity, then from the triangle inequality we have for the elements of V:

$$1/N(1+|\rho|) < |V_i| < 1/N(1-|\rho|) \qquad\qquad (3.6)$$

Thus we can write,

$$P_{out} = \langle yy^* \rangle = \frac{2}{3}N^{-1}(M-1)^{-2}(1-|\rho|)^{-2} \qquad\qquad (3.7)$$

Simulation experiments show that this result is usually bettered, especially when the quantisation is coarse. The reason, pre- sumably, is that the feedback loop can find better quantised weight coefficients than those with purely random errors. Thus we are justified in treating Eq.3.7 as a worst case result and can expect results one or two dB better but exact gain is very sensitive to interference direction.

4.   CONCLUSIONS

Analysis of the effects of signal and weight vector quantisation in adaptive processors shows that both can degrade the depth of nulls which the processor can steer toward a discrete interference source. In the case of signal quantisation there is no difficulty in obtaining good results from modest conversion accuracy and eight bits should be sufficient to give null depths greater than 40 dB. Obtaining sufficiently accurate weight coefficients may be more difficult especially in the radio antenna situation since 4 to 5 bits of accuracy are required to obtain 40 dB null depths with a ten element array. The situation is alleviated by the ability of the processor to take care of inaccuracy in the coefficients, and it is only necessary to provide high resolution. Exact calibration of the coefficients is not necessary.

ACKNOWLEDGEMENT

This work is supported by the Ministry of Defence, U.K.

REFERENCES

1.   J.W.R.Griffiths and J.E.Hudson, An introduction to adaptive processing in a passive sonar, Nato Signal Processing Conference, La Spezia 1976.
2.   B. Widrow, P.E.Mantey, L.J.Griffiths and B.B.Goode, Adaptive antenna systems, Proc.IEEE Vol 55 No.12 (Dec 1967).
3.   R.T.Lacoss, Adaptive combining of wideband data for optimal reception, IEEE Trans. Vol GE-6 pp 78-86 (May 1968).
4.   O.L.Frost, An algorithm for linearly constrained adaptive array processing, Proc.IEEE Vol 60 No.8 pp 926-935 (Aug 1972).
5.   B.Liu, Effect of finite word length on the accuracy of digital filters, IEEE Trans CT-18 (Nov 1971) pp 670-677.
6.   R.T.Lacoss, Data adaptive spectral analysis methods, Geophysics Vol 36 No.4 pp 661-675 (August 1971).

ADAPTIVE BEAMFORMING IN AN ACTIVE DOPPLER SONAR

J. N. Maksym

Defence Research Establishment Atlantic
Research & Development Branch
Department of National Defence
Dartmouth, Nova Scotia, Canada

ABSTRACT.   The paper considers reverberation reduction in an
active sonar system transmitting long C.W. pulses from a moving
platform. Somewhat different from the MTI radar situation in
which clutter can be considered as arriving from a specific
bearing within a given doppler cell, the reverberation in the
active doppler processing sonar is somewhat spread in angular
extent largely because the scatterers, waves and the like, are
moving. Reverberation suppression, therefore, requires rather
wide nulls in the directivity pattern of the receiver.

The basic array optimization problem will be discussed from the
point of view of minimization of the total beamformer output
power subject to a mainlobe maintenance constraint. The
covariance matrix of the reverberation noise field will be
modelled based upon empirical results for spreading due to wave
motion and the geometry of the receiving transducer. From these
idealized models the optimized array response patterns are
computed and will be shown for various doppler frequency cells,
platform velocities and sea state conditions.

Next the physical transducer is measured and the actual
performance of adaptive beamforming, taking into account the
transducer limitations is examined. Conclusions are made
concerning the effect of the non-ideal transducer response
patterns on the degree to which the overall response can be
optimized. These conclusions are used to indicate potential
design improvements in active transducers which are necessary
to full utilization of optimum and adaptive techniques in
active sonar.

G. Tacconi (ed.), Aspects of Signal Processing, Part 2, 429-441. All Rights Reserved.
Copyright © 1977 by D. Reidel Publishing Company, Dordrecht-Holland.

INTRODUCTION

This paper describes the results of current research at DREA in
which techniques of optimum array processing are being applied to
active sonar. We are presenting these results at the Advanced
Study Institute in order to illustrate some actual applications
for such processing and to point out some of the practical
considerations which arise in real systems. In particular, the
paper concerns the problems which arise when the individual
sensor elements have a complicated directivity pattern themselves.
This is a common phenomenon in active systems where the receiving
sensors are complex resonant structures and are housed in a dome
or towed body presenting various baffling and diffraction effects.
Most treatments of array processing consider ideal elements
which have well behaved directivity properties and are transparent
to the field. The results of this paper show that where these
properties are not met, careful in situ array measurements are
required, and even with such measurements practical array gains
may not be as good as predictions based on ideal sensors.

The active sonar problem that is considered is optimization of the
receiving array response in the presence of a directional
reverberational background. This condition is produced when
narrowband signals and filtering are employed to maximize the
detectability of moving targets. When the sonar platform itself
is moving the reverberation is doppler shifted as a function of
angle with respect to the direction of motion. This effect
combined with a frequency spread due to motion of the scatterers
and the finite transmitted pulse length give the reverberation a
directional characteristic when viewed through a particular
doppler filter. The objective of optimum array processing here
is to design a beam pattern which strongly discriminates in the
directions of highest reverberation.

The study was carried out with a research sonar transmitting and
receiving at 10 kHz with a 24 stave cylindrical transducer
having a nominal radius of 28 cm and height approximately twice
this. Each stave is considered as an array element for beamforming
in the horizontal plane. This is a reasonable simplification for
a duct sonar, for which signals and reverberation are confined
to a narrow region of angles near the horizontal. A sketch of
the overall system is shown in Figure 1. In the figure ten of the
24 staves form a receiving beam at the doppler channel at
frequency f. In a typical application of this transducer in a
doppler processing sonar 24 contiguous beams are formed at each
doppler frequency.

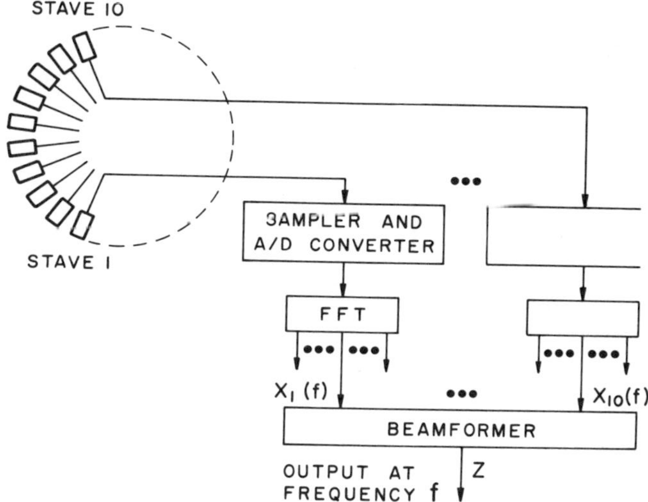

Fig. 1. Doppler-Processing Sonar

The response pattern of a beam depends not only on the array
geometry and the beamforming coefficients but also on the
directional response of each stave comprising the array. In the
research sonar the same transducer is used for both transmission
and reception and is enclosed in a variable depth sonar (VDS)
body. Under these conditions the stave response patterns are
not smooth functions of angle but are similar to that in
Figure 2, which displays the power response of a typical stave.
Shown also in Figure 2 is a cardioid function used as an example
of an idealized stave response. This is used later to compare
the relative effects on beamforming of a smooth stave response
and that obtained from calibration measurements of the
transducer mounted in the VDS body.

MODEL FOR REVERBERATION NOISE

In order to determine the maximum improvement to be gained by
adaptive beamforming it is necessary to model the directional
properties of the reverberation field at each doppler frequency.
For this purpose it is assumed that reverberation arises from
reflections off scattering points uniformly distributed in the
horizontal plane. For reflections from a stationary point at

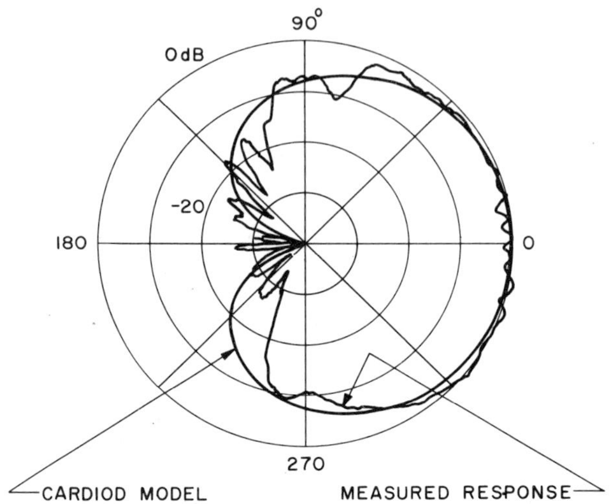

Fig. 2. Power response pattern of a typical stave

bearing angle θ the echo will be shifted in frequency by

$$f = \frac{2v}{\lambda} \cos \theta \tag{1}$$

where, v is the platform velocity and, λ is the acoustic wavelength
at the transmitted frequency. The scattering points are normally
in random motion so that the doppler frequency Δf from a
scattering point at bearing angle θ is a random variable whose
probability density function was found empirically[1] in the case of
a one-second CW transmission to be proportional to

$$p(\Delta f) = \begin{cases} e^{-\gamma |\Delta f|^2} & , \quad |\Delta f| \leq f_c \\ \\ \alpha e^{-\beta |\Delta f|} & , \quad |\Delta f| > f_c \end{cases} \tag{2}$$

where the parameters α, β, γ, and f_c are defined by Table I for
three windspeed conditions. From equations (1) and (2) and
Table I the reverberation power can be determined as a function

WINDSPEED (KNOTS)	α	β	γ	f_c (Hz)
<3	1.122	0.253	0.046	5.0
∿10	1.778	0.216	0.018	8.0
∿20	3.981	0.1445	0.00375	17.5

Table I   Parameters for Reverberation Frequency Spread

of direction for any given doppler frequency. An example is given
in Figure 3 which shows the resulting patterns at three selected
frequencies for a 5 knot platform velocity, a 10 knot wind
velocity, and a one-second CW transmitted pulse.

BEAMFORMING

In order to use notation which has become more or less standard in

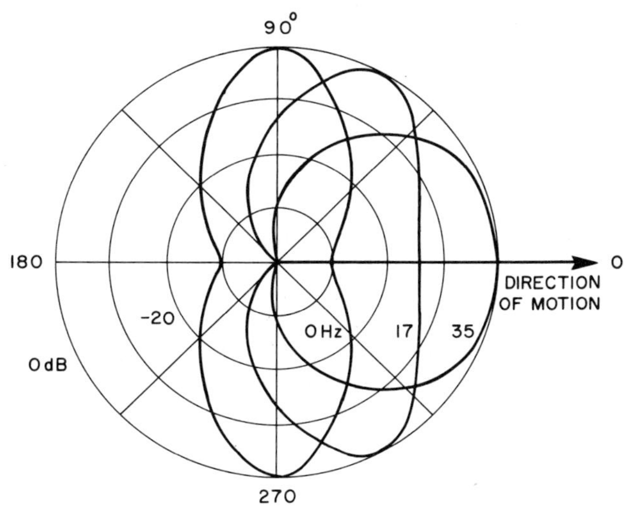

Fig. 3. Power direction patterns of reverberation as a
         function of doppler frequency

the literature on optimum and adaptive beamforming[2] the beamformer
inputs $x_1(f),\ldots,x_{10}(f)$ are represented by a complex vector $\underline{X}$.
Beamforming may then be represented by the complex inner product
given by

$$z = \underline{W}^{\dagger}\underline{X} \tag{3}$$

where, $\underline{W}$ is a vector of complex numbers which are the beamformer
coefficients, and the symbol $\dagger$ denotes the complex conjugate
transpose operation.

When a unit amplitude plane wave is incident upon the array at a
bearing angle θ the response is denoted by a complex vector $\underline{S}(\theta)$
with i^{th} element given by

$$S_i(\theta) = g_i(\theta)\, e^{jkR\cos(\theta-\theta_i)} \tag{4}$$

where, $k = 2\pi/\lambda$ = wavenumber,

  $g_i(\theta)$ is the complex response of the ith stave,

  θ_i is the bearing of the ith stave, and

  R is the transducer radius.

The vector $\underline{S}(\theta)$ is known as the "signal-direction vector", and
expresses the amplitude and phase relationships of the stave
output relative to that of the incident wave at the origin. A
conventional phased-to-a-plane beamformer may be expressed by

$$\underline{W} = \frac{1}{M}\,\underline{S} \tag{5}$$

where, M is the number of array elements. Figure 4 is a polar
plot of the beam power patterns obtained using the phasing of
equation (5) and shading the amplitudes of the elements of $\underline{W}$ by
the numbers {3, 7, 11, 13, 15, 15, 13, 11, 7, 3}. Shading
similar to this is generally used to reduce the level of the
sidelobes. Figure 4(a) is the beam power pattern when each stave
response is real and modelled by a cardioid function

$$g_i(\theta) = \frac{1 + \cos(\theta-\theta_i)}{2} \tag{6}$$

where, θ_i is the bearing angle of the ith stave. Figure 4(b)

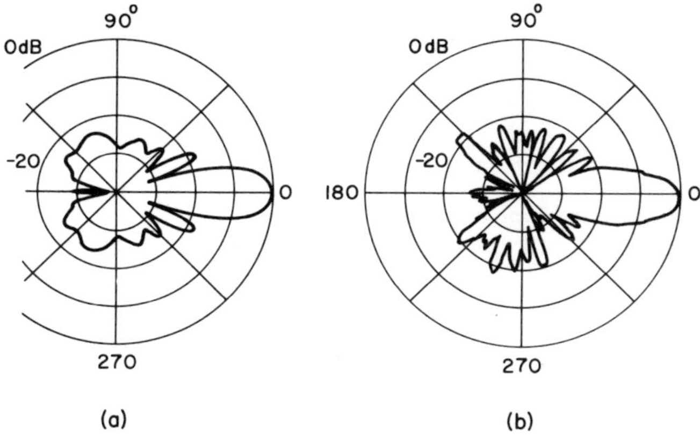

Fig. 4   Conventional shaded beam power pattern

    (a)   Computed using cardioid stave response

    (b)   Computed using transducer calibration data

shows the power response pattern for the forward-looking beam computed using transducer calibration data. The response in this case is more random in appearance, and is in fact different in detail from one beam to another as a consequence of the differences among the stave responses themselves.

When the noise field is directional such as the reverberation fields shown in Figure 3 the patterns of Figure 4 do not provide optimum rejection. One may consider the possibility of forming response nulls in directions of maximum reverberation. A beamformer which does this is the minimum variance beamformer[2] whose weight vector is given by

$$\underline{W} = \frac{\underline{Q}^{-1}\underline{S}(\theta)}{\underline{S}^{\dagger}(\theta)\underline{Q}^{-1}\underline{S}(\theta)} \tag{7}$$

where, $\underline{Q}$ is the covariance matrix of the noise, defined by

$$\underline{Q} = \mathcal{E}\{\underline{XX}^{\dagger}\} \tag{8}$$

where, $\mathcal{E}\{\cdot\}$ denotes the statistical expectation, and $\underline{X}$ is the

input to the beamformer when reverberation alone is present.
Equation (7) minimizes the total output power while ensuring
unity power output for a unit amplitude plane wave incident in
the desired steering direction.

COVARIANCE MATRIX FOR REVERBERATION

With a reverberation power function $N(\theta)$ and known stave response
an expression for the covariance matrix can be derived from the
equality

$$\underline{W}^\dagger \underline{Q} \underline{W} = \int_0^{2\pi} N(\theta) \, |\underline{W}^\dagger \underline{S}(\theta)|^2 \, d\theta. \tag{9}$$

The elements of Q can be derived from the above and the expression
for $S_i(\theta)$ in equation (4). The element q_{mn} is then expressed by

$$q_{mn} = \int_0^{2\pi} N(\theta) g_m(\theta) g_n^*(\theta) \; e^{jkR[\cos(\theta-\theta m)-\cos(\theta-\theta n)]} \, d\theta \tag{10}$$

where, * denotes complex conjugate.

The covariance matrix for the actual transducer placed in a
reverberation field with power function $N(\theta)$ may be computed from
calibration measurements taken at a sufficient number of
orientations of the transducer relative to a source of plane waves
at 10 kHz center frequency. The measurements were made possible
through the construction at DREA of real-time hardware capable of
simultaneous complex sampling, analogue to digital conversion,
digital filtering, buffering and recording on magnetic tape, the
output of the 24-stave transducer. The totality of these
measurements is referred to here as a coherent calibration of
the transducer. Mathematically a typical measurement, of the ith
stave at orientation θ_ℓ consists of

$$X_i(\theta_\ell) = g_i(\theta_\ell) \; e^{jkR \cos(\theta_\ell-\theta_i) + j\psi_\ell} \tag{11}$$

where, $g_i(\theta_\ell)$ is the measured complex response of the ith
stave at orientation θ_ℓ, and ψ_ℓ is an unknown phase term
associated with the travel distance and projector phase. At
each θ_ℓ an array response vector $\underline{X}_\ell$ can be formed from the above
stave responses, and an array response matrix may be computed
according to the complex outer product operation

$$\underline{M}_\ell = \underline{X}_\ell \underline{X}_\ell^\dagger \; . \tag{12}$$

This matrix is independent of the unknown phase term ψ_ℓ which
disappears in the course of taking the complex outer product.
The totality of array response matrices for equispaced θ_ℓ values
in the interval 0° to 360° can be used to compute the covariance
matrix for the array according to

$$\underline{Q} = \sum_{\theta_\ell=1°}^{360°} N(\theta_\ell) \, \underline{M}_\ell \quad . \tag{13}$$

Equation (13) is a simple-minded implementation of the integration
given by equation (10), and may be replaced by more sophisticated
forms such as Simpson's rule if desired.

RESULTS

The optimized beamformer response patterns will have minimum
response in those directions in which reverberation power is
greatest. This can be seen in Figures 5 and 6 where typical
optimized beam patterns are shown for sea-state conditions
corresponding to a 10 knot wind, with the sonar moving through

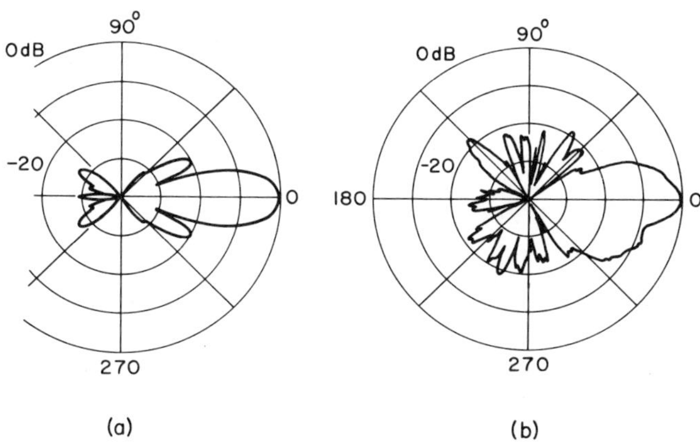

Fig. 5. Power response for a front-looking beam at zero
         Hz doppler

         (a) Computed using cardioid stave response

         (b) Computed using transducer calibration data

                                                                J. N. MAKSYM

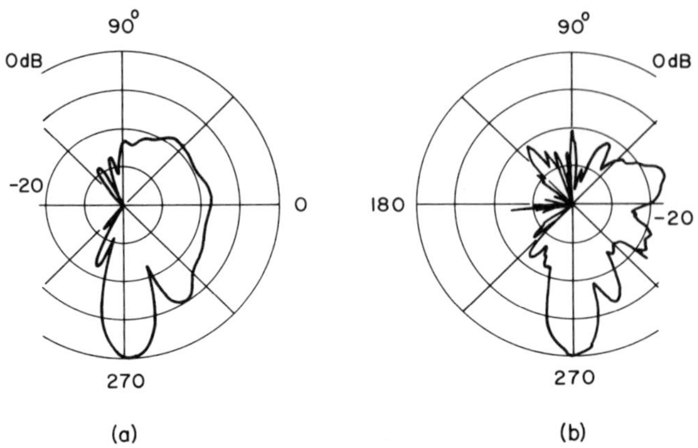

(a)   (b)

Fig. 6.   Power response for a side-looking beam at -35 Hz
          doppler

          (a)   Computed using cardioid stave response

          (b)   Computed using transducer calibration data

the water at 5 knots. The optimizations were carried out by
computing the covariance matrix according to equations (10), or
(13), and solving for the beamforming coefficients using
equation (7). In all cases self noise with a power 30 dB below
reverberation as seen by an omnidirectional transducer was added
by augmenting the diagonal elements of the $\underline{Q}$ matrix. The
addition of self noise acts to constrain the level of sidelobes
in those directions where the reverberation is low. Part (a)
of Figures 5 and 6 shows the pattern obtained with the cardioid
model for each stave, while part (b) is computed using actual
data from the transducer calibrations together with the same
$N(\theta)$ as part (a). Figure 5 is for a beam looking directly
ahead at 0 Hz doppler while Figure 6 is for a side-looking beam
at -35 Hz doppler. In the first case maximum reverberation
comes from the sides and in the second case it is from the
rear. It appears from these typical results that the actual
response patterns are not as effective as those obtained from
the idealized cardioid model.

Figures 7 and 8 illustrate the relative performance of 1) The
conventional beam based on measured stave response illustrated in
Figure 4(b), 2) Optimized beams based on transducer calibration,
and 3) Optimized beams using cardioid stave response functions.
Two situations were chosen for the sake of illustration, a beam
looking directly forward in Figure 7, and one looking to the
side in Figure 8. The performance parameter in each case is

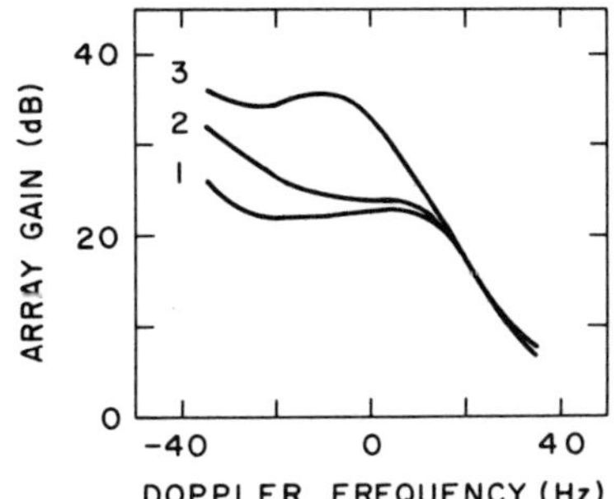

Fig. 7. Array gain vs reverberation for a front-looking beam, 5 knot platform velocity, and a 10 knot wind velocity

(1) Shaded beamformer as computed using calibration data

(2) Optimized using calibration data

(3) Optimized using cardioid stave model

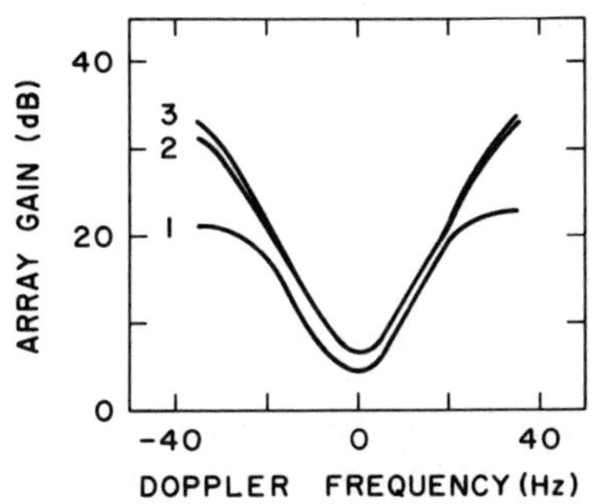

Fig. 8. Array gain vs reverberation for a side-looking beam, 5 knot platform velocity, and 10 knot wind velocity

(1) Shaded beamformer as computed using calibration data

(2) Optimized using calibration data

(3) Optimized using cardioid stave model

array gain versus the reverberation noise power achieved by the
array relative to an omnidirectional receiver. It should be
noted that a high value of array gain represents good rejection
of the reverberation noise, and provides improved detectability
of an echo. The highest array gain, and also the greatest
potential for its improvement by optimum processing, is seen
to occur at those doppler frequencies for which the reverberation
occurs outside the main lobe of the beam.

DISCUSSION

This paper has pointed out some of the practical considerations
in the design of optimum beamformers, and because adaptive
systems attempt to approach optimality the same conclusions on
the attainable performance apply to them as well. The results
emphasize the importance of carrying out simultaneous measurements
of the complex response of the array elements.

The process of beamforming can be regarded as the formation of
a desired response by a linear combination of complex functions
of angle comprising the individual element responses modified by
a geometric phase function due to their locations relative to
the center of the array. In real systems such as the active
sonar considered here local scattering of sound waves from
transducers and nearby structures contributes to rather complicated
response patterns for the individual elements. The variation of
response as a function of angle can be described by a spatial
bandwidth parameter. When this bandwidth is low as in the
cardioid model it is possible to find a linear combination (beam
pattern) which has a reasonably low response over the region of
angles where the reverberation has a high power level. If
however the spatial bandwidth is high as was found during the
calibrations no linear combination of array weights may be found
to sufficiently minimize response over the required range of
angles.

A number of techniques to improve the performance may be
considered. One might try, for example, to improve individual
element responses by increasing sonar body transparency. If
the transmit and receive functions of an active sonar were
separated, point elements for reception might simultaneously
increase the array transparency and reduce the problems
associated with the scattering of waves in the near field.

The process of forming optimized beams can be regarded as the
application of constraints on the beam pattern at a number of
angular locations. For example with M staves it is well known
that the response pattern can be constrained at M-1 points, say
to form M-1 nulls. Having done this the response at all other

angles depends on the variations in the individual element response functions. It might be possible with more elements comprising a beam to form a better pattern than those examined in this paper. However, the use of more elements must be weighed against the increased hardware and computational cost, and may not be a practical alternative.

The research reported in this paper originated as part of a wider study of the application of adaptive beamforming algorithms to active sonar. When adaptive beamforming was applied to actual reverberation data recorded at sea it was found that a discrepancy existed between predicted performance based upon an idealized transducer and that actually experienced.

Our results presented here have shown that much of the discrepancy can be attributed to limitations imposed by the responses of individual transducer elements.

REFERENCES

1. R. S. Thomas, J. C. Moldon, J. M. Ross, "Shallow Water Acoustics Related to Signal Processing", Proc. NATO Advanced Study Institute on Signal Processing, Loughborough, U.K. (Aug 1972). (The raw data on spectral spreading of surface reverberation obtained by Thomas et al was used to obtain Table I.)

2. H. Cox, "Sensitivity Considerations in Adaptive Beamforming", Proc. NATO Advanced Study Institute on Signal Processing, Loughborough, U.K. (Aug 1972).

ADAPTIVE ARRAY PROCESSING EXPERIMENTS AT HF*

Lloyd J. Griffiths

Department of Electrical Engineering
University of Colorado
Boulder, Colorado, USA 80309

ABSTRACT. This paper summarizes experimental results which demonstrate that significant improvements in signal-to-noise ratio SNR can be achieved by utilizing adaptive array methods in a 15 MHz, electromagnetic, high frequency (HF), radar system. The experiments were conducted with a bistatic radar which employed adaptive beamforming at the receive array elements only. Two adaptive algorithms were studied and in both cases it was shown that signal-to-noise ratio improvements of 10 to 15 dB are readily achieved when adaptive beamforming is compared with conventional, Dolph-taper beamforming methods using identical received data in an HF backscatter environment. It was also demonstrated that the time scale of coefficient variation in an adaptive processor operating in this environment is the order of one second. Successful tracking of the adaptive algorithm under these conditions was demonstrated. The use of MTI clutter suppression filters at the subarray outputs, prior to adaptation, was investigated. No significant improvement was observed with the use of these filters on experimental data. Finally, it was shown that the presence of fading nulls can significantly affect the determination of optimal subarray location and spacing in an HF environment. In general, the adaptive beamformer performance was found to be less dependent upon array geometry than was the case for conventional processing.

---

*This work has been supported by the Defense Advanced Research Projects Agency of the Department of Defense and by the Electronic Systems Command (ESD) of the Air Force Systems Command. The Wide Aperture Research Facility (WARF) which was used to obtain the experimental results is operated by the Remote Measurements Laboratory at Stanford Research Institute, Stanford, California.

1. INTRODUCTION AND RADAR SYSTEM DESCRIPTION

The results summarized briefly in this paper are based on experimental observations which have been conducted during the past three years. Complete and detailed descriptions of these observations have been prepared and accepted for publication in the IEEE Transactions on Antennas and Propagation. This work will appear in the September and November issues for 1976 (see References [1] and [2]) and the data presented below are abstracted from these papers. The purpose of the present paper is to briefly summarize all pertinent results and to comment on the applicability of these results to the sonar environment.

Data for the study were obtained using the Wide Aperture HF Radio Research Facility (WARF) which is located near Los Banos in the central valley of California and is operated by Stanford Research Institute, California. Active transmitters at both Lost Hills, California and Bearden, Arkansas were used during the investigation with the latter repeater transmitter simulating a target having a discrete range/doppler return. Figure 1 illustrates the relative locations of these facilities and summarizes their parameters. All tests were conducted with both the Los Banos and the Lost Hills arrays steered in the 90° true (eastward) direction. The WARF was configured to operate as an OTH backscatter radar for the purpose of detecting targets at one-hop ionospheric distances. Thus, signals transmitted from Lost Hills were reflected by the ionosphere to the ground near Bearden, Arkansas -- a distance of about 2600 km -- and scattered back toward California. The signals were then received at Los Banos after a second ionospheric reflection. Most of the received energy consisted of ground backscatter at zero doppler with a small component representing the doppler-shifted energy generated by the Bearden repeater.

A sawtooth, sweep-frequency CW transmit signal similar to that shown in Fig. 2 was employed for all experimental tests with a center frequency f_c chosen to provide one-hop ionospheric propagation between Lost Hills, California and Bearden, Arkansas. Typical values for the tests reported here were in the range 12-15 MHz. Note that the sweep repetition rate was 60 Hz, a value chosen for purposes of minimizing the effects of power-line related interference.

The receiving system at Los Banos utilized the outputs of eight spatially separated adjacent subarrays. Each output consisted of the sum of 32 vertical monopole elements with a 10 meter spacing between elements. The array is linear and oriented in the North-South direction. Thus, a given subarray aperture was 320 meters and the total aperture spanned was 2.56 km. Figure 3 illustrates the receiving and recording system used in this work. The

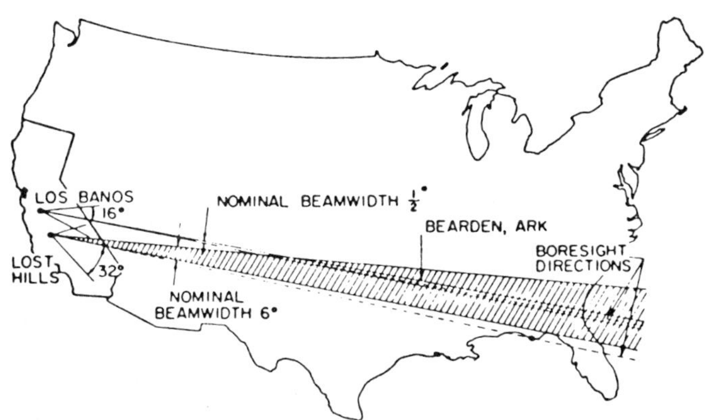

Fig. 1. Transmitter/receiver location map for WARF facility.
Lost Hills Transmitter: 30 kW average power; 18 element antenna
array oriented in true north/south direction; frequency range
9-27 MHz; beamwidth at 20 MHz, 6°; steerable boresight ±32° in
4° steps. Bearden Repeater: 1 kW average power. Los Banos
Receiving System: 256 vertical monopole elements; array length
2.5 km; array oriented in true north/south direction; steerable
boresight ±16° in ° steps.

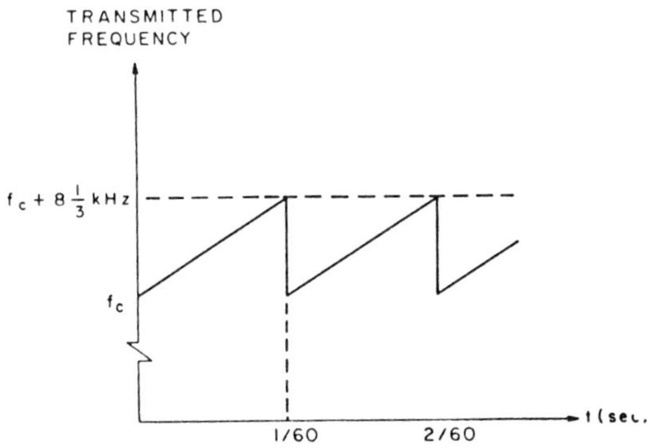

Fig. 2. Instantaneous frequency of transmitted waveform used
during experimental tests.

subarray output signals were demodulated using eight identical mixers driven by a common local oscillator. Manual, fixed gain control was employed at each receiver during the tests and a common calibrate signal was used prior to each recording session for purposes of phase and gain alignment of the receivers. An IF bandwidth of 960 Hz was employed.

The receiver output signals generated by a stationary target at range r_d consist of a constant-frequency sinusoid which may be set to a prescribed value f_d through appropriate choice of the local oscillator waveform. An example is presented in Fig. 4. For the results reported in this paper, the receiver passband shown in this figure was between 0 and 980 Hz. Note that for a brief portion of the sweep (τ_d sec.) the beat frequency falls outside of the receiver passband. The time waveform observed at the receiver output for this signal is shown in Fig. 5. Dashed lines are used to indicate portions of the sweep in which the frequency is outside the passband of the receiver. For the parameters used in this study, 980 Hz receiver bandwidth and 8 1/3 kHz sweep range, the maximum allowable time period outside the passband is 1/(8 1/3) of the sweep period or about 2 milliseconds.

Inspection of Fig. 5 shows that the receiver output waveform for a stationary target consists of a sinusoid during each sweep with the phase of the sinusoid held constant with respect to the start time of the sweep. A moving target at the same range would produce a similar waveform except that the phase reference of the sinusoid would drift with respect to the start time of the sweep. In Fig. 6, successive sweeps are arranged vertically to illustrate this property. Note that the above discussion is based on the assumption that the doppler frequency caused by target motion produces only a small frequency change during one sweep. This assumption is consistent with the targets of interest in the HF environment.

The signal observed at the output of an FM/CW radar during any one sweep may thus be approximated by a sinusoid at frequency f_d , determined by the target range, and at phase ϕ_d , determined by the target range and doppler, and by the sweep number. The transient observed after frequency flyback consists of both an amplitude change and a phase shift.

In the experiments described here, the detected audio output signals from the eight channels were digitized using simultaneous sample and hold circuits and a multiplexed 12 bit analog-to-digital converter. The effective per channel sampling rate was 1.92 kHz, a value intentionally chosen to be a multiple of 60 Hz so that power-line related spurious signals would fold into zero doppler in the final display. This sampling rate provided a total of 32 digital samples per transmitted sweep at each of the eight

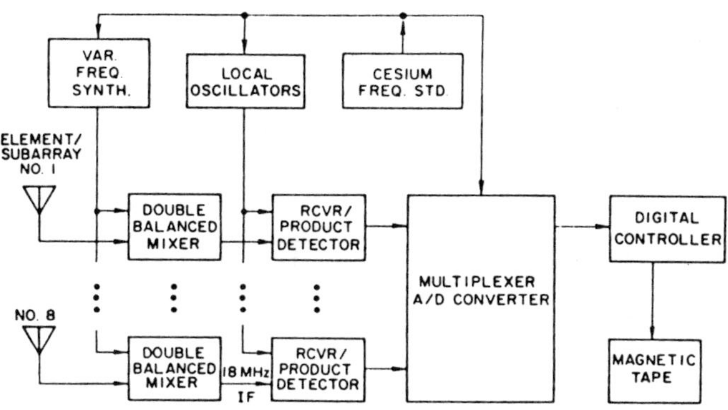

Fig. 3. Block diagram of eight-channel receiving and digital recording system for Los Banos array.

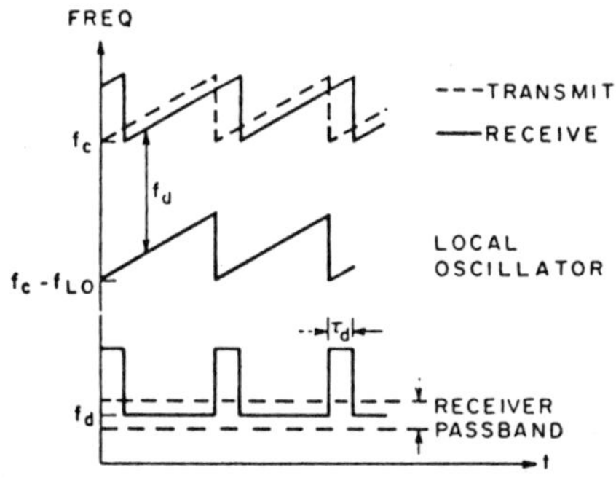

Fig. 4. Receiver output frequency generated by stationary target in FM/CW system.

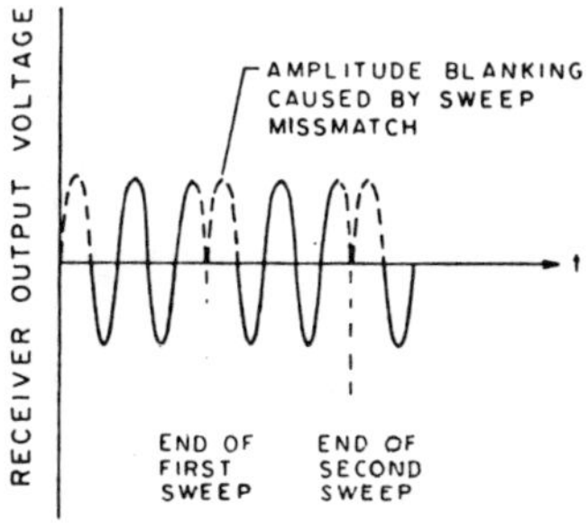

Fig. 5. Receiver output time waveform for stationary target shown in Fig. 4.

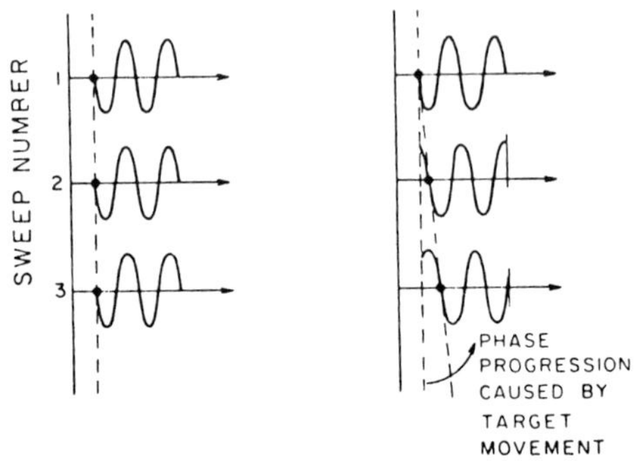

a) stationary target   b) moving target

Fig. 6. Receiver output waveforms observed for single targets
in FM/CW system.

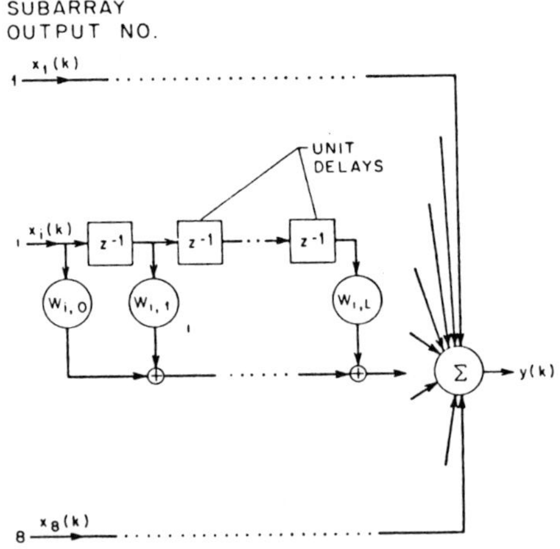

Fig. 7. Tapped-delay line array processing configuration used
during experimental tests.

received subarray outputs. At this sampling rate, the maximum
number of samples in any given sweep which could occur during
the time τ_d shown in Fig. 4 was four samples. Thus, the desired
output signal for a discrete doppler shifted target, after digi-
tizing is a sinusoid with amplitude and phase discontinuity oc-
curing at known 32 sample increments. In the adaptive processing
work described here, the frequency of this sinusoid was assumed
known, consistent with the search mode of an operational radar.
The effects of a mismatch between the assumed and actual target
frequencies are discussed below.

Digitized on-line recordings from the eight receiver channels
were taken at Los Banos, shipped to the University of Colorado,
and processed using a CDC 6400 general purpose computer. The
general structure of the tapped-delay-line beamforming system
used in the present study is shown in Fig. 7. Conventional processing consisted of using fixed weighting coefficients at each
received subarray output, prior to beamforming, with Dolph taper
[3] coefficients. Two adaptive algorithms were studied, the P-
vector algorithm [4] and a constrained least-squares procedure
proposed by Frost [5].

2. DATA PROCESSING AND EXPERIMENTAL RESULTS

One advantage of an FM/CW radar is that target range/doppler maps
are readily generated from the received output signal using a
two-dimensional FFT transform. Specifically, if $q(k;n)$ is used
to denote the kth sample (k = 1, 2,...32) of the nth received
sweep (n = 1, 2, ...64) from a particular receiver, we form a
64 x 32 matrix by arranging the samples from the nth sweep in the
n^{th} row of the matrix. In real time, this matrix would take
64/60 seconds to collect, representing about a one second integration for the radar. It is easily shown [1] that if the rows
and columns of this matrix are Fourier transformed, then the
resulting matrix contains 16 resolvable range cells and 64 resolvable doppler cells.

Figure 8 shows the range-doppler display obtained using the
WARF on March 15, 1974. Eastward transmissions at 12.37 MHz were
employed and a conventional beamformer was in use. The plot has
been normalized to provide an average noise floor level which is
15 dB below the dynamic range of the display. Signals which
exceed this range, such as the strong clutter returns discernible
at zero doppler and all ranges, were clipped at the maximum plot
value. A weak target having a negative doppler frequency of
about -15 Hz can be seen centered in range cells 6 and 7.

This target signal is much more evident in Figures 9 and 10
which show the processing improvement provided with the use of

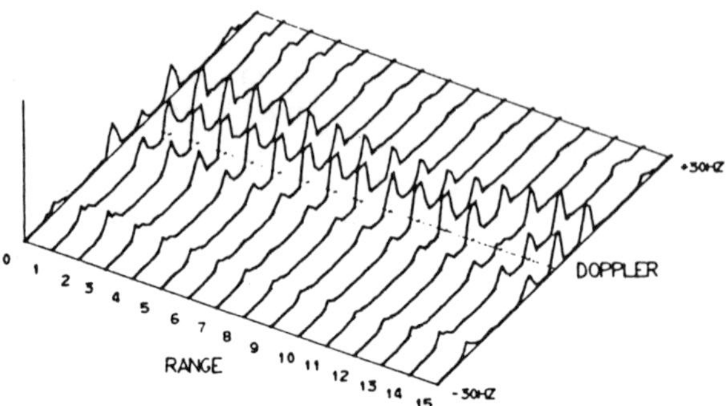

Fig. 8.   Range-doppler map generated using conventional beamforming.

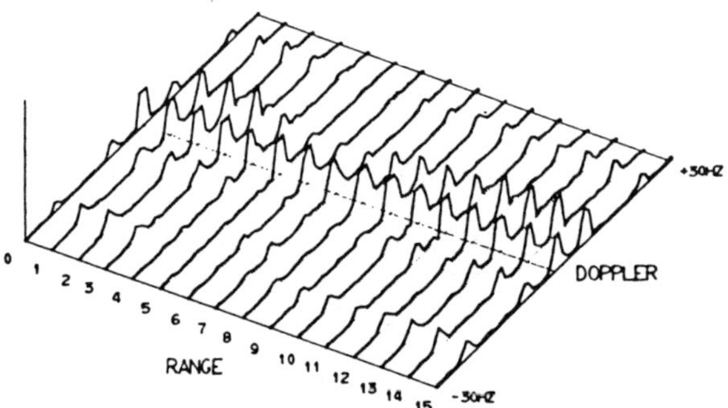

Fig. 9.   Range-doppler map generated using adaptive beamforming
with Frost algorithm.

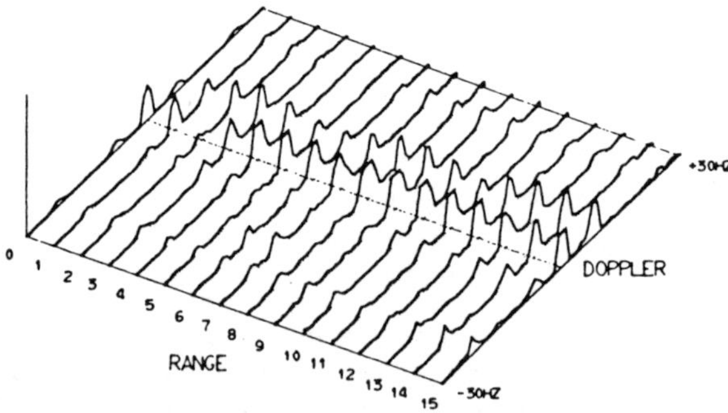

Fig. 10.   Range-doppler map generated using adaptive beamforming
with P-vector algorithm.

adaptive beamforming. The data set used to produce these figures
was identical to that used for Fig. 8. However, a tapped-delay-
line beamformer of the type illustrated in Fig. 7 was used to
produce the beamformed output prior to range/doppler computation.
In both cases, five adaptive weights per subarray were used and
the adaptive proportionality constant was $\alpha = 0.1$ (see Ref. [1]).
Figure 9 was obtained with Frost's constrained least-squares
procedure [4] and Fig. 10 resulted from the use of the P-vector
technique [5]. The strong bands of energy observed in these plots
at -21, -9, +7, and +23 Hz doppler frequencies are interference
lines. This test was intentionally conducted under conditions of
strong HF interference to illustrate the capabilities of the
adaptive processor and the advantages are readily apparent in
these figures.

A comparison of these results illustrates that the received
target signal-to-noise and signal-to-interference ratio is generally greater for either type of adaptive processing than for conventional beamforming. In addition, P-vector adaptation offers a
small improvement over that provided by the constrained, least-
squares procedure. Further evidence of this effect is shown in
Fig. 11 which shows the amplitude values in the RD map as a function of range, taken through the target located at a doppler frequency of -15 Hz. These results clearly indicate that adaptive
processing provides an overall reduction of the noise floor in
addition to the suppression of strong interference lines such as
that present at doppler frequencies -21 Hz and -9 Hz.

Many additional examples and comparisons of this type are
presented in References [1] and [2] and the interested reader is
referred to these sources for further information. In addition,
experiments relating to the time variation of the beamformer coefficients during adaptation, the use of clutter suppression
filters (MTI) prior to adaptation, the effect of array aperture
on performance, and a comparison with quadrature receivers have
all been conducted and reported in these references.

3. DISCUSSION AND CONCLUSIONS

This paper has presented results relating to the use of time-
domain, continuously adaptive beamformers in an HF bistatic FM/CW
radar system. The particular adaptive algorithms investigated
were originally designed for use in baseband receiving arrays
such as those found in passive sonar and seismic systems. The
results presented herein have shown that these techniques are
well suited to modulated systems which employ complex modulation
formats. Specifically, it has been demonstrated that substantial
signal-to-noise and signal-to-interference improvements can be
achieved in HF backscatter radar systems which employ time-domain

adaptive beamforming methods. It has been shown that interference
levels in the processed range/doppler maps can be reduced as much
as 20 dB over that provided by conventional beamforming methods.
In addition, the overall noise floor, exclusive of those areas
containing interference, is substantially lowered. Under every
condition investigated to date, the adaptive beamformer has provided performance levels superior or equal to that offered by
conventional processing. No conditions have been found in which
the use of adaptive processing induced a degradation in performance.

Although the results presented are based on analysis of
adaptive beamforming characteristics using a relatively small
sample of field-recorded data, detailed studies which test these
procedures over longer time periods and under a variety of ionospheric and interference conditions have been conducted at both
the University of Colorado and Stanford Research Institute [6]
using the WARF experimental facility. The latter effort has also
included a fixed point, real time implementation of the P-vector
algorithm. These studies have substantiated the improvements
reported herein and the specific examples chosen for the present
paper were selected as being representative of the total data set.

In conclusion, this study has demonstrated that a significant
processing advantage may be achieved in an HF backscatter radar
system through the use of adaptive beamforming methods. The techniques presented are relatively simple to implement and represent
a minimal addition to the overall system cost. Because the HF
environment involves the propagation of waveforms through a random, turbulent media -- i.e. the ionosphere -- which allows a
variety of multipath structures, and because the HF noise and
interference environment typically contains many more sources than
there are degrees of freedom in the array processor, the overall
evnironment shares many similarities with that found in sonar.
As a result, the conclusions reached from this study may well
reflect comparable improvements available in sonar systems with
the use of a simple adaptive beamformer.

REFERENCES

1. L.J. Griffiths, "Time-domain adaptive beamforming of HF back-
    scatter radar signals," IEEE Trans. Antennas and Propagation,
    (AP-24), 5, Sept. 1976.
2. L.J. Griffiths, "A comparison of quadrature and single-channel
    receiver processing in adaptive arrays," IEEE Trans. Antennas
    and Propagation, (AP-24), 6, Nov. 1976.
3. C.L. Dolph, "A current distribution for broadside arrays which
    optimizes the relationship between beam width and side-lobe
    level," Proceedings IRE, (34), 335-348, June 1946.

4. L.J. Griffiths, "A simple adaptive algorithm for real-time
    processing in antenna arrays," Proc. IEEE, (57), 10, 1696-
    1704, Oct. 1969.
5. O.L. Frost, III, "An algorithm for linearly constrained adap-
    tive array processing," Proc. IEEE, (60), 8, 926-935, Aug.
    1972.
6. T.W. Washburn and L.E. Sweeney, Jr., "An on-line adaptive
    beamforming capability for HF backscatter radar," IEEE Trans.
    Antennas and Propagation, (AP-24), 5, Sept. 1976.

SEPARATION AND REPRESENTATION OF SOURCES AND SIGNALS BY A LINEAR ANTENNA

Hubert Debart

CIT/ALCATEL, France

INTRODUCTION. Among the various functions that can be filled by an antenna, specially attractive is the reconstruction of the pattern of signal emitting sources. If the antenna receives monochromatic coherent signals, this function can be performed rather easily, a lot of Fourier transforms is used so as to reconstruct the pattern on concentric spheres surrounding the antenna ; the operation is well-known as acoustic holography.

Such a result is much more difficult to attain if the sources are emitting completely unknown and wide-band signals. However, it is possible to reconstruct the source pattern at infinity (on a very remote sphere) if the sources are mutually incoherent. This operation can be achieved at the price of an important amount of numerical computation ; the basic treatment consists of slicing the useful spectrum and performing correlations between signals on sensors (for no reference is available). The performances of this reconstruction are limited by imperfections of the sound propagation in sea-water ; contrary to astronomy, no long time is available to superimpose many measurements ; on the other hand, the features of propagation are worse in sea-water than in atmosphere (this will be precised).

.1. PRINCIPLES

.1.1. Mutual coherence function (E. WOLF)

Let us consider two quasi-monochromatic sources, located in two remote points, designated by their bearings (θ_1 θ_2). They are

emitting signals of central angular frequency ω ; their bandwidth extends from

$$\omega - \frac{\Delta\omega}{2} \qquad\qquad \text{to} \qquad\qquad \omega + \frac{\Delta\omega}{2} \quad ;$$

if these signals are named $s_1(t)$, $s_2(t)$ and their signals-in-
quadrature

$$\sigma_1(t), \ \sigma_2(t), \ S_1(t) = s_1(t) + i\,\sigma_1(t) \qquad \text{and}$$
$$S_2(t) = s_2(t) + i\sigma_2(t)$$

are the associated analytic signals. The time average
$S_1(t)\,S_2{}^{*}(t)$ calculated on a long time versus
$2\,\pi\big/\Delta\,\omega$ is called Mutual Coherence Function and

$$\frac{\overline{S_1(t)\ S_2{}^{*}(t)}}{(|S_1|^2\,|S_2|^2)} \qquad \text{Mutual coherence Coefficient}$$

of the sources.

Such functions can be defined for a couple of acoustical sources, located in (θ_1, θ_2) if the spectrum of their signals is
sliced in pieces of bandwidth $\Delta\omega$.

.1.2. Signal decomposition

We are just considering the case (the most usual one) in which
the sources (θ_1, θ_2) have a coherence distance (i.e. the
angular distance for which the M.C.C. falls to $\frac{1}{e}$) very short,
i.e. very smaller than the angular distance defined by
the highest spatial frequency transmitted by the antenna. Thus,
the spectrum is to be sliced in pieces of width $\Delta\omega$, defined
as follows : $\Delta\omega$ is short enough, so as to avoid appreciable
variation of spectrum modulus in an interval

$\Delta\omega$ is long enough, so as to make $\Delta\omega \gg \frac{1}{T}$, if T
is the duration of the measured signal.

The slicing is achieved as follows : the spectrum of signals
of limited duration T, can be sampled with steps of

$\frac{1}{2\,T}$, and reconstructed by Shannon

interpolation from the sampling. The spectrum is sliced in
adjacent pieces of bandwidth $\Delta\omega$; every so defined elementary signal has as Fourier transform the whole set of

components inside the corresponding piece of spectrum. An
elementary analytic signal is derived by adjunction of the
signal in quadrature.

.1.3. Result of the correlation of elementary analytic signals
on 2 sensors.

The whole observable space is filled by sources ; the elementary analytic signal sent by the source located in θ can be
represented by

$$A(\theta, t)\, e^{i\,\phi(\theta, t)}$$

Let γ be $\sin\theta$

   c celerity of sound

Let us consider two sensors (a is their distance) and achieve,
on a duration T, the correlation of two elementary analytical
signals (ω their central angular frequency).

The result is :

$$\int_0^T dt \int_{-\pi/2}^{\pi/2} S(\theta_1, t)\, S^*(\theta_2, t)\, e^{-\,i\,\frac{\omega\gamma_1 x}{c}}\, e^{i\omega\gamma_2\,\frac{(x+a)}{c}}\, d\theta_1\, d\theta_2$$

(if x, x+a are abscissal of sensors).

$$\overline{S(\theta_1, t)\, S^*(\theta_2, t)} = 0 \quad \text{if} \quad \theta_1 \neq \theta_2$$

and it remains :

$$T \int_{-\pi/2}^{\pi/2} \overline{S(\theta, t)^2}\, e^{i\,\frac{\gamma\omega a}{c}}\, d\theta$$

As a result of the correlation, we obtain a sample of the
Fourier transform of $\overline{S(\theta)^2}$.

.2. RECONSTRUCTION FROM A LINEAR ARRAY

All the couples of sources of the array, and all the elementary
signals included in the raw signal can be used so as to reconstruct this Fourier transform from samples, hence the power
repartition $\overline{|S(\theta)|^2}$ in the whole space.

It can be noticed that, from this point of view, the notion of

"noise" (outside of the antenna) is evanescent, for it is
considered as signal from sources, exactly as any "useful"
signal.

If the linear array includes n + 1 sensors, and if the useful
spectrum includes m elementary pieces, mn distinct samples of
the Fourier transform can be evaluated, and even, if the mutual
distances of the elements are arbitrary :

$$m \left[(n - 1) + (n - 2) + \ldots + 2 + 1 \right] = \frac{mn \, (n - 1)}{2} \quad \text{distinct}$$

samples.

This operation is simple if the spectrum is flat ; i.e.
$|S \, (\theta)|^2$ does not depend on ω in the useful band ; if it is
not flat, the measurements result in m distinct Fourier transforms, each of them being defined by

$$\frac{n \, (n - 1)}{2} \quad \text{samples.}$$

It can be seen very easily that the spatial frequency spectrum
is not homogeneously sampled. Let us call ω and Ω the extreme frequencies of the spectrum.

Samples are available at spatial frequencies :

$$\frac{a_1 \, \omega}{c} \; , \; \frac{a_1 \, (\omega + \Delta\omega)}{c} \; , \ldots , \; \frac{a_1 \, (\omega + (m - 1) \, \Delta\omega)}{c} = \frac{a_1 \Omega}{c}$$

$$\frac{a_2 \omega}{c} \, \ldots \ldots \ldots \ldots \, \frac{a_2 \, \Omega}{c}$$

$$\frac{a_n \, \omega}{c} \, \ldots \ldots \ldots \ldots \, \frac{a_n \, \Omega}{c}$$

So as to find no gap in the sampling, it is necessary that
$a_2\omega < a_1\Omega$ or $\frac{\Omega}{\omega} > \frac{a_2}{a_1}$ . . . . and so on. For a regular disposition, the relative bandwidth of the whole
spectrum must be at least 1 octave (the other conditions are
fulfilled a fortiori).

A diagram, for example corresponding to the case $\Omega = 2 \, \omega$, and a regular disposition, can be drawn (we have represented its lower part)

$$a\omega/c \quad (nm) \quad a\Omega/c$$

$$2a\omega/c \quad (n-1)m \quad 2a\Omega/c$$

$$3a\omega/c \quad (n-2)m \quad 3a\Omega/c$$

It can be noticed that :

1°) This repartition can be modulated, in modifying the relative distances of the elements in the array, and the relative bandwidth of the spectrum.

2°) The extreme spatial frequencies are given by : $\dfrac{a_1\omega}{c}$, $\dfrac{a_n\Omega}{c}$

3°) The loosest sampling in spatial frequencies is located in the last part of the spectrum ; its step is

$$\frac{a_n}{c}\,\Delta\omega.$$

As the conjugated variable γ goes from -1 to $+1$, the condition

$$\frac{\Delta\omega\, a_n}{c} < \pi \quad \text{or} \quad \Delta\omega < \frac{\pi c}{a_n}$$

must be fulfilled so as to build a sufficient sampling in Shannon sense.

As the duration of the signal T must be larger than $\dfrac{2\pi}{\omega}$:

$$T \gg \frac{2a_n}{c}$$

This sampling is to be transformed into a regular sampling (uniform step) by any method of smoothing if the Fourier transform technique is to be applied to it in order to obtain a panoramic view of the sources.

If the power spectrum of signals is not flat, as many Fourier transforms as elementary signals (with, of course, a worse definition) are obtained and a diagram (θ, ω) can be drawn, just as in focal planes of lenses in optical treatment of acoustical signals.

.3. ERRORS

Their origin lays essentially in the imperfections of propagation. For an angular frequency ω and a source located in the direction θ , it is actually evaluated :

$$\overline{S\,(\theta)^2}\left[\alpha\,(x)\,\alpha\,(x + a)\,e^{i\beta\,(x)- i\,\beta\,(x + a)}\right]$$

the factor $\alpha\,e^{i\beta}$ represents the random transfer function of the medium-statistically independant on the signal ; no $\overline{}$ (time averaging) is added, for, during the measurement, this transfer function does not undergo appreciable modifications.

Hence appears an error, that can be characterized by the factor

$$\alpha\,(x)\,\alpha\,(x + a)\left[1 + i\left[\beta\,(x)-\beta\,(x + a)\right]\right]$$

if the angular difference is small (which is actually the case).

The real random term $\alpha(x)\,\alpha(x + a)$ has a mean value : if

$$\overline{\alpha\,(x)} = m + \varepsilon \qquad (\varepsilon ; \eta \quad \text{centered random variables})$$
$$\alpha\,(x + a)\,m + \eta$$

$$\overline{\varepsilon^2} = \overline{\eta^2} = b^2 \qquad\qquad \overline{\varepsilon\eta} = b^2\,c\,(a)$$

$$\overline{\alpha\,(x)\,\alpha\,(x + a)} = m^2 + \quad b^2\,c\,(a)$$

and a variance :

$$m^4 + 2\,b^2\,m^2 + \overline{\varepsilon^2\,\eta^2} - \left[m^2 + b^2\,c\,(a)\right]^2$$

If the random variables ε , η are gaussian, it reduces to :

$$2\,b^2\,m^2\left[1 - c\,(a)\right] + b^4\left[1 + c^2\,(a)\right]$$

The imaginary term is :

$$i\alpha\,(x)\,\alpha\,(x + a)\left[\beta\,(x) - \beta\,(x + a)\right]$$

It is, as a rule, centered, but its variance is of a very difficult evaluation, because the irregularities of the medium are not large versus wavelengths. This term is responsible for angular errors ; it is desirable that progresses in the field of random propagation could give a good idea of its mean-square-value.

.4. CONCLUSION

The possibility of reconstructing the spatial location of the
acoustical sources from an array is not bound to the existence
of monochromatic and coherent signals ; the operation can be
made, in rather loose conditions, from unknown signals, at the
price of a larger amount of calculation.

The irregularities of the propagation are limiting the possibilities of analysis especially as the time of measurement is
strictly limited.

REFERENCE

J.C. DAINTY, Diffraction limited imaging of stellar objects using
telescopes of low optical quality, OPTICS COMMUNICATIONS,
Vol. 7(2), pp. 129-134 (Feb. 1975).

COMPLEMENTARITY OF PROPAGATION MODEL DESIGN WITH ARRAY PROCESSING[*]

H. Mermoz

L.S.D.M. - D.C.A.N. - Toulon - France
Attaché aux Services Techniques de l'Armée Française.

1. INTRODUCTION

Most generally, the specialists of propagation models and those
of array processing choose to work with rather loose connections,
trying to improve their particular technique, namely:

- to improve the representation of the acoustic field in order to
  predict its value at point P for a source at point S.

- to refine spatial processing in order to discriminate different
  sources, and to determine their individual signal together with
  their individual position.

This paper is intended to give some clues to increase the amount
of interaction between the two techniques and to suggest an
approach towards more powerful imaging processings at the cost
of much increased real-time computing capabilities.

We shall here restrict ourselves to the case of discrete noise
sources transmitting independent noises.

2. CORRELATION MATRIX FOR ONE SOURCE

Consider an array made of N sensors in a medium which fluctuates slowly in time, where slowly means with a time constant T
that is much larger than the inverse of the useful bandwidth.

---

* *Translated from French.*

G. Tacconi (ed.), *Aspects of Signal Processing, Part 2*, 463-468. *All Rights Reserved.*
Copyright © 1977 *by D. Reidel Publishing Company, Dordrecht-Holland.*

Under these conditions we may be able, with a sufficient real-
time computing power, to plot the N spectral densities on the
sensor outputs and the N(N - 1)/2 cross-spectral densities out
of every pair of sensors. Doing so, we build, in the frequency
domain, a short-term estimation of the correlation matrix. This
estimation is valid for a time delay roughly equal to T, then
needs to be refreshed by a new estimation. Computing continually,
we can follow the slow fluctuations of the correlation matrix.

As long a there is only one source in the medium, and whatever
the fluctuations are, it is expected that the rank of the correlation matrix will remain unity, in other words that only one
eigenvalue will remain significant compared with the others, at
any frequency. This is to assume, of course, some "short term"
perfect coherence between the different ray paths leading from
the source to a sensor.

If this is true, by exploring the rank of the matrix continually,
we have a test of the one source presence. Practically, we have
to compute very fast all the eigenvalues of the matrix at many
frequencies, if we can afford that computing power. If along the
slow fluctuations, only one of the eigenvalues remains significant, it will be clear that only one source is present. This
result is independent of the propagation model — of the medium
structure — as well as the other following result.

The non-zero eigenvalue, as a function of frequency, represents
the spectral density of the source (more precisely the apparent
spectral density available at the array, the true spectral density
being inaccessible as the source itself).

It can also be shown that all the information about the position
of the source lies in the normalized eigenvector associated with
the non-zero eigenvalue. This vector is made of N complex
components, which are yielded by the fast processing of the
matrix.

3. MODEL

We have now to use this vector and try to find the position of
the source. It depends obviously on the propagation model. The
model is, in essence, the expression of the transfer functions
from the source to the sensors just as they are shaped by the
whole medium structure. Consequently it is a set of N slowly-
varying filters. Considering their frequency responses to be
valid over a time delay T, we have a vector in the frequency
domain. Basically, the components of this vector are known
functions of the unknown position of the source (three coordinates)
and of all the parameters involved in the medium description.

Some of these parameters are well known and can be expressed in numerical values in a rather accurate way. We do not mention them any more. Some other parameters are very poorly extrapolated or completely unknowable (especially in real time).

Model designers choose to worry about the last category. They fight to eliminate these unreliable parameters or to have a rough numerical estimate stand for them in the model. This leads either to oversimplified models, very unable to represent the real world, or to models representing a kind of "mean value" of every possible situation, with so large a possible dispersion that it is hardly significant of any particular occurrence. We would need, on the contrary, very rich models with many physical parameters, if only we could always give them the right value.

On the other hand, the array processing specialists are generally shy at fully using the N-dimensional space available to them at the outputs of the array.

4. ONE-SOURCE SITUATION: SATURATED "MODEL"

Let us try to tie everything together and accept to design a model with a set of p parameters of unknown numerical values, the ensemble being represented by P. The model is now an N-dimensional complex vector depending on: (frequency being no longer mentioned)

- the position of the source represented by m, which stands for the three coordinates,

- the set of p unknown parameters represented by P

$$Z(m, P) \ .$$

In the one-source case it happens that this vector is equal to the eigenvector of the correlation matrix (associated to the non-zero eigenvalue) except by an arbitrary phase.

It can be expressed by:

$$Z(m, P) = E \exp\{i\alpha\}$$

where

      Z is a set of N known functions of unknown parameters m and P,

      E is the known eigenvector computed from the matrix,

      α is an arbitrary phase.

The previous relationships stands for   N   scalar equations with
obviously   p + 4   unknown parameters.   In principle it can be
solved for:

$$N = p + 4$$

which means for   p   up to   (N - 4).   If we think of an array
with many more than 10 sensors, we can afford a "saturated" model
with   (N - 4)   unknown parameters whose values will be eventually
yielded by the array processing as well as the coordinates of the
source.

In other words, with a rich array we can afford a rich model,
even widely unknown at the beginning.   Such a model is more
likely to represent the real medium, even a fluctuating one since
some of the parameters may fluctuate in time.

5. THE TWO-SOURCES SITUATION

When several sources are present simultaneously in the medium,
things are a bit different but the general conclusion still holds.
The two-sources situation is significant of the way to cope with
their mutual interaction in the correlation matrix.

Remember that the two sources are assumed to transmit independent
noises.   The correlation matrix, measured at the sensor outputs,
is consequently of rank 2;   the processing yields two non-zero
eigenvalues and the two associated eigenvectors which are orthogonal.   Unfortunately we cannot assigned the first one to source
one and the second to source two.   Both sources contribute to
both eigenvalues and vectors.

Looking in detail into how far we can discriminate between the
two sources, with the correlation matrix only, the situation can
be sketched as follows.

All the qualities of a source can be summarized in a "source-
vector":

$$V = d^{\frac{1}{2}} U$$

where

    d   is the spectral density (apparent),
    U   is the normalized vector built with the frequency responses
       of the filters "inserted" between the source and the
       sensors, these filters reflecting the medium properties.

So the question really is: can we derive V_1 and V_2 from the
matrix?
The answer is: "not quite".

More precisely, the source-vectors V_1 and V_2 , are, of course,
in the N-dimensional complex space, and indeed localized into the
"plane" of the two known eigenvectors. Furthermore, we can derive
three relationships between their four components, referred, in
this plane, to the (orthogonal) eigenvectors. But, in the end,
we are short of one equation to fully determine V_1 and V_2.

Nevertheless this means a possibility to express V_1 and V_2
as <u>known</u> functions of <u>one unknown</u> and well-chosen parameter p.
But this is the furthest we can go with the matrix and we are
left with $V_1(p)$ and $V_2(p)$ as matrix-vectors.

Now what about the model?
Basically the model gives an expression of vector U as known
function of the unknown parameters m and P.

Equaling the model-vector and the matrix-vector, we get two
vector equations, which means 2N scalar relationships with the
following unknown parameters

$(d_1)^{\frac{1}{2}} \exp\{i\alpha\}$
$(d_2)^{\frac{1}{2}} \exp\{i\beta\}$ $\Big\}$ where α and β are arbitrary phases

m_1 : 3 coordinates

m_2 : 3 coordinates

P : p parameters

p : 1 parameter

which, in totals, give p + 9 parameters.

The system can be solved for

$$2N = p + 9$$

which means for p up to (2N - 9).
We now can afford 2N - 9 "free" parameters in a saturated model,
compared to the (N - 4) in the one-source case.
There is a net gain of (N - 5) free parameters, very significant
for a practical number of sensors between 10 and 100.

6. MORE THAN TWO SOURCES

These results can be easily generalized. For N sensors and n
sources we can afford

$$p = n\left[N - \frac{n + 7}{2}\right]$$

free parameters in a saturated model.

Nevertheless, this relationship should not be taken for an
accurate one, but only as a rough estimate of the degree of
freedom for the "array + model" system. Other conditions could
be added, for example the fact that some parameters — the
coordinates — have to be real at any frequency.

Still, it appears that for a given N, the model can be more
and more enriched in free parameters when more and more sources
"light" the medium, up to a limit settled by N itself. Within
this limit, the array is able both to discriminate the sources
and to "describe" the medium, through the solution of parameters
P. This is a way to bring model and array to cooperate. It is
also clear that the same principles apply in pure experimental
modeling, using a number of sources in perfectly known positions.
With two sources we could then afford some (2N - 3) free parameters to describe the medium.

It is necessary to remember that all this computer work should be
made at several frequencies, which may help to solve eventual
ambiguities on a source position.

7. CONCLUSION

This is an oversimplified presentation of some wide possibilities
which lie behind the real-time computation and resolution (eigenvectors and values) of the correlation matrix. It is also an
insight into a way of enriching propagation models with unknown
parameters so that only the general structure of the model has
to be trusted. It is a way to use the full capacity of the
N-dimensional space provided by an N-sensor array. It is an
approach of "imaging" the sources (indicating the position and
the spectral density) and, at the same time, of partly describing
the medium through a set of a *priori* unknown parameters, regardless of the way they may vary with frequency or fluctuate in time.

REFERENCE

1.   Mermoz, H. Imagerie, corrélation et modèles.
      Annales des Télécommunications 31(1-2) 1976 : 17-36.

SPECTRAL SIGNAL SET EXTRACTION*

Norman L. Owsley

Naval Underwater Systems Center, New London Laboratory,
New London, CT 06320

ABSTRACT. A technique is proposed for the extraction of spectral
signals for which the components within a signal set are characterized by complex envelopes which exhibit correlation. The
technique is based on an orthogonal decomposition of the cross
(discrete) frequency correlation matrix describing the set in the
presence of additive uncorrelated noise. Certain advantages of
the technique relative to those based on Fourier analysis with
either estimated power spectrum or minimum variance processing
are presented. Specifically, it is illustrated how the detection
performance of the technique is established by the total energy
in the spectral set rather than by the levels of individual spectral elements in the set. Furthermore, it is shown how minimum
variance spectrum analysis actually can suppress components in
such a set.

I.   INTRODUCTION

In many spectral signal detection systems the detector implementation embodies a model for the signal which assumes that the
signal is characterized by a random phase parameter. Furthermore,
it is typically assumed that the behavior of amplitude and phase
of one spectral component is uncorrelated with the amplitude and
phase of a spectral component which is occurring simultaneously
in another region of the analysis band. As such, phase information
is ignored and detection is performed on the basis of estimated
power spectrum. More recently [1, 2, 3] sequential adaptive

---

*This work has been supported by U. S. Naval Sea Systems Command,
Codes 06H1 and 06H2, at the Naval Underwater Systems Center, New
London, Connecticut, U. S. A.

G. Tacconi (ed.), Aspects of Signal Processing, Part 2, 469-475. All Rights Reserved.
Copyright © 1977 by D. Reidel Publishing Company, Dordrecht-Holland.

schemes have been considered which exploit stationary phase estimators which yield coherent detection algorithms that exhibit improved performance. However, very little consideration has been given to exploiting potential correlation between the complex envelopes of multiple spectral components with the objective of improving the multiple component extraction process. Intuitively, one feels that the best set extraction scheme should perform simultaneous coherent detection and component relating. In the following section, such an approach to coherent component relation which occurs simultaneous with detection is presented. The proposed scheme is based on a principle component analysis of the random Fourier transformed data vector (for example see [4]). Specifically, an orthogonal decomposition of the cross frequency correlation (CFC) matrix into its eigenvectors indicates that the frequency information on spectral components with coherent envelopes appears in a single eigenvector. Furthermore, the corresponding eigenvalue contains energy from all spectral components. Accordingly, it is proposed that spectral sets be extracted by identification of the appropriate eigenvalues and eigenvectors.

II.   THE CROSS FREQUENCY CORRELATION (CFC) MATRIX

Let the deterministic complex N-dimensional vector $\underline{E}_k$ define the position (frequency) of a spectral component in an N-dimensional complex sample space. A random premultiplier s_k, termed the complex envelope, accounts for identical multiplicative random amplitude disturbances and additive phase perturbations on each element of the position vector $\underline{E}_k$. Consider an observed N-dimensional random data vector $\underline{X}$ consisting of $K \leq N$ spectral components and additive noise

$$\underline{X} = \sum_{k=1}^{K} s_k \underline{E}_k + \underline{N} \tag{1}$$

where $\underline{N}$ is a sample vector from an identically distributed and statistically independent vector random process with zero mean and variance σ_0^2. If the matrix E with kth column $\underline{E}_k$ and the matrix P with i-jth element $p_{ij} = \overline{s_i s_j^*}$ are introduced, then the CFC matrix

$$R = \overline{\underline{X}\,\underline{X}^*} \tag{2}$$

$$= EPE^* + \sigma_0^2 I_N \tag{3}$$

is defined wherein ( )* is the matrix complex conjugate transpose operator and I_N is an N by N identity matrix. Assuming that the vectors $\underline{E}_k$ are linearly independent a matrix Φ can be found such that

$$E = \Phi B^{-1} \tag{4}$$

where Φ has the property $\Phi^*\Phi = I_K$ and B is a K-dimensional upper triangular matrix. The matrix $B^{-1} P B^{-1*}$ is now written in diagonal form as

$$B^{-1} P B^{-1*} = \gamma \sum \gamma^* \tag{5}$$

where γ is a K by K matrix consisting of the orthonormal eigenvectors of $B^{-1}PB^{-1*}$ and $\sum$ is a K by K diagonal matrix consisting of the eigenvalues $\sigma_1^2 \geq \sigma_2^2 \geq \cdots \sigma_K^2$ of $B^{-1}PB^{-1*}$.

Using (4) and (5) in (3) and post-multiplying by $\Phi\gamma$ yields

$$R\Phi\gamma = \Phi\gamma\lambda_K \tag{6}$$

where

$$\lambda_K = \sum + \sigma_0^2 I_K \tag{7}$$

and

$$M = \Phi\gamma \tag{8}$$

$$= EB\gamma \tag{9}$$

give the first K eigenvalues and corresponding eigenvectors of the CFC matrix. All remaining (N-K) eigenvalues are identically σ_0^2. Equation (9) indicates that the eigenvectors $\underline{M}_\ell$ of R (columns of M) are linear combinations of the spectral position vectors $\underline{E}_k$ where the combination coefficients are functions of (a) the inner products of the position vectors $\underline{E}_i^*\underline{E}_j$ through B and (b) the complex envelope factor correlation characteristics through γ.

To realize the significance of (7) and (9) consider the case of K spectral components with orthogonal position vectors, i.e.

$$\underline{E}_i^* \underline{E}_j = G_i \delta_{ij} \quad (\delta_{ij} = \text{dirac delta}) \tag{10}$$

and complex envelope power

$$S_i = \overline{|s_i|^2} \tag{11}$$

Let the K spectral components consist of L sets where the ℓth set $\{i \in \Omega_\ell\}$ contains components for which the magnitude squared coherence

$$|C_{ij}|^2 = \frac{\overline{|s_i \, s_j^*|}^2}{\overline{|s_i|^2} \; \overline{|s_j|^2}} = \begin{cases} 1 \text{ for } i, j \in \Omega_\ell \\[2ex] 0 \text{ for } i \in \Omega_\ell \text{ and } j \in \overline{\Omega}_\ell \end{cases} , \quad (12)$$

that is, envelope coherence exists only for components within a set. It can be shown from (7) and (9) that

$$\lambda_\ell = \sum_{i \, \in \, \Omega_\ell} G_i \, S_i + \sigma_0^2$$

$$\underline{M}_\ell = \frac{1}{\left[\sum_{i \, \in \, \Omega_\ell} G_i \, S_i \right]^{\frac{1}{2}}} \sum_{i \, \in \, \Omega_\ell} (G_i \, S_i)^{\frac{1}{2}} \, \underline{E}_i \text{ for } \ell \leq L$$

$$(13)$$

and

$$\lambda_\ell = \sigma_0^2 \quad \text{for} \quad N-L < \ell \leq N . \tag{14}$$

Thus, detecting the ℓth spectral set is accomplished by comparing an estimated λ_ℓ to σ_0^2 and relating components within the ℓth spectral set is accomplished by interrogating $\underline{M}_\ell$ to determine which components of the set $\{\underline{E}_k\}_{k=1}^{K}$ are present in the linear combination $\underline{M}_\ell$. The above case pertains to the situation where each of the spectral sets is disjoint, i.e., $\Omega_k \cap \Omega_\ell = 0$ with $\ell \neq k$. For the case $\Omega_k \cap \Omega_\ell \neq 0$, which includes partially coherent complex envelopes, leakage between sets occurs. However, inclusion of a spectral component within a given set will tend to be in proportion to the amount of coherence which exists between that component and the components within the set.

III. THE RELATION BETWEEN ORTHOGONAL CROSS-FREQUENCY-CORRELATION (CFC) MATRIX DECOMPOSITION AND MINIMUM VARIANCE SPECTRUM ANALYSIS

A minimum variance estimate, y_{MV}, of the complex envelope for a possible component of the random data vector $\underline{X}$ with spectral position vector $\underline{D}$ requires the filter

$$\underline{H}_{MV} = (G/\underline{D}^* \, R^{-1} \, \underline{D}) \, R^{-1}\underline{D} \tag{15}$$

where $\underline{D}^*\underline{D} = G$ (see [7] for example). The expected output power for the minimum variance filter is

$$\overline{|y_{MV}|}^2 = (G^2/\underline{D}^* \, R^{-1} \, \underline{D}) , \tag{16}$$

whereas the expected output power for a conventional filtering operation is

$$\overline{|y_C|^2} = \underline{D}^* R \underline{D} \ .$$ (17)

In the preceeding section it was shown that the CFC matrix can be written as

$$R = \sum_{\ell=1}^{L} \sigma_\ell^2 \ \underline{M}_\ell \underline{M}_\ell^* + \sigma_0^2 \ I_N$$ (18)

Using (18) in (16) and (17) gives

$$\overline{|y_{MV}|^2} = G^2 \sigma_0^2 / (G - \sum_{\ell=1}^{L} \psi_\ell |\underline{D}^* \underline{M}_\ell|^2)$$ (19)

and

$$\overline{|y_C|^2} = \sum_{\ell=1}^{L} \sigma_\ell^2 \ |\underline{D}^* \underline{M}_\ell|^2 + G \sigma_0^2$$ (20)

where $\psi_\ell = \sigma_\ell^2 / (\sigma_\ell^2 + \sigma_0^2)$. Consider the special case of two spectral components for both coherent and incoherent envelopes. Given orthogonal position vectors with $\underline{D}^* \underline{E}_i = \alpha_i \ \exp(j\phi_i)$ such that for real α_i $(0 \leq \alpha_i \leq G)$ there results from (19)

$$\overline{|y_{MV}|^2} = \begin{cases} G\sigma_0^2 / \left[1 - \dfrac{(SNR)_1 \alpha_1^2}{1 + (SNR)_1} - \dfrac{(SNR)_2 \alpha_2^2}{1 + (SNR)_2} \right] & \text{for } |C_{12}|^2 = 0 \\[4ex] G\sigma_0^2 / \left[1 - \dfrac{1}{1 + (SNR)_1 + (SNR)_2} \right. \\ \left. [(SNR)_1 \alpha_1^2 + (SNR)_2 \ \alpha_2^2 + \right. \\ \left. 2(SNR)_1^{\frac{1}{2}} (SNR)_2^{\frac{1}{2}} \ \alpha_1 \alpha_2 \ \cos(\phi_1 - \phi_2) \right] \\ \qquad \text{for } |C_{12}|^2 = 1 \end{cases}$$ (21)

and from (20)

$$\overline{|y_C|^2} = \begin{cases} G\sigma_0^2 \ [(SNR)_1 \ \alpha_1^2 + (SNR)_2 \ \alpha_2^2 + 1] \ \text{for} \ |C_{12}|^2 = 0 \\[2ex] G\sigma_0^2 \ [(SNR)_1\alpha_1^2 + (SNR)_2\alpha_2^2 + \\[1ex] \qquad 2(SNR)_1(SNR)_2\alpha_1\alpha_2 \ \cos(\phi_1-\phi_2) + 1] \\[1ex] \qquad \text{for} \ |C_{12}|^2 = 1 \end{cases} \qquad (22)$$

where $(SNR)_i = GS_i/\sigma_0^2$ is the post-filter signal-to-noise ratio for $\underline{D} = \underline{E}_i$ with only the ith spectral component present. It is noted from (21) for $|C_{12}|^2 = 1$ that the filter output power for $\underline{H}_{MV}$ estimating s_1 ($\alpha_1 \simeq G$ and $\alpha_2 \simeq 0$) is reduced as $(SNR)_2$ increases. This represents a suppression of the filtered signal even though $\underline{D}$ is well matched to $\underline{E}_1$. This is because the minimum variance filter assumes that s_1 and s_2 are uncorrelated. This can be seen by considering the sensitivity expression

$$\frac{\partial \overline{|y_{MV}|^2}}{\partial \alpha_1^2}\Bigg|_{\alpha_2=0} = [1 + (SNR)_1 + (SNR)_2] [G\sigma_0^2(SNR)_1]/$$

$$[1 + (1 - \alpha_1^2)(SNR)_1 + (SNR)_2]^2 \qquad (23)$$

which is a monotonically decreasing function of $(SNR)_2$. No such signal suppression effect is present in either conventional filtering or signal set extraction using eigenvalue(-vector) component relating.

IV. CONCLUSION

A spectrum analysis technique based on an orthogonal decomposition of the cross frequency correlation matrix has been proposed and compared to both conventional linear and minimum variance spectrum analysis. Orthogonal component spectrum analysis exhibits a capability for simultaneous extraction of all spectrally disjoint components characterized by coherence between the envelopes of the spectrally disjoint components. It has been shown that minimum variance spectrum analysis for this class of signals can actually lead to worse performance than conventional analysis. Orthogonal decomposition, on the other hand, exhibits detection performance governed by the ratio of the total signal power in an envelope coherent spectral set to the noise level in a single spectral resolution cell. As such, orthogonal spectrum analysis would have application to the extraction of a limited,

but perhaps interesting, class of signals in both frequency
and spatial wavenumber analysis [5, 6]. Finally, adaptive
algorithms based on gradient search techniques are available
for realizing the orthogonal component spectrum analysis technique described herein [7]. These algorithms will circumvent
the necessity of first estimating and storing the CFC matrix
with subsequent orthogonal decomposition.

ACKNOWLEDGMENTS

The author wishes to thank Edward Eby, Ronald Kneipfer,
and Joseph Wolcin of the Naval Underwater Systems Center,
New London, Connecticut, U. S. A., for their many helpful discussions and criticisms regarding the contents of the work
reported herein.

REFERENCES

1.   Roberts, R. A., "On the Detection of Signals Known Except
     for Phase," IEEE Trans. on Information Theory, Vol. IT-11,
     No. 1, January 1965, p. 76.

2.   Owsley, N. L., "The Parametric Sequential Classification of
     Spectral Patterns with Application to Signal Detection and
     Extraction," IEEE EASCON Proceedings, pp. 585-592, 1968.

3.   Nuttall, A. H., "Detection Capabilities of Several Phase-
     Processing Receivers," NUSC Report No. 4529, New London,
     Conn., 13 July 1973.

4.   Okamoto, M., "Optimality of Principle Components," Multi-
     variate Analysis, Proceedings of an International Symposium,
     Editor P. R. Krishnaia, Academic Press, 1967, pp. 673-685.

5.   Cox, H., "Resolving Power and Sensitivity to Mismatch of
     Optimum Array Processors," Signal Processing (Proceedings
     of NATO Institute), McGraw Hill, 1973.

6.   Owsley, N. L., "Noise Cancellation in the Presence of
     Correlated Signal and Noise," NUSC Report 4639, New London,
     Conn., 11 January 1974.

7.   Owsley, N. L., "A Recent Trend in Adaptive Array Processing:
     Constrained Adaptation," Signal Processing (Proceedings of
     NATO Institute), Academic Press, 1973.

A CONSTRAINED ADAPTIVE BEAMFORMER TOLERANT OF ARRAY GAIN AND PHASE ERRORS

J. M. McCool

Fleet Engineering Department, Naval Undersea Center,
San Diego, California

ABSTRACT. This paper describes a new method of constrained adaptive beamforming whose performance is relatively unaffected by array gain and phase errors. The method consists of setting certain of the adaptive parameters of a conventional adaptive beamformer to zero. It provides a means of rejecting large signals incident on the side lobes of an antenna array while preserving small broadband and narrowband and large broadband signals incident in the steering direction. In exchange for error tolerance broadband superdirectivity and large narrowband signal reception are lost.

Figure 1 shows a broadband adaptive antenna system consisting of an antenna array of K elements; a conventional time-delay-and-sum beamformer; a bank of K adaptive transversal filters, one corresponding to each array element, whose tap gains or "weights" are governed by an adaptive algorithm; and a differencer. The outputs of the array elements are routed in parallel to the conventional beamformer, which steers a beam in the desired "look" direction, and to the bank of adaptive filters. The conventional beamformer output forms the desired response d_j of the adaptive process, where j is the time index. The sum y_j of the adaptive filter outputs, a linear combination of delays of the element waveforms, is subtracted from the desired response to form the system output z_j, which also serves as the "error" signal ε_j for the adaptive algorithm.

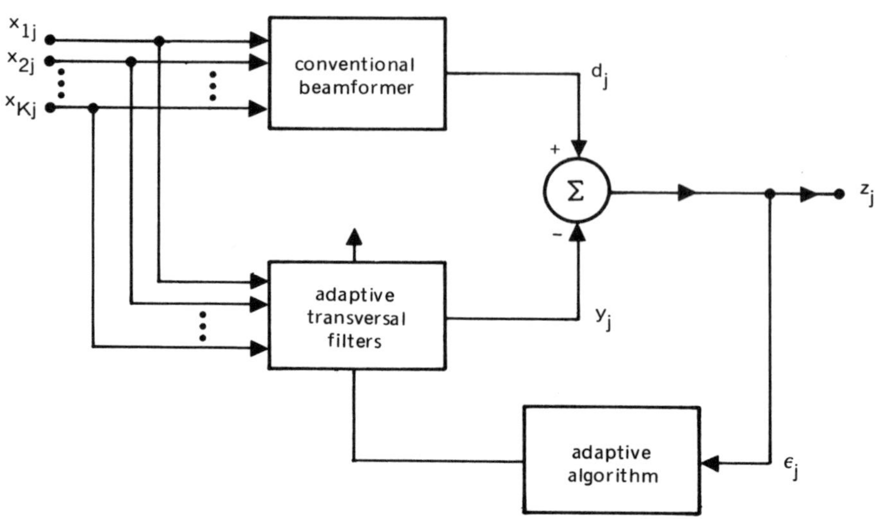

Figure 1. Broadband adaptive antenna system.

This system, first described in reference 1, minimizes the
mean square of the difference between the conventional beamformer
output d_j and the sum y_j of the adaptive filter outputs. Array
performance in the presence of directional interference is en-
hanced as long as the conventional beamformer and adaptive filter
inputs are derived from different array elements. If the inputs
are derived from the same elements, however, the adaptive filter
output forms an exact model of the conventional beamformer output, resulting in a mean square difference of zero and no response. This problem can be avoided by the method described in
references 2 and 3, which is to constrain the sum of each column
of filter weights to zero. Such a procedure guarantees that the
filter output due to a signal of the same magnitude and phase at
each element will be zero. That is, for an ideal array it guarantees that the response in the steering direction will be that
of the conventional beamformer. In an actual array, however, the
directional response will be a function of array gain or phase
errors. This is an example of the well known sensitivity of constrained algorithms to array tolerances.

An alternate method of avoiding the exact model problem,
previously described in reference 4 for broadband signals and
interference, is to remove from the adaptive system the nonzero
weights of the exact model. This procedure eliminates sensitivity to element gain errors. Sensitivity to phase errors is eliminated by also removing the weights within the delay tolerance
associated with each element. The result is a constrained
adaptive beamformer whose performance is relatively unaffected
by array gain or phase errors.

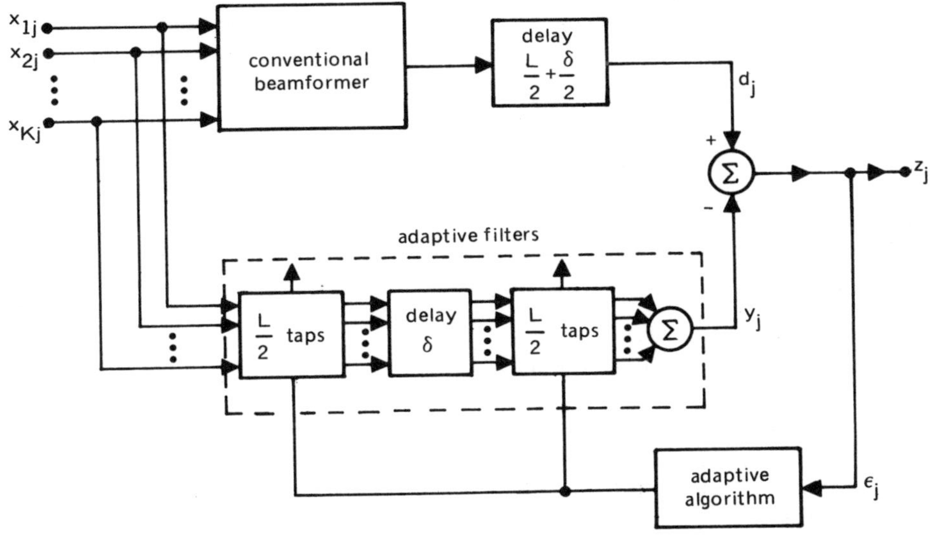

Figure 2. Constrained adaptive beamformer tolerant of array
         gain and phase errors.

Figure 2 is a block diagram of the error-tolerant constrained beamformer as applied to a line array with its beam
steered in the direction normal to the array's axis.* This system is identical to the system shown in figure 1 except in two
respects. First, the adaptive filters are divided into two
halves, each of length L/2 taps, separated by a delay of δ samples.+ Second, the desired response is delayed by an amount
equal to half the total adaptive filter delay or L/2 + δ/2
samples.

The operation of the beamformer system of figure 2 can be
understood by examining the following relationships. The system
output or error signal ε_j, from reference 1, is given by

$$\varepsilon_j = d_j - \underline{W}^T \underline{X}_j, \tag{1}$$

---

*For steering angles other than normal to the array's axis, as
 indicated in reference 4, the adaptive filter inputs are derived
 from the conventional beamformer delays rather than directly
 from the element outputs.
+This delay is effected by inserting columns of zeros in the
 adaptive filter weight coefficient matrix (ref. 4).

where $\underset{\sim}{W}$ is the weight vector of the adaptive filters and $\underset{\sim}{X}$ is the
signal vector. The weight vector that minimizes the expected
value of the square of the error $E[\varepsilon^2]$ is

$$\underset{\sim}{W} = \underset{\sim}{R}^{-1}\underset{\sim}{P}, \tag{2}$$

where $\underset{\sim}{R}$ is the autocorrelation matrix of $\underset{\sim}{X}$ and $\underset{\sim}{P}$ is the vector of
cross correlations between the components $x_{1,j},\ldots,x_{K*L,j}$ of the
signal vector $\underset{\sim}{X}$ and the desired response d_j defined by

$$\underset{\sim}{P} = (E[x_{1,j}d_j],\ldots, E[x_{K*L,j}])^T. \tag{3}$$

If a broadband signal with a white spectrum is incident on
the array, no correlation will exist between any component of $\underset{\sim}{X}$
and the desired response until the angle of incidence is such
that $\Delta > \delta - 1$, where Δ is the signal delay across the array's
aperture. In the region $0 < \Delta < \delta - 1$, therefore, no component
of $\underset{\sim}{P}$ will be other than zero, and hence $\underset{\sim}{P}$ and the solution to
equation (2) will be zero. Under these circumstances the error
signal is equivalent to the desired response, and equation (1)
reduces to $\varepsilon_j = d_j$. When $\Delta > \delta - 1$, $\underset{\sim}{P}$ becomes other than zero,
and the possibility of rejection exists. The effect of array
gain or phase errors is to alter the relationship between source
angle and delay Δ.

Figure 3 is a plot of the broadband directional response of
the adaptive beamformer system of figure 2. This plot, obtained
by solving equations (1) and (2) numerically, shows the gain of
the beamformer as a function of steering angle for an incident
broadband signal with a white spectrum.* The array comprises ten
uniformly weighted elements in a line configuration; element
spacing is one-half wavelength at the sampling frequency, and the
end elements are offset one-quarter of a wavelength from the
array's axis. Independent broadband noise with a white spectrum
is incident on each element. The signal-to-noise ratio is 14 dB.
The conventional beamformer, whose response pattern is shown by a
dotted line, is steered in the direction normal to the array's
axis. Note that the peak gain of the two beams is approximately
the same but that beyond 20 degrees the sidelobe rejection of
the adaptive beam is superior.

---

*The terms "signal" and "signal-to-noise ratio" are used in this
 and the following example for convenience. It should be noted,
 however, that in both examples the "signal" is "interference"
 when it is not incident in the steering direction.

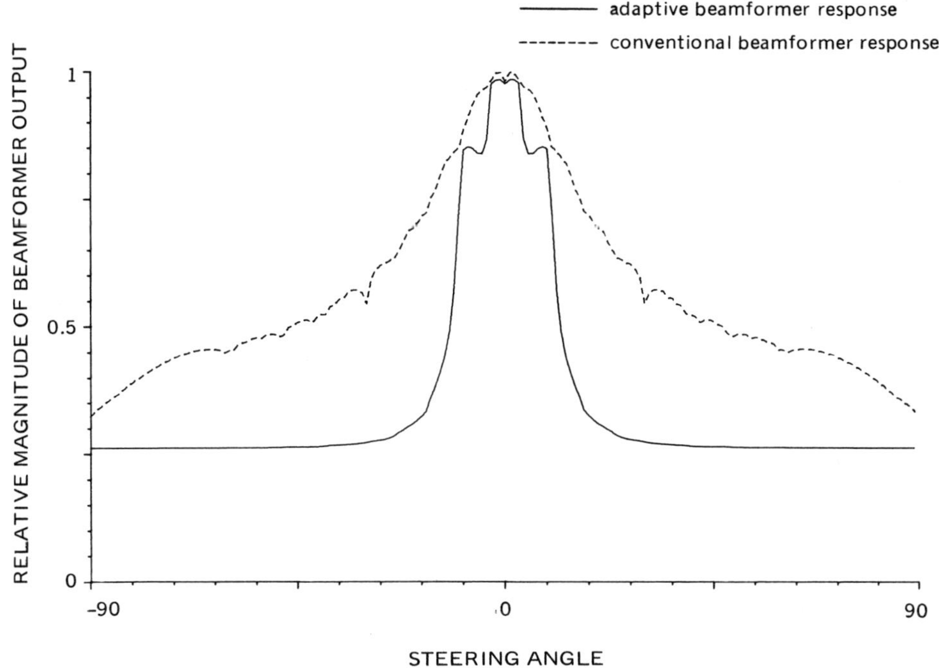

Figure 3. Broadband directional response of constrained adaptive
           beamformer tolerant of array gain and phase errors.

For narrowband directional signals the above argument is
invalid because all components of X are correlated. In the case
of a directional source of frequency f, however, with independent
white noise at each element of the receiving array, equation (2)
can be solved analytically by the method of reference 5 to obtain

$$\underset{\sim}{W}_{\ell+k*L} = \underset{\sim}{W}_0 \cos\ [-2\pi(f/f_s)\ell - \phi_k],\ 1 \leq \ell \leq (L/2),\ 1 \leq k \leq K,$$

$$= \underset{\sim}{W}_0 \cos\ [-2\pi(f/f_s)(\ell + \delta) - \phi_k],$$

$$(L/2) \leq \ell \leq L,\ 1 \leq k \leq K, \tag{4}$$

where

$$\underset{\sim}{W}_0 = \frac{S/N}{\frac{K*L}{2}(S/N) + 1}. \tag{5}$$

In these expressions ℓ is one of L adaptive filter taps; k is one
of K array elements or corresponding adaptive filters; f_s is the
sampling frequency; and ϕ_k is the phase of the k<u>th</u> element relative to the desired response.

Figure 4 shows the frequency response of the adaptive beamformer of figure 2 in the steering direction in the presence of a
narrowband signal with a frequency of 0.22 relative to the sampling frequency. This plot, which shows the magnitude of the
digital Fourier transform of the square of the error signal ϵ^2 or
output power as a function of frequency, was obtained by solving
equations (1) and (4) numerically. The array is the same as in
the previous example except that it consists of eight elements;
each adaptive filter has 64 taps or weights. Independent white
noise is incident on each element, and the signal-to-noise ratio
is -24 dB. Note the null at the frequency of the narrowband signal. The output signal-to-noise ratio of the adaptive system at
this frequency is given by

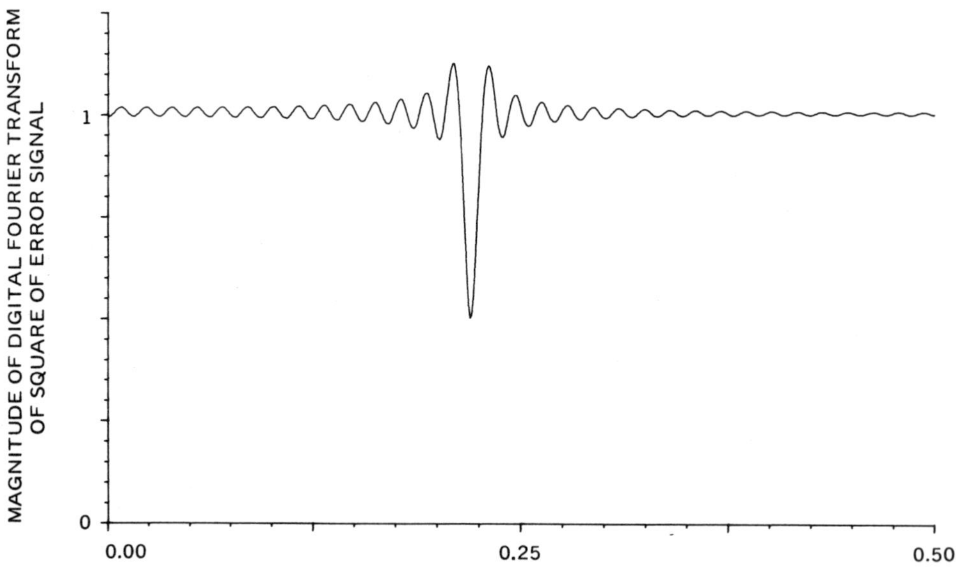

Figure 4. Frequency response of constrained adaptive beamformer
           in steering direction in the presence of a narrowband
           signal.

$$(S/N)_{out} = \frac{K(S/N)}{K^2(L/2)(S/N)^2 + 2K(L/2)^{1/2}(S/N) + 1}. \tag{6}$$

For small signal-to-noise ratios, S/N << 1, this expression simplifies to

$$(S/N)_{out} = K(S/N), \tag{7}$$

the output signal-to-noise ratio of the conventional beamformer; for large signal-to-noise ratios, S/N >> 1, it becomes

$$(S/N)_{out} = \frac{1}{K*L(S/N)}. \tag{8}$$

At frequencies well removed from that of the interference the gain is nearly equal to that of the conventional beamformer output. The frequency resolution can be shown to be the maximum realizable with a finite-impulse-response (FIR) filter of L taps.

The adaptive beamformer system described in this paper provides a means of rejecting large signals incident on the side lobes of an antenna array regardless of array gain or phase errors. The system preserves small narrowband and large or small broadband signals incident in the steering direction. In exchange for error tolerance, however, broadband superdirectivity and large narrowband signals incident in the steering direction are lost.

REFERENCES

1. B. Widrow, P. Mantey, L. Griffiths, and B. Goode, "Adaptive antenna systems," _Proc. IEEE_, vol. 55, pp. 2143-2156, Dec. 1967.
2. O.L. Frost III, "An algorithm for linearly constrained adaptive array processing," _Proc. IEEE_, vol. 60, pp. 926-935, Aug. 1972.
3. L.J. Griffiths, "Adaptive Monopulse Beamforming," _Proc. IEEE_ (forthcoming).
4. B. Widrow and J.M. McCool, "A comparison of adaptive algorithms based on the methods of steepest descent and random search," _IEEE Trans. Antennas and Propagation_ (forthcoming).
5. B. Widrow, J.R. Glover, J.M. McCool, _et al._, "Adaptive noise cancelling: Principles and Applications," _Proc. IEEE_, vol. 63, pp. 1692-1716, Dec. 1975.

SIGNAL EXTRACTION ALGORITHMS FOR ADAPTIVE PROCESSING OF ARRAY DATA

D.R. Farrier* and T.S. Durrani**

*Atomic Energy Establishment, Winfrith, U.K.
**Department of Electronic Science, University of
Strathclyde, Glasgow, U.K.

ABSTRACT. This paper presents a set of adaptive algorithms for
signal power estimation and signal extraction of array data based
on a MMSE criterion.   The algorithms are particularly useful for
processing low frequency signals, and have the advantage of requiring no a priori signal statistics.   Computer simulation results
are included which verify the theoretical derivations and compare
the performance of the algorithms under practical conditions of
interest.

1. INTRODUCTION

It is well known that for low frequency data arriving at an array
of sensors the conventional phased array beam former such as the
DIMUS[1] processor gives poor resolution due to the very large
array beam width, and leads to poor output signal to noise gains.
Devising an adaptive processor based on a minimum mean square error
(MMSE) criterion for beam forming can be difficult as it requires
a priori signal statistics[2,3].   These are seldom available.   In
this paper we present a technique for estimating signal power from
different directions.   An adaptive MMSE processor is derived for
beam forming in several directions simultaneously and adapting
sequentially until the beam formed outputs are consistent with the
received sensor data.   Such a processor gives an improved gain as
compared to the Griffiths algorithm[3], and is less sensitive to
statistical variations in the data.   Moreover it does not need any
a priori signal statistics.   Several versions of the processor are
developed here, in particular a set of algorithms are proposed which
allow fast convergence for low frequency data.

2. SIGNAL MODEL

Consider an array of n omnidirectional sensors receiving directional
signals from the environment.   Let $y_i(\omega)$ be the frequency domain
representation of the output at the ith sensor obtained by filtering
or after performing block F.F.T.   Let $x_i(\omega)$; i=1,2, ... m be
signals from m discrete directions incident on the array.   Then

$$\underline{y}(\omega) = \underline{H}(\omega)\ \underline{x}(\omega) + \underline{v}(\omega) \tag{1}$$

where
$$\underline{y}^T(\omega) = \begin{bmatrix} y_1(\omega) & y_2(\omega) & \ldots & y_m(\omega) \end{bmatrix}$$
$$\underline{x}^T(\omega) = \begin{bmatrix} x_1(\omega) & x_2(\omega) & \ldots & x_n(\omega) \end{bmatrix}$$
$$\underline{v}^T(\omega) = \begin{bmatrix} v_1(\omega) & v_2(\omega) & \ldots & v_n(\omega) \end{bmatrix}$$

T denotes transpose, and the vector $\underline{v}(\omega)$ is the frequency domain
representation of the additive Gaussian distributed measurement
noise and the matrix $\underline{H}(\omega)$ is a function of the array geometry and
the signal directions.   In general, $\underline{y}$, $\underline{x}$ and $\underline{v}$ will be complex
valued.   Figure 1 shows a simple arrangement where n = m = 2 and
$\underline{H}(\omega)$ is given by

$$\underline{H}(\omega) = \begin{bmatrix} 1 & e^{-j\omega\tau_{12}} \\ e^{-j\omega\tau_{21}} & 1 \end{bmatrix} \tag{2}$$

A typical beamforming problem is to extract $\underline{x}(\omega)$ from $\underline{y}(\omega)$[3,4].
The phased array beamformer is given by $\underline{x}^\dagger(\omega)$ where

$$\underline{x}^\dagger(\omega) = \frac{1}{m}\underline{H}^{*T}\underline{y}(\omega) \tag{3}$$

* denotes complex conjugate.

For low frequency processing the beamwidth of any beamformer
is wide and so the total number of beamform directions (n) may be
chosen as small.   Although the signals incident on the array may
arise from any direction, the processor may model them as coming
from n discrete directions.   Most processors use this property,
since only a finite number of beams are desired.

Since signals and measurement noise are independent, we have

$$E\begin{bmatrix} \underline{y}\ \underline{y}^{*T} \end{bmatrix} = \underline{H}\ E\begin{bmatrix} \underline{x}\ \underline{x}^{*T} \end{bmatrix} \underline{H}^{*T} + E\begin{bmatrix} \underline{v}\ \underline{v}^{*T} \end{bmatrix} \tag{4}$$

i.e. $\quad \underline{R}_y = \underline{H}\ \underline{R}_x\ \underline{H}^{*T} + \underline{R}_v \tag{5}$

As the measurement noise depends on the sensors, $\underline{R}_v$ will often
be known, a priori.   If the measurement noise is independent from
sensor to sensor and the sensors are of similar construction, then
$\underline{R}_v = \sigma_v^2\ \underline{I}$, where $\underline{I}$ is an identity matrix.

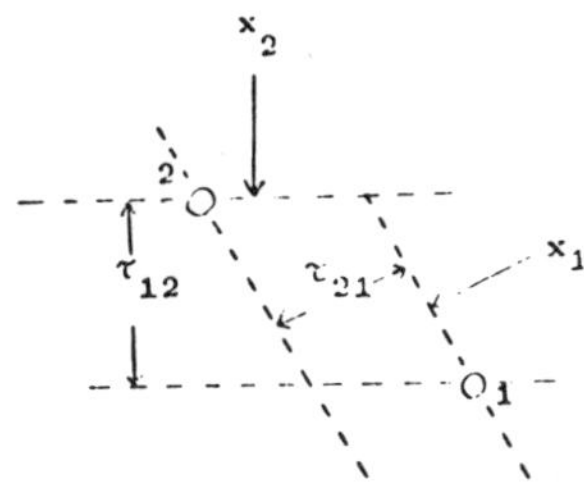

Fig. 1.   A signal model. (τ = time delay.)

Two distinct problems will be considered in this paper:

(a) Deriving recursive estimators for $\underline{R}_x$.   If the signal power from each direction is only required, then all that is necessary is $\underline{R}_x$.

(b) Obtaining a MMSE estimator for the signal vector $\underline{x}$.   It may be shown [6] that that the MMSE estimator is given by

$$\underline{G} = \underline{R}_x \, \underline{H}^{*T} \, (\underline{H} \, \underline{R}_x \, \underline{H}^{*T} + \underline{R}_v)^{-1} \tag{6}$$

where
$$\hat{\underline{x}} = \underline{G} \, \underline{y} \tag{7}$$

3. SIGNAL POWER ESTIMATION ALGORITHM

Consider $\hat{\underline{R}}_x$ to be an estimate of $\underline{R}_x$.   An estimate $\hat{\underline{R}}_y$ may then be obtained from Eq. 5, if R_v is known

$$\hat{\underline{R}}_y = \underline{H} \, \hat{\underline{R}}_x \, \underline{H}^{*T} + \underline{R}_v \tag{8}$$

The estimate $\hat{\underline{R}}_x$ may be improved by the following adaptation:

$$\hat{\underline{R}}_{x_{k+1}} = \hat{\underline{R}}_{x_k} + \gamma_k \, \underline{A}(\hat{\underline{R}}_y - \hat{\underline{R}}_y)\underline{B} \tag{9}$$

Where γ_k (> o) determines the step size at each iteration and where $\underline{A}$ and $\underline{B}$ are chosen to ensure convergence of $\hat{\underline{R}}_{x_k}$.   This algorithm adjusts $\hat{\underline{R}}_x$ if $\hat{\underline{R}}_y$ does not equal $\underline{R}_y$.

If $\underline{A} = \underline{H}^{*T}$, $\underline{B} = \underline{H}$ and R_y is replaced by the instantaneous covariance estimate, $\underline{y}_{k+1} \, \underline{y}^{*T}_{k+1}$; Eq. 9 becomes

Algorithm 1
$$\hat{\underline{R}}_{x_{k+1}} = \hat{\underline{R}}_{x_k} + \gamma_k \, \underline{H}^{*T} \, (\underline{y}_{k+1} \, \underline{y}^{*T}_{k+1} - \underline{H} \, \hat{\underline{R}}_x \, \underline{H}^{*T} - \underline{R}_v)\underline{H} \tag{10}$$

It may be noted that the stability of the mean is ensured since $\underline{H}^{*T}\,\underline{H}$ is non-negative definite[5].

Eq. 8 may be simplified by using the property that signals from different directions are uncorrelated. In this case R_x is a diagonal matrix, thus a reduction in computing effort may be obtained if $\hat{R}_x$ is also taken to be diagonal.

4. EXTENSIONS FOR LOW FREQUENCY DATA ANALYSIS

For low frequency data (i.e. wavelength, $\lambda \gg$ largest array dimension) any incident signal only suffers a very small phase change across the array. In this case $\underline{H}^{*T}\,\underline{H}$ will have one eigenvalue much larger than all others. The eigenvector corresponding to this eigenvalue will be $(11 \ldots 1)^T/\sqrt{n}$. In this section, we will make use of this eigenvector to develop an algorithm particularly useful for low frequency processing. A choice for matrix $\underline{A}$(c.f. Eq. 9) which is more suitable than $\underline{H}^{*T}$ will be sought to ensure faster convergence.

$$\text{Let } \underline{H}^{*T}\,\underline{H} = \underline{Q}\,\underline{\Lambda}\,\underline{Q}^{*T} \tag{11}$$

where $\underline{\Lambda}$ is a diagonal matrix of eigenvalues and
$\underline{Q}$ is a matrix of eigenvectors,

$$\text{and} \quad \underline{H}^{*T}\,\underline{H} = (\sigma_1^2 - \sigma^2)\,\frac{\underline{1}\,\underline{1}^T}{n} + \underline{Q}\,\underline{\Lambda}_1\underline{Q}^{*T} \tag{12}$$

where σ_1^2 is the largest eigenvalue; $\underline{1}^T = (11 \ldots 1)$, $(1\times n)$ vector $\underline{\Lambda}_1$ is the diagonal matrix $\underline{\Lambda}$, with σ_1^2 replaced by σ^2, for any $\sigma^2\,(\geqslant 0)$. Now consider a matrix T, such that

$$\underline{T} = (\sigma_1^2 - \sigma^2)\,\frac{\underline{1}\,\underline{1}^T}{n} + \sigma^2\,\underline{Q}\,\underline{I}\,\underline{Q}^{*T}$$

$$= (\sigma_1^2 - \sigma^2)\,\frac{\underline{1}\,\underline{1}^T}{n} + \sigma^2\,\underline{I} \quad (\text{Since } \underline{Q}\,\underline{Q}^{*T} = \underline{I})$$

By the matrix inversion Lemma[7]

$$\underline{T}^{-1} = \frac{1}{\sigma^2}\,(\underline{I} - \frac{\sigma_1^2 - \sigma^2}{\sigma_1^2\,n}\,\underline{1}\,\underline{1}^T) \tag{13}$$

$\underline{T}^{-1}\,\underline{H}^{*T}\,\underline{H}$ equals $\underline{Q}\,\underline{\Lambda}_1\,\underline{Q}^{*T}/\sigma^2$, whereas $\underline{H}^{*T}\,\underline{H}$ is $\underline{Q}\,\underline{\Lambda}\,\underline{Q}^T$.
$\underline{T}^{-1}\,\underline{H}^{*T}\,\underline{H}$ is a better approximation to the identity matrix $\underline{I}$ than $\underline{H}^{*T}\,\underline{H}$ (the eigenvalues of $\underline{T}^{-1}\,\underline{H}^{*T}\,\underline{H}$ have a smaller dynamic range than $\underline{H}^{*T}\,\underline{H}$).

The advantage of this is that γ_k can be made larger than previously while still ensuring stability. The smaller eigenvalues will then converge faster (see Widrow et al[2]). $\underline{T}^{-1}\,\underline{H}^{*T}$ is easy to compute because of Eq. 13. For all $\sigma^2(\geqslant 0)$, $\underline{T}^{-1}\,\underline{H}^{*T}\,\underline{H}$ is non-negative definite and therefore correct values for σ^2 are not necessary for convergence. However if necessary, values for σ^2 may be stored for various frequencies, and σ_1^2 may be obtained from $\underline{1}^T\,\underline{H}^{*T}\,\underline{H}\,\underline{1}/n$.

For processing low frequency data the algorithm may be recast as

Algorithm 2

$$\hat{\underline{R}}_{x_{k+1}} = \hat{\underline{R}}_{x_k} + \gamma_k \underline{T}^{-1} \underline{H}^{*T} (\underline{y}_{k+1} \underline{y}^{*T}_{k+1} - \underline{H} \hat{\underline{R}}_{x_k} \underline{H}^{*T} - \underline{R}_v) \underline{H} \underline{T}^{-1} \tag{14}$$

The implementation of the algorithm is only marginally more costly in computing time than Eq. 10 but its convergence is faster (see computed results). The ease of implementation is due to the matrix $\underline{1}\,\underline{1}^T$ and the matrix inversion lemma. In this section an approximation of $\underline{H}^{-1}\,\underline{H}$ as $\underline{T}^{-1}\,\underline{H}^{*T}\,\underline{H}$ has been used. The exact version would be both computationally expensive and illconditioned for low frequency operation. For other situations of interest, different approximations may be available (for instance, the technique is similar to nulling a known interference source[8]).

5. A FEEDBACK PROCESSOR

In figure 2(a) a diagram of a signal power estimator using a conventional phased array processor (Eq. 3) is shown. This processor is an open-loop system, with no error correcting facility (the averager merely smoothes out the noise), and the estimates are not compared with the data, $\underline{y}_k$.

Comparison with the proposed processor shows that a feedback loop has been included (see figure 2(b)). The estimate is also passed through a second filter (to ensure stability) before being compared with the data for an error measure. Hence the adaptive processor employs incoming data to correct errors in $\hat{\underline{R}}_x$. The averaging circuit in (a) corresponds to the updating processor in

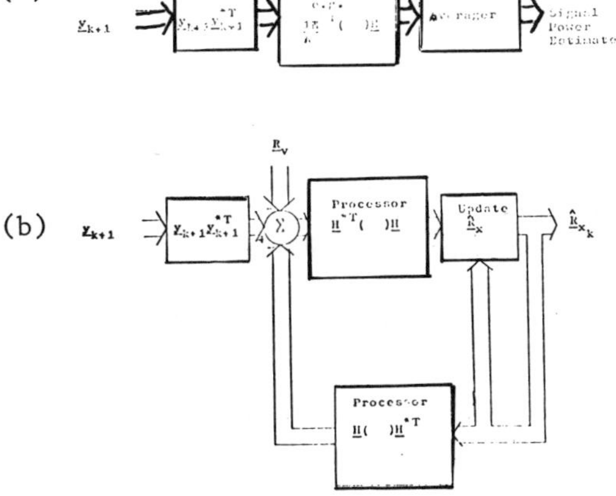

Fig. 2. Block diagrams for processors. (a), Conventional processor; (b), Adaptive signal power estimator.

(b).   Both the conventional processor ($=\underline{H}^{*T}(\ )\ \underline{H}$), and the updating processor are within the feedback loop.   This ensures that any errors are corrected.

Furthermore, it may be noticed that no processor in (a) (even $\underline{H}^{-1}(\ )\ \underline{H}^{*T^{-1}}$) ensures an unbiased output.   Only a feedback processor is capable of giveing an unbiased estimate.

6. A FILTER FOR SIGNAL EXTRACTION

In this section a recursive algorithm for forming a MMSE estimator for signal extraction will be developed for the observation model of Eq. 1(c.f. Eq. 6-7).

If $\underline{R}_x\ \underline{H}^{*T}$ was known a priori the filter could be found iteratively by

$$\underline{G}_{k+1} = \underline{G}_k + \gamma_k\ \underline{\Delta G}_k \tag{15}$$

where

$$\underline{\Delta G}_k = \underline{R}_x\ \underline{H}^{*T} - \underline{G}_k\ \underline{R}_y \tag{16}$$

Since $\underline{R}_x\ \underline{H}^{*T}$ is not usually known and only $\underline{R}_y$ may be estimated Eq. 15 cannot be implemented.   Even recasting the definition of $\underline{R}_y$ as $\underline{R}_y - \underline{R}_v = \underline{H}\ \underline{R}_x\ \underline{H}^{*T}$ does not allow evaluation of $\underline{R}_x$ as $\underline{H}^*$ may be ill-conditioned.   However noting that the algorithm of Eq. 15 may converge if we premultiply $\underline{\Delta G}_k$ by any non-negative definite matrix, and following the arguments of the previous section, we obtain

Algorithm 3

$$\underline{G}_{k+1} = \underline{G}_k + \gamma_k\ \underline{H}^{*T}(\underline{y}_{k+1}\ \underline{y}_{k+1}^{*T} - \underline{R}_v - \underline{H}\ \underline{G}_k\ \underline{y}_{k+1}\ \underline{y}_{k+1}^{*T}) \tag{17}$$

To operate on the measurements, the individual terms in Eq. 17 may be implemented as follows:

$$\hat{\underline{x}}_k = \underline{G}_k\ \underline{y}_k\ ;\ \hat{\underline{y}}_k = \underline{G}_k\ \hat{\underline{x}}_k\ ;\ \underline{e}_k' = \underline{y}_k - \hat{\underline{y}}_k$$

$$\underline{G}_{k+1} = \underline{G}_k + \gamma_k\ \underline{H}^{*T}\ (\underline{e}_{k+1}'\ \underline{y}_{k+1}^{*T} - \underline{R}_v) \tag{18}$$

It may be noted that Eq. 18 only consists of operations similar to those required to find $\hat{\underline{x}}_k$ when $\underline{G}$ is known (Eq. 7).   No inversions or matrix-matrix multiplications are necessary.   For m sensors and n beamform directions only (5m+1)n complex multiplications are required to implement Eq. 18.   By a careful choice of γ_k, this may be further reduced.   This represents approximately 5m complex multiplications for each beamform direction.   For low frequency data (wavelength, $\lambda \gg$ largest array dimension) a similar approach to that of section 4 may be employed by using $\underline{T}^{-1}\ \underline{H}^{*T}$ instead of $\underline{H}^{*T}$.   Then the algorithm for $\underline{G}_k$ becomes:

Algorithm 4

$$\underline{G}_{k+1} = \underline{G}_k + \gamma_k\ \underline{T}^{-1}\ \underline{H}^{*T}\ (\underline{e}_{k+1}'\ \underline{y}_{k+1}^{*T} - \underline{R}_v) \tag{19}$$

Again there is only a slight increase in complexity.   The computational advantages mentioned in section 4 also apply to Eq. 19.

If an estimate $\underline{R}_x$ is already available (i.e. from the signal power estimation algorithm - Eq. 10 or 14 - a simpler method may be used.   This approach is similar to that employed by Griffiths [3] :

<u>Algorithm 5</u>

$$\underline{G}_{k+1} = \underline{G}_k + \gamma_k \, (\underline{\hat{R}}_k \, \underline{H}^{*T} - \underline{G}_k \, \underline{y}_k \, \underline{y}_k^{*T}) \tag{20}$$

7. RESULTS OF COMPUTER SIMULATION

The algorithms proposed in this paper were tested using simulated data.   Though the techniques are not restricted to any particular array geometry, we have used a circular array of 8 equi-spaced sensors for our computations.

8 complex Gaussian variables were generated and given appropriate phase shifts to represent signals arriving at the sensors from different directions in the medium (see figure 3).   Complex Gaussian additive measurement noise on each sensor was also generated.   To test the signal power estimation algorithm, signals of power 2 (in arbitrary scaled units) were generated by each signal source except in direction "3" where the signal power was taken as 4.   Each sensor was assumed to be very noisy, generating a noise power of unity.   Since low frequency signals were of direct interest, the ratio of signal wavelength to array radius was chosen to be 15:1.

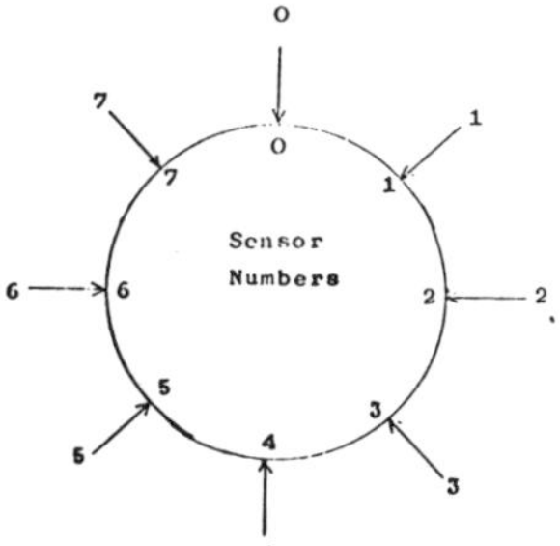

Fig. 3.   Array model. Beamform directions and nominal signal directions.

To make the simulation representative of practical situations, the beamform directions did not coincide with the signal directions. Figure 4 illustrates the output from Algorithms 1 and 2, corresponding to two different directions.   Note that the estimates obtained from Algorithm 1 are very poor even after 200 iterations. None of the outputs are representative of the signal power present, and it is very difficult to recognise that a signal of power 4 is

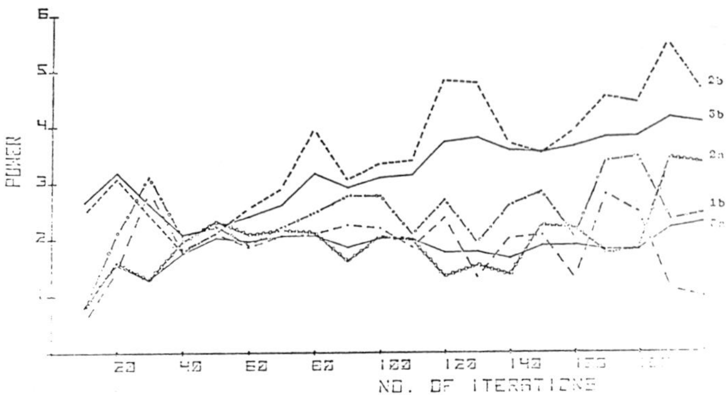

Fig. 4.   Convergence of Algorithms 1 and 2.
          a = Output from direction 7.
          b = Output from direction 3.
          1: γ_k = 0.025   Algorithm 1.
          2: γ_k = 0.4
          3: γ_k = 20/(40+k)   Algorithm 2.

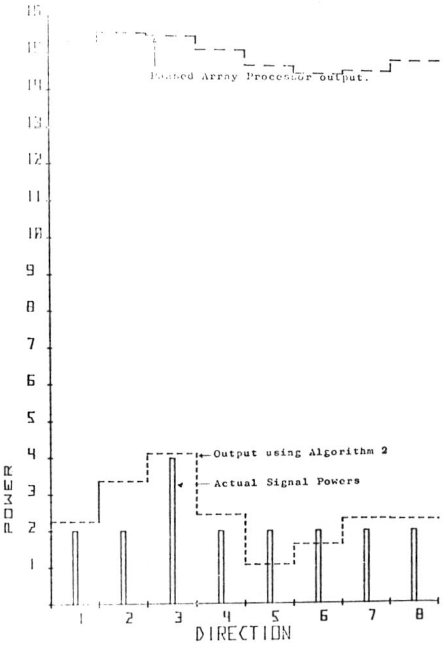

Fig. 5.   Power estimates from different processors.

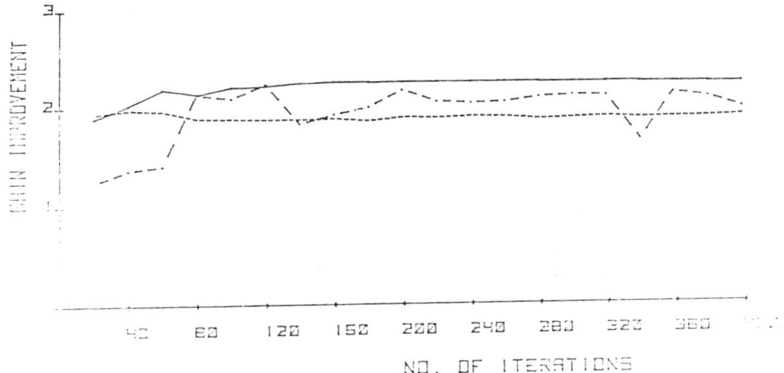

Fig. 6.   Improvements in S/N gains using algorithms 3-5.
          -----   Algorithm 3,
          _______ Algorithm 4,
          -·-·-   Algorithm 5.

present in direction 3.   Considerable improvement is obtained when
Algorithm 2 is used on the same data.   Figure 4 consists of the
algorithm output for two values of γ_k (Eq. 14).   For fixed γ_k, the
algorithm converges to the appropriate signal power, however the
fluctuations are large.   Using a stochastic approximation sequence
($\gamma_k = {}^{20}/_{40+k}$) it may be seen that the two signal sources can be
clearly distinguished after only 100 iterations.

Figure 5 shows the actual signal power output as determined
by a phased array processor and by the use of Algorithm 2.   Also
plotted on the diagram are the 'true' values of the signal power as
used in the simulation.   It is immediately obvious that the phased
array processor cannot identify the presence of any particular
source, such as the strong peak from direction 3, whereas the peak
is obvious for the latter case.   The graphs are obtained after 200
iterations of each processor.

The adaptive filtering algorithms (algorithms 3-5) were tested
by generating signals of power 2 in each direction except direction
"5" where the signal power was 8.   Measurement noise of power unity
was again included, while the ratio of signal wavelength to array
radius was taken as 14:1.   To illustrate the improvement due to
the processor, the ratio of processor gain to the gain of a phased
array is plotted in figure 6.   Here beamform directions
were not considered as coincident with signal directions.

Figure 6 shows the gain improvements for algorithms 3,4 and 5
respectively for beamforming in direction 5.   Note that Algorithm
5 has much larger variations about the desired gain.   This is one
of the disadvantages of a Griffiths-type algorithm.

The performance of Algorithm 4 is seen to be best and it leads
to very small variations about the desired gain.   In all three

cases the initial processor was phased array processor (gain improvement = 1).   Note the very fast response to all three algorithms.   For Algorithms 3 and 5 considerable improvement in gain is apparent after only 30 iterations.

9. ACKNOWLEDGEMENTS

This research was conducted under a U.K. Ministry of Defence grant. The authors would like to acknowledge many useful discussions with Mr. G.A. Lawson and Professor J.M. Nightingale.

10. REFERENCES

1.   V.C. Anderson, _Digital Array Phasing_, J.A.S.A.,Vol.32,pp.867-870, 1960.
2.   B. Widrow, _Adaptive Filters I: Fundamentals_, Stanford Electronics Labs.,Stanford,Calif.,Rept.SEL-66-126(Tech.Rept.6764-6), 1966.
3.   L.J. Griffiths, _Minimum Mean Square Error Array Processing: An Iterative Approach_, Symposium on computer processing in Communications,Polytechnic Institute of Brooklyn, 1969.
4.   O.L. Frost, _An algorithm for linearly constrained adaptive array processing_, Proc. IEEE,Vol.60,pp.925-936, 1972.
5.   J. La Salle and S. Lefchetz, _Stability by Liapinov's Direct Method_, New York,Academic Press, 1961.
6.   D.R. Farrier, _Adaptive algorithms for array processing_, Ph.D. Thesis, University of Southampton, 1975.
7.   F.A. Graybill, _Introduction to matrices with applications in statistics_, Wadsworth,Belmont,California, 1969.
8.   V.C. Anderson and P. Rudnick, Rejection of a coherent arrival at an array, J.A.S.A.,Vol.41,pp.406-410, 1969.

PROPERTIES OF THE DISCRETE FOURIER TRANSFORM (DFT)

B.W. Conolly

University of London, Chelsea College
London, England

ABSTRACT. This paper will be divided into two parts. The first
is intended as a companion to a tutorial session on those basic
properties of the DFT which lead to Fast Fourier Transform
algorithms. The second part will range more widely, in particular
considering ways in which certain less well-known properties of
the DFT could be turned to practical use. The two parts are
independent.

PART I.

THE MATHEMATICAL BASIS OF THE FAST FOURIER TRANSFORM (FFT)
ALGORITHMS

1. The harmonic analysis of a signal sampled at finite equally
spaced intervals leads to the reciprocal formulae

$$f_r = \sum_{n=-N}^{+N} c_n w^{+rn} \tag{1.1}$$

$(r = 0,1,2,\ldots,2N-1)$ $(w = \exp 2\pi i/2N$, the $2N^{th}$ fundamental root
of unity), and

$$c_n = \frac{1}{2N} \sum_{r=0}^{2N-1} f_r w^{-rn}. \tag{1.2}$$

These are analogues of the continuous time pair:

G. Tacconi (ed.), Aspects of Signal Processing, Part 2, 495-516. All Rights Reserved.
Copyright © 1977 by D. Reidel Publishing Company, Dordrecht-Holland.

$$f(t) \sum_{-\infty}^{\infty} c_n e^{int}, \tag{1.3}$$

$$c_n = \frac{1}{2\pi} \int_0^{2\pi} e^{-int} f(t)\, dt. \tag{1.4}$$

By suitable shift of origin and change of notation we derive from
(1.1) and (1.2) the following definition of the one-dimensional
mod t DFT of the sequence $(a_0, a_1, \ldots, a_{t-1})$ of t real numbers:
This is another sequence of t (complex) numbers a_s^* defined by

$$a_s^* = \sum_{r=0}^{t-1} a_r\, w^{rs}, \tag{1.5}$$

$(s = 0,1,2,\ldots,t-1)$
where $w = \exp(2\pi i/t)$ is the tth fundamental root of unity. The
inversion formula is

$$a_r = \frac{1}{t} \sum_{s=0}^{t-1} a_s^*\, w^{-rs}, \tag{1.6}$$

$(r = 0,1,2,\ldots,t-1).$

Other properties will be mentioned in Part II. Here we
concentrate on the DFT property that leads to FFT algorithms.
This is that a one-dimensional transform can, if t is composite,
be converted into a multi-dimensional transform. To follow the
argument we proceed to define a multi-dimensional DFT.

2.   The following is modelled on the exposition of [4]. A multi-
dimensional mod $(t_1, t_2, \ldots, t_n)$ DFT of the n-dimensional array of
$\tau = t_1 \cdot t_2 \cdots t_n$ real numbers $a_{\underline{r}}$ is the n-dimensional array of
complex numbers $a_{\underline{s}}^*$ defined by

$$a_{\underline{s}}^* = \sum_{\underline{r}} a_{\underline{r}}\, w_1^{r_1 s_1}\, w_2^{r_2 s_2} \cdots w_n^{r_n s_n} \tag{2.1}$$

$(r_j, s_j = 0,1,2,\ldots,t_j-1;\ j = 1,2,\ldots,n),$

with $w_j = \exp(2\pi i/t_j)$ the t_j^{th} fundamental root of unity. Here
$\underline{r} = (r_1, r_2, \ldots, r_n)\ (r_j = 0,1,\ldots,t_j-1;\ j = 1,2,\ldots,n)$
with a similar definition for the vector $\underline{s} = (s_1, s_2, \ldots, s_n).$

To give a simple concrete example of an $n = 2$ dimensional transform
we take $t_1 = 2$, $t_2 = 3$. The array $a_{\underline{r}}$ can be organised lexico-
graphically (if this is felt to help) as a vector, thus:
$a_{\underline{r}} = (a_{00},\ a_{01},\ a_{02},\ a_{10},\ a_{11},\ a_{12},)$, (commas between the vector
subscripts having been omitted), and similarly $a_{\underline{s}}^*$. We have, in
full,

$$a_{\underline{s}}^* = \sum_{r_1=0}^{1} \sum_{r_2=0}^{2} a_{\underline{r}}\, w_1^{r_1 s_1}\, w_2^{r_2 s_2}$$

and for example,

$$a_{01}^* = \sum_{r_1=0}^{1} \sum_{r_2=0}^{2} a_{\underline{r}}\, w_2^{r_2}$$

$$= a_{00} + a_{01}\, w_2 + a_{02}\, w_2^{2} + a_{10} + a_{11}\, w_2 + a_{12}\, w_2^{2}.$$

3.   The reduction of one-dimensional to multi-dimensional depends
on establishing unique correspondences between $\underline{r}$ and $\underline{s}$ and their
components $\{\ r_i\ \}$ and $\{\ s_i\ \}$. One such set of correspondences
leads to the procedure on which the Cooley-Tukey algorithm is
based. A second leads to a possibly less known procedure, though
in fact predating the publication of CT.

The first set of correspondences is congruential. The s-
correspondence derives from the so-called Chinese Remainder
Theorem (see [5]), and has the limitation that the numbers
t_1, t_2,...,t_n must be mutually prime in pairs (comprime). Then if

$$s \equiv s_1 \ (\mathrm{mod}\ t_1),\ \ .... \ \ s \equiv s_n \ (\mathrm{mod}\ t_n), \tag{3.1}$$

the theorem provides the unique mod τ congruence

$$s \equiv \frac{\tau s_1}{t_1}\, R_1 + ... + \frac{\tau s_n}{t_n}\, R_n \tag{3.2}$$

$(\mathrm{mod}\ \tau;\ 0 \le s < \tau\ ;)$
where R_j is the mod t_j reciprocal of τ/t_j i.e.

$$R_j \left(\frac{\tau}{t_j}\right) \equiv 1\ (\mathrm{mod}\ t_j), \tag{3.3}$$

which is unique (not obvious). As an example we take $t_1 = 3$,
$t_2 = 5$, $\tau = 15$. Then R_1 and R_2 are such that $5R_1 \equiv 1\ (\mathrm{mod}\ 3)$ and
$3R_2 \equiv 1\ (\mathrm{mod}\ 5)$ giving $R_1 = 2$, $R_2 = 2$, whence we have the unique

$$s \stackrel{=}{} 10s_1 + 6s_2 \ (\text{mod } 15).$$

The (1-1) relationship implied by the congruence is displayed in the following table:

s_1 \ s_2	0	1	2	3	4
0	0	6	12	3	9
1	10	1	7	13	4
2	5	11	2	8	14

Thus the pair $s_1 = 1$, $s_2 = 2$ $s = 7$, uniquely.

The reader is invited to contemplate the diagonal ("bishop-like", to use a chess analogy) pattern resulting from tracing successive values of s through the Table. The companion correspondence for r is given by

$$r \stackrel{=}{} \frac{\tau}{r_1} r_1 + \ldots + \frac{\tau}{t_n} r_n \ (\text{mod} \tau). \qquad (3.4)$$

The previous example is again used. The unique correspondence of $r \ (= 0, 1, \ldots, 15)$ to $r_1 \ (= 0, 1, 2)$ and $r_2 \ (= 0, 1, 2, 3, 4)$ is shown in the next Table. It corresponds to

$$r \stackrel{=}{} 5r_1 + 3r_2 \ (\text{mod } 15).$$

r_1 \ r_2	0	1	2	3	4
0	0	3	6	9	12
1	5	8	11	14	2
2	10	13	1	4	7

Again the pattern is interesting. In chess terminology it is
"knight-like". From any value of r one preceeds two steps to the
right in the row, and then one step upwards. If this carries
one out of the Table one assumes that the top row can be placed
below the bottom, thus permitting space to be filled with infinite
biperiodic repetitions. The reader may feel challenged to find
other correspondences and to follow their consequences as FFT
producing algorithms.

4.   We now show that the bishop and knight correspondences reduce
a one-dimensional DFT to a multi-dimensional one. We start with
the one-dimensional

$$a^*_s = \sum_{r=0}^{\tau-1} a_r w^{rs},\tag{4.1}$$

$(s = 0, 1, 2, \ldots \tau-1)$
where $w = \exp(2\pi i/\tau)$ and $\tau = t_1 \cdot t_2 \ldots \ldots t_n \cdot$ all t_j being
mutually prime in pairs. It is first clear from (3.2) and (3.4)
that

$$r \equiv (\tau/t_j)r_j \pmod{t_j}, \quad s \equiv (\tau s_j/t_j)R_j \pmod{t_j}.\tag{4.2}$$

It follows that

$$rs \equiv \left(\frac{\tau}{t_j}\right)r_j \left(\frac{\tau s_j}{t_j}\right)R_j = r_j s_j \frac{\tau}{t_j} \pmod{t_j}$$

and the Chinese Remainder Theorem then gives

$$rs \equiv \frac{\tau}{t_1} r_1 s_1 \frac{\tau}{r_1} R_1 + \ldots + \frac{\tau}{t_n} r_n s_n \frac{\tau}{t_n} R_n \pmod{\tau}$$

$$= \sum_j r_j s_j \, \tau/t_j \pmod{\tau}.$$

Hence

$$w^{rs} = \exp 2\pi i\left[\frac{r_1 s_1}{t_1} + \ldots + \frac{r_n s_n}{t_n}\right] = w_1^{r_1 s_1} \ldots w_n^{r_n s_n},$$

and therefore (4.1) reduces to an n-dimensional transform of form
(2.1). We show in the next paragraph by a particular example how
this decomposition reduces computing labour.

5.   Consider the elementary case $t_1 = 2$, $t_2 = 3$, $\tau = 6$. To
evaluate the vector a^*_s $(s = 0,1,2,\ldots 5)$ defined by

$$a^*_s = \sum_{r=0}^{5} a_r w^{rs} \tag{5.1}$$

requires 6 operations (multiplications and additions) for each s, $6^2 = 36$ operations in all. But if, using the bishop-knight decomposition, we write with obvious notation

$$a^*(s_1, s_2) = \sum_{r_1=0}^{1} \sum_{r_2=0}^{2} a(r_1, r_2) w_1^{r_1 s_1} w_2^{r_2 s_2}, \tag{5.2}$$

a saving becomes apparent.

If as a first stage we evaluate the summation with respect to r_1 we can form a quantity $a_1(r_2, s_1)$ defined by

$$a_1(r_2, s_1) = \sum_{r_1=0}^{1} a(r_1, r_2) w_1^{r_1 s_1} \tag{5.3}$$

requiring 2 operations of multiplication (including by unity) and addition for each fixed r_2 (0,1,2) and s_1 (0,1). In the second and final stage we form

$$a^*(s_1, s_2) = \sum_{r_2=0}^{2} a_1(r_2, s_1) w_2^{r_2 s_2}$$

requiring for each fixed s_2 (0,1,2) and s_1 (0,1) 3 operations of multiplication and addition. Thus for each s_1 and s_2 5 operations are needed, and since there are 6 combinations 6(3+2) = 30 operations are needed altogether instead of 6^2 as would be the case with a straight forward evaluation of

$$a^*_s = \sum_{r=0}^{5} a_r w^{rs}.$$

Table 1 which follows shows the successive stages. At each stage we need to store only 2 x 3 = 6 items. Organisation of an algorithm is left as an exercise to the interested reader.

TABLE 1

Stages in the first FFT Procedure

__STAGE 0.__   6 storage locations are required for the data a_r $(r = 0, 1, \ldots, 5)$. The r and s correspondences are:

$$r = 3r_1 + 2r_2 \ (\mathrm{mod}\ 6); \qquad\qquad s = 3s_1 + 4s_2 \ (\mathrm{mod}\ 6)$$

viz.

r_1 \ r_2	0	1	2
0	0	2	4
1	3	5	1

s_1 \ s_2	0	1	2
0	0	4	2
1	3	1	5

The basic data are stored in "dictionary" order of the pair (r_1, r_2). In the first stage we depart from and calculate

$$a_1 \ (r_2, s_1) = a \ (0, r_2) + a \ (1, r_2) \ w_1^{s_1}.$$

__STAGE 0.__   __STAGE 1.__

r_1	r_s	a_r	s_1	r_2	$a_1 \ (r_2, s_1)$	
0	0	a_0	0	0	$a(0,0) + a(1,0)$	$= a_0 + a_3$
0	1	a_2	0	1	$a(0,1) + a(1,1)$	$= a_2 + a_5$
0	2	a_4	0	2	$a(0,2) + a(1,2)$	$= a_4 + a_1$
1	0	a_3	1	0	$a(0,0) + a(1,0)w_1$	$= a_0 + a_3 \, w_1$
1	1	a_5	1	1	$a(0,1) + a(1,1)w_1$	$= a_2 + a_5 \, w_1$
1	2	a_1	1	2	$a(0,2) + a(1,2)w_1$	$= a_4 + a_1 \, w_1$

In the final stage we calculate

$$a^* \ (\underline{s}) = a^* \ (s_1, s_2) = a_1 \ (0, s_1) + a_1 \ (1, s_1) \ w_2^{s_2} + a_1 \ (2, s_1) w_2^{2s_2}$$

$$\underline{\text{STAGE} \quad 2.}$$

$$s_1 \quad s_2 \quad s \quad \underline{a^*(s_1,s_2) = a_s^*}$$

$$0 \quad 0 \quad 0 \quad a_1(0,0)+a_1(1,0)+a_1(2,0) = a_o+a_3+a_2+a_5+a_4+a_1$$

$$0 \quad 1 \quad 4 \quad a_1(0.0)+a_1(1,0)w_2+a_1(2,0)w_2^2 = a_o+a_3+(a_2+a_5)w_2+(a_4+a_1)w_2^2$$

$$0 \quad 2 \quad 2 \quad a_1(0,0)+a_1(1,0)w_2^2+a_1(2,0)w_2^4 = a_o+a_3+(a_2+a_5)w_2^2 (a_4+a_1)w_2^4$$

$$1 \quad 0 \quad 3 \quad a_1(0,1)+a_1(1,)+a_1(2,1) = a_o+a_3w_1+a_2+a_5w_1+a_4+a_1w_1$$

$$1 \quad 1 \quad 1 \quad a_1(0,1)+a_1(1,1)w_2+a_1(2,1)w_2^2 = a_o+a_3w_1+(a_3+a_5w_1)w_2+(a_4+a_1w)w_2^2$$

$$1 \quad 2 \quad 5 \quad a_1(0,1)+a_1(1,1)w_2^2+a_1(2,1)w_2^4 = a_o+a_3w_1+(a_2+a_5w_1)w_2^2+(a_4+a_1w_1)w_2^4$$

The final a_s^* are also in scrambled order of s.

6.   By use of the bishop-knight (perhaps more succinctly "chess")
correspondence a one-dimensional DFT has been reduced to an
n-dimensional one. Symbolically, the reduction from

$$a_s^* = \sum_{r=0}^{r-1} a_r \, K\,(r,s) \tag{6.1}$$

to

$$a_s^* (s_1,s_2,\ldots,s_n) = \sum_{r_1=0}^{t_1-1} \cdots \sum_{r_n=0}^{t_n-1} a(r_1,r_2,\ldots,r_n)$$

$$K_1(r_1,s_1)\ldots K_n(r_n,s_n) \tag{6.2}$$

has been achieved. In the first stage we evaluated from the
$\tau = t_1 t_2 \ldots t_n$ values of a_r the τ quantities

$$a_1 (r_2,r_3,\ldots r_n, \ s_1) = \sum_{r_1=0}^{t_1-1} a(r_1,r_2,\ldots,r_n) K_1(r_1,s_1) \tag{6.3}$$

each of which requires t_1 operations of multiplication and addition.
Next we evaluate the τ quantities

$$a_2 (r_3,r_4,\ldots r_n, \ s_1,s_2) = \sum_{r_2=0}^{t_2-1} a_1 (r_2,r_3,\ldots r_n, \ s_1) K_2(r_2,s_2) \tag{6.4}$$

each requiring t_2 operations, and so on until finally we obtain

$$a^* (s_1,s_2,\ldots,s_n) = \sum_{r_n=0}^{t_n-1} a_{n-1} (r_n,s_1,s_2,\ldots,s_{n-1}) K_n(r_n,s_n) \quad (6.5)$$

in the τ storage cells. The job is complete and a total of $\tau (t_1 + t_2 + \ldots + t_n)$ operations has been needed instead of τ^2.

It is relevant to the Cooley-Tukey procedure to note that a similar saving in effort can be achieved if the unidimensional transform (6.1) can be expressed in the multi-dimensional form

$$a^* (s_1,s_2,\ldots,s_n) = \sum_{r_1=0}^{t_1-1} \sum_{r_2=0}^{t_2-1} \cdots \sum_{r_n=0}^{t_n-1} a(r_1, r_2,\ldots r_n) \text{ X}$$

$$\text{X } K_1 (r_1,s_1) K_2(r_2,s_1,s_2)\ldots\ldots K_n(r_n,s_1,s_2,\ldots,s_n). \quad (6.6)$$

The first summation with respect to r_1 gives with t_1 operations

$$a_1 (r_2,\ldots r_n, s_1) = \sum_{r_1=0}^{t_1-1} a(r_1,\ldots,r_n) K_1(r_1,s_1). \quad (6.7)$$

Next we get

$$a_2 (r_3,\ldots,r_n,s_1,s_2) = \sum_{r_2=0}^{t_2-1} a_1 (r_2,\ldots,r_n,s_1) K_2(r_2,s_1,s_2)$$

$$(6.8)$$

requiring t_2 operations and the same number τ of storage locations. Continuation in this way achieves the transformation, again in $\tau (t_1 + t_2 + \ldots + t_n)$ operations instead of τ^2. Next we show how the Cooley-Tukey algorithm is an example of this kind of decomposition.

7.   The key likes in the unique representation of $\underline{r}$ and $\underline{s}$ in terms of their n components. In both cases a mixed radix representation is employed. τ is supposed factorised into n components t_1, $t_2,\ldots,t_n$ which do not in this case have to be comprime.

For r we write

$$r = t_1 \, t_2, \ldots, \, t_{n-1} \, r_1 + t_1 \, t_2, \ldots, \, t_{n-2} \, r_2 + \ldots + t_1 \, r_{n-1} + r_n$$

$$(7.1)$$

where

$$r_1 = 0,1,\ldots,t_n-1; \; r_2 = 0,1,\ldots,t_{n-1}-1,\ldots,r_n = 0,1,\ldots,t_1-1:$$

for s we write

$$s = t_2 \, t_3 \ldots \, t_n \, s_n + t_3 \, t_4 \ldots \, t_n \, s_{n-1} + \ldots + t_n \, s_2 + s_1$$

$$(7.2)$$

where

$$s_n = 0,1,\ldots,t_1-1; \; s_{n-1} = 0,1,\ldots,t_2-1; \; \ldots;$$

$$s_2 = 0,1,\ldots,t_{n-1}-1; \quad s_1 = 0,1,\ldots, \, t_n-1.$$

Then

$$w^{rs} = \exp \frac{2\pi i}{\tau} \left[t_1 \, t_2, \ldots t_{n-1} r_1 \, s_1 + t_1 \, t_2, \ldots, t_{n-2} \, r_2 \, (t_n \, s_2 + s_1) \right.$$

$$+ t_1 \, t_2, \ldots, t_{n-3} r_3 (t_{n-1} \, t_n \, s_3 + t_n \, s_2 + s_1) + \ldots$$

$$\left. + r_n \, (t_2 \, t_3 \ldots \, t_n \, s_n + t_3 \, t_4 \ldots \, t_n \, s_{n-1} + \ldots + t_n \, s_2 + s_1) \right].$$

$$(7.3)$$

Thus the one-dimensional transform

$$a^*_s = \sum_{r=0}^{\tau-1} a_r \, w^{rs}$$

can be reduced to the form (6.6) (though note that

$$r_1 = 0,1,\ldots, \, t_1-1; \; r_2 = 0,1,\ldots,t_2-1; \; \text{etc.})$$

and the claimed computational saving ensues.

8.   For illustration, and comparison with the previous method, we write out the steps in full for the $n = 2$-dimensional case $t_1 = 2$, $t_2 = 3$, $\tau = 6$.

The r and s correspondences are:

$$r = 2r_1 + r_2 \quad , \qquad (r_1 = 0,1,2, \ r_2 = 0,1) \ ;$$

$$s = 3s_2 + s_1 \quad , \qquad (s_1 = 0,1,2; \ s_2 = 0,1) \ ;$$

specifically we have

r_1 \ r_2	0	1
0	0	1
1	2	3
2	4	5

s_1 \ r_2	0	1
0	0	3
1	1	4
2	2	5

The one-dimensional transform

$$a^*_s = \sum_{r=0}^{5} a_r \, w^{rs} \qquad (s = 0,1,\ldots,5)$$
$$(w = \exp(2\pi i/6) \qquad\qquad (8.1)$$

is reduced to

$$a^*(s_1, s_2) = \sum_{r_1=0}^{2} \sum_{r_2=0}^{3} a(r_1,r_2) \, w^{(2r_1+r_2)(3s_2+s_1)}$$

$$= \sum_{r_1=0}^{2} \sum_{r_2=0}^{1} a(r_1,r_2) \, w^{2r_1 s_1} \, w^{r_2(3s_2 + s_1)} \qquad (8.2)$$

$(s_1 = 0,1,2; \ s_2 = 0,1)$.

Calculation is shown in Table 2.

TABLE 2.

Stages in Second FFT Procedure

The data are stored in natural order of r thus

r_1	r_2	a_r
0	0	a_0
0	1	a_1
1	0	a_2
1	1	a_3
2	0	a_4
2	1	a_5

Then

$$a_1(r_2, s_1) = \sum_{r_1=0}^{2} a(r_1, r_2)\, w^{2r, s_1}$$

$$= a(0, r_2) + a(1, r_2)\, w^{2s_1} + a(2, r_2)\, w^{4s_1}$$

giving

STAGE 1.

r_2	s_1	$a_1(r_2, s_1)$	
0	0	$a(0,0) + a(1,0) + a(2,0)$	$= a_0 + a_2 + a_4$
0	1	$a(0,0) + a(1,0)\, w^2 + a(2,0)\, w^4$	$= a_0 + a_2 w^2 + a_4 w^4$
0	2	$a(0,0) + a(1,0)\, w^4 + a(2,0)\, w^8$	$= a_0 + a_2 w^4 + a_4 w^8$
1	0	$a(0,1) + a(1,1) + a(2,1)$	$= a_1 + a_3 + a_5$
1	1	$a(0,1) + a(1,1)\, w^2 + a(2,1)\, w^4$	$= a_1 + a_3 w^2 + a_5 w^4$
1	2	$a(0,1) + a(1,1)\, w^4 + a(2,1)\, w^8$	$= a_1 + a_3 w^4 + a_5 w^8$

Then we have $a^*(s_1, s_2) = \sum_{r_2=0}^{1} a_1(r_2, s_1)\, w^{r_2(3s_2 + s_1)}$

$$= a_1(0, s_1) + a_1(1, s_1)\, w^{(3s_2 + s_1)}$$

STAGE 2.

s_1	s_2	s	$a*(s_1,s_2)$
0	0	0	$a_1(0,0) + a_1(1,0) = a_0 + a_2 + a_4 + a_1 + a_3 + a_5$
0	1	3	$a_1(0,0) + a_1(1,0)w^3 = a_0 + a_2 + a_4 + (a_1 + a_3 + a_5)w^3$
1	0	1	$a_1(0,1) + a_1(1,1)w = a_0 + a_2 w^2 + a_4 w^4 + a_1 w + a_3 w^3 + a_5 w^5$
1	1	4	$a_1(0,1) + a_1(1,1)w^4 = a_0 + a_2 w^2 + a_4 w^4 + a_1 w^4 + a_3 w^6 + a_5 w^8$
2	0	2	$a_1(0,2) + a_1(1,2)w^2 = a_0 + a_2 w^4 + a_4 w^8 + a_1 w^2 + a_3 w^6 + a_5 w^{10}$
2	1	5	$a_1(0,2) + a_1(1,2)w^5 = a_0 + a_2 w^4 + a_4 w^8 + a_1 w^5 + a_3 w^9 + a_5 w^{13}$

Notice: (1) The basic data a_r are stored in "natural" order of
ascending r;
      (2) In the course of the stages the order is scrambled
and, as in the first algorithm, the storage location
corresponding to the original a_r does not contain $a*_r$.

The values of $a*_s$ may be cross checked with those given in Table 1.

From the above we have

$$a*_3 = a_0 + a_2 + a_4 + (a_1 + a_3 + a_5)w^3$$

$$= a_0 + a_2 + a_4 + (a_1 + a_3 + a_5)e^{\pi i}.$$

From Table 1

$$a*_3 = (a_0 + a_3 w_1) + (a_2 + a_5 w_1) + (a_4 + a_1 w_1)$$

where $w_1 = e^{\frac{2\pi i}{2}} = e^{\pi i}$. Similarly, from Table 1

$$a*_1 = a_0 + a_3 w_1 + (a_2 + a_5 w_1) w_2 + (a_4 + a_1 w_1) w_2^2$$

$$= a_0 + a_3 e^{\pi i} + a_2 3^{\frac{2\pi i}{3}} + a_5 e^{\frac{5\pi i}{3}} + a_4 e^{\frac{4\pi i}{3}} + a_1 e^{\frac{7\pi i}{3}}.$$

From Table 2, rearranging,

$$a*_1 = a_0 + a_3 e^{\frac{6\pi i}{6}} + a_2 e^{\frac{4\pi i}{6}} + a_5 e^{\frac{10\pi i}{6}} + a_4 e^{\frac{8\pi i}{6}} + a_1 e^{\frac{2\pi i}{6}}$$

which is plainly the same. The reader may wish to confirm the
identity of the remaining results.

9.   To conclude it is worth recalling that in both procedures
the results at each stage require only τ locations for storage.
Although the C-T procedure does not require it, it is nevertheless
usual to use sample sizes τ which are a power of $2, \tau = 2^N$ say.
Instead of $\tau^2 = 2^{2N}$, the number of operations, in this case
merely additions and subtractions, is only $2N\tau$, or $2\tau \log \tau$
a notable reduction.

REFERENCES

1.   J.W. Cooley and J.W. Tukey, <u>An algorithm for the machine
      calculation of comples Fourier series</u>, Maths. Computation,
      <u>19</u>, 297-301, (1965).
2.   I.E.E.E. Trans. Audio Electroacoust. Vol. AU-15, June 1967.
3.   I.E.E.E. Trans. Audio Electroacoust. Vol. AU-15, June 1969.
4.   I.J. Good, <u>The relationship between two Fast Fourier
      Transforms</u>, I.E.E.E. Transactions on Computers C-20, 310-317,
      1971.
5.   G.H. Hardy and E.M. Wright, <u>An Introduction to the theory
      of numbers</u>, Oxford: **Clarendon** Press, (1938).

PART II.

SOME INTRINSIC PROPERTIES OF THE DISCRETE FOURIER TRANSFORM (DFT)

1.   It is less well-known that the DFT has a theory in its own
right with fundamental properties closely analogous to those of
Fourier series and the Fourier integral. For present purposes
we define the one dimensional mod t DFT, with which we shall be
exclusively concerned, as a linear transformation applied to the
array a_r (r = 0, 1, ...t-1) of real or complex numbers to produce
an array a* (s = 0, 1, ...t-1) of complex numbers. Let
$w = \exp 2\pi^s i/t$ with t a positive integer. Then the transformation
is

$$a^*_s = \sum_{r=0}^{t-1} a_r w^{rs},$$ (1.1)

(s = 0, 1, ...t-1).

It is easy to verify the inversion formula

$$a_r = \frac{1}{t} \sum_{s=0}^{t-1} a^*_s w^{-rs}$$ (1.2)

(r = 0, 1,...,t- 1).

From the point of view of symmetry it is often better to define

$$a^*_s = \frac{1}{\sqrt{t}} \sum_{r=0}^{t-1} w^{rs} a_r, \text{ so that } a_r = \frac{1}{\sqrt{t}} \sum_{s=0}^{t-1} w^{-rs} a^*_s.$$

The DFT as defined in (1.1) and (1.2) possesses a "Rayleigh-
Parseval" formula

$$\frac{1}{t} \sum_{s=0}^{t-1} |a^*_s|^2 = \sum_{r=0}^{t-1} a_r^2,$$ (1.3)

and the convolution relation that $a^*_s b^*_s$ is the DFT of

$$\sum_r a_r b_{\tau-r}.$$

An analogue for multidimensional transforms to the Poisson
summation formula is developed in [1]. The DFT has not found a
place like that of its integral counterpart with rich and versatile

fields of application to problem solving, but it occupies the
central role in digital harmonic analysis, and the existence of
FFT algorithms endows it with a proper distinction. In this
paper we are going to comment on other properties of the DFT
which seem to us to deserve wider attention.

CONSEQUENCES OF SAMPLING THEOREM: TERMINATING SERIES.

2.   Digital analysts are well aware that the Fourier coefficients
of a real band-limited signal $f(2\pi t/T)$ observed over a time
interval T can be calculated exactly by summation over a sufficient
number of equidistant sample points, and that the result is
equivalent to integration. Another way of looking at this is to
suppose that $f(\Theta)$ has period 2π and possesses the Fourrier
expansion

$$f(\Theta) = \sum_{n=-N}^{+N} c_n e^{in\Theta} \, , \tag{2.1}$$

from which the mathematician, knowing $f(\Theta)$, can in principle
calculate c_n by integration from the Fourier series inversion
formula

$$c_n = \frac{1}{2\pi} \int_0^{2\pi} f(\Theta) e^{-in\Theta} d\Theta \, . \tag{2.2}$$

If he can not perform the integration the mathematician may resort
to numerical methods and then he knows, provided that $f(\Theta)$ has
suitable properties that

$$c_n = \lim_{t\to\infty} \frac{1}{t} \sum_{r=0}^{t-1} f\left(\frac{2\pi r}{t}\right) w^{-nr} \, , \tag{2.3}$$

where t is an integer, r is the index of the sample point, and
$w = \exp(2\pi i/t)$. Now it also follows from (2.1) that for any
fixed positive integer t,

$$\frac{1}{t} \sum_{r=0}^{t-1} f\left(\frac{2\pi r}{t}\right) w^{-nr} = \sum_m c_{mt+n} \, , \tag{2.4}$$

$(n = 0, 1, \ldots t - 1),$

where the summation extends over positive and negative integer m, providing a kind of Poisson summation formula for Fourier series. Clearly if $t \geq 2N + 1$ (2.4) reduces to the exact equality

$$c_n = \frac{1}{t} \sum_{r=0}^{t-1} f\left(\frac{2\pi r}{t}\right) w^{-nr} \qquad (2.5)$$

$(|n| \quad N)$. t can be less for small values of $|n|$. This is equivalent to the sampling theorem and has interesting consequences.

3.   The first is that for functions $f(\Theta)$ with finite term Fourrier expansion there will for large enough t be the following <u>exact</u> equality for series and integral:

$$\frac{1}{2\pi} \int_0^{2\pi} f(\Theta) e^{-in\Theta} d\Theta = \frac{1}{t} \sum_{r=0}^{t-1} f\left(\frac{2\pi r}{t}\right) w^{-nr}. \qquad (3.1)$$

A striking example has been given by Good [2].   Consider

$$f(\Theta) = [a\, e^{-i\Theta} + b + a\, e^{i\Theta}]^N$$

Then, by (3.1), for $|n| \quad N$,

$$c_n = \frac{1}{2\pi} \int_0^{2\pi} e^{-in\Theta}[ae^{-i\Theta} + b + ae^{i\Theta}]^N d\Theta$$

$$= \frac{1}{t} \sum_{r=0}^{t-1} w^{-nr}[aw^{-r} + b + aw^r]^N$$

certainly for $t \geq 2N + 1$.   In the case that $b = x$, $a = \frac{1}{2}(x^2-1)^{\frac{1}{2}}$, $n = 0$, the integral defines the Legendre polynomial $P_N(x)$, and accordingly we then have the integral and series equivalent

$$P_N(x) = \frac{1}{2\pi} \int_0^{2\pi} [x + \sqrt{(x^2-1)}\, \cos\Theta]^N d\Theta$$

$$= \frac{1}{t} \sum_{r=0}^{t-1} [x + \sqrt{(x^2-1)}\, \cos\frac{2\pi r}{t}]^N,$$

in this case for $t \geq N + 1$.   It also follows that $P_N(x)$ is the term independent of Θ in the Fourier expansion of $[x + (x^2-1)^{\frac{1}{2}}\cos\Theta]^N$

in exponentials. Computational savings can be achieved if the series is used rather than the integral.

INFINITE SERIES: ACCURACY OF DFT APPROXIMATION

4. It is of interest to be able to assess the accuracy of a DFT approximation to a harmonic amplitude c_n in the case that $f(\theta)$ possesses an infinite trigonometric expansion. To illustrate we consider the sawtooth function

$$f(\theta) = \theta - \pi , \quad (0 < \theta < 2\pi) \tag{4.1}$$

Let

$$f(\theta) = \sum_{n=-\infty}^{\infty} c_n e^{in\theta} \tag{4.1a}$$

and then we have

$$c_n = \frac{1}{2\pi} \int_0^{2\pi} e^{-in\theta} (\theta - \pi)\,d\theta = \frac{i}{n} , \quad (n \neq 0), \tag{4.2}$$

and $c_0 = 0$. To check this it is instructive to reconstruct $f(\theta)$ by summing the series in (4.1a) directly. Choose $n \neq 0$ and define for some positive integer t,

$$R_n(t) = \frac{1}{t} \sum_{r=1}^{t-1} w^{-rn} \left(\frac{2\pi r}{t} \right) - \pi). \tag{4.3}$$

Then $R_n(t) \to c_n$ as $t \to \infty$ since it approximates the Riemann integral in (4.2). We may now evaluate (4.3) by direct summation. Indeed for $r = 1, 2, \ldots t - 1$ and fixed t define

$$a_r = \frac{2r}{t} - 1. \tag{4.4}$$

The DFT is

$$a_s^* = \sum_{r=1}^{t-1} \left(\frac{2r}{t} - 1 \right) w^{rs} \quad (s = 0, 1, \ldots t-1 \tag{4.5}$$

and $a_0^* = 0$). Direct summation and use of $w^t = 1$ gives

$$a_s^* = \frac{1}{i} \cot \frac{\pi s}{t}, \quad (s = 1, 2, \ldots t-1), \tag{4.6}$$

$$a_o^* = 0 .$$

Then

$$R_n(t) = \frac{\pi}{t} a_{-n}^* = \frac{\pi i}{t} \cot \frac{\pi n}{t}.$$

For fixed n and t >> n we have

$$R_n(t) \doteq \frac{\pi i}{t} \frac{(1 - \dfrac{\pi^{+2} n^2}{2t^2})}{\dfrac{\pi n}{t} (1 - \dfrac{\pi^2 n^2}{6t^2})}$$

so that

$$\left| \frac{R_n(t) - c_n}{c_n} \right| = 1 - \frac{\pi^2 n^2}{3t^2} + 0 \left(\frac{1}{t^4}\right), \tag{4.7}$$

indicating the manner of convergence of $R_n(t)$ to c_n.

USE OF TEST PAIRS FOR CHECKING & TESTING

5.   It is certainly not a new idea to seek test functions for
the purpose of checking the accuracy and speed of algorithms on
the same computer, and for comparing implementations of a given
algorithm on different computers (benchmark tests, for instance).
The idea is to fix on one or more representative pairs a_r and
a_s^*. Then one generates $\hat{a}_s^*$, say and the first check can be a
comparison of $\hat{a}_s^*$ with the known a^* for each s. This will be
most telling near singular values of s (e.g. small values of
π s.t in $a_s^* = i^{-1} \cot(\pi s/t)$. An overall measure of accuracy
might be the mean square error taken over all s, viz.

$$\delta^2 = \frac{1}{t} \sum_{s=0}^{t-1} |\hat{a}_s^* - a_s^*|^2 \tag{5.1}$$

Then one can transform back with the same algorithm (adapted of
course to deal with complex data) from $\hat{a}_s^*$ to $\hat{a}_r$ (say) and make a
similar accuracy comparison, repeating as often as required.

The application to the timing of an algorithm is obvious. It may in addition be worth while attempting to develop exact formulae to test a more elaborate package. In digital analysis one thinks naturally of spectral analysis. If we define the autocovariance $g(\tau)$ by

$$g(\tau) = \sum_{r=0}^{t-1} a_r \, a_{r+\tau} \tag{5.2}$$

$(\tau = 0, 1, \ldots t-1)$

with the convention that $a_{r+\tau}$ is to be interpreted mod t, then the DFT $g^*(s)$ of $g(\tau)$ is given by

$$g_s^* = a_s^* \, a_{-s}^* . \tag{5.3}$$

For the test pair of para. 4 we then have

$$g_s^*(s) = \cot^{-2} \frac{\pi s}{t} \tag{5.4}$$

and test of a spectral density package might be to generate samples of $a_r = \frac{2r}{t} - 1 (r \neq 0)$ and hence $g(\tau)$. Then one would calculate $g^*(s)$ by machine and compare it with $g^*(s)$: then go back to $g(\tau)$, and so on. For cross-correlation and convolution additional pairs are useful and can be generated straightforwardly from the commoner polynomials. For example,

$$a_s^* = \sum_{r=0}^{t-1} \binom{t-1}{r} w^{rs} = (-)^s e^{-\frac{\pi i s}{t}} \left[2 \cos \frac{\pi s}{t} \right]^{t-1} \tag{5.5}$$

has the image $a_r = \binom{t-1}{r}$, and it may not be at all obvious that

$$\binom{t-1}{r} = \frac{1}{t} \sum_{s=0}^{t-1} (-)^s e^{-\frac{\pi i s}{t}(2r+1)} \left[2 \cos \frac{\pi s}{t} \right]^{t-1} . \tag{5.6}$$

Disbelievers will find it instructive to evaluate (5.6) by hand for, say, $t = 4$.

OTHER METHODS FOR GENERATING TEST PAIRS

6.   The writer and I.J. Good have compiled a short dictionary of basic DFT pairs. The results given were generated by polynomial, observation and derivation from Fourier series. For example if

$$f(\theta) = \sum_{-\infty}^{\infty} c_n \, e^{\,in\theta}, \tag{6.1}$$

where $f(\theta)$ has period 2π, then subject to suitable conditions

$$\frac{1}{t} \sum_{r=0}^{t-1} f(\tfrac{2\pi r}{t}) \, w^{rs} = \sum_{m=-\infty}^{\infty} c_{mt-s}$$

giving the DFT pair

$$a_s^* = \sum_{m=-\infty}^{\infty} c_{mt-s} \;, \quad a_r = \frac{1}{t} f(\tfrac{2\pi r}{t}) \;. \tag{6.2}$$

This pair can be generalised. An easy application can be seen by recovering essentially the pair (4.4) and (4.6). Starting from $f(\theta) = \theta - \pi \; (0 < \theta < 2\pi)$ we obtained $c_n = \frac{1}{n} \; (n \neq 0)$ and consequently we have the pair

$$a_r = \frac{\pi}{t} (\tfrac{2r}{t} - 1) \; (r \neq 0) \text{ and } a_s^* = i \sum_m \frac{1}{(mt-s)} \; (s \neq 0)\,.$$

Now by combining terms we get $(s \neq 0)$

$$-\,i\,a_s^* = -\frac{1}{s} + 2s \sum_{m \geq 1} \frac{1}{m^2 t^2 - s^2} \;.$$

We use the result (e.g. [3])

$$-\frac{1}{2a} + \frac{\pi}{2\sqrt{(ab)}} \cot h \, (\pi \sqrt{\tfrac{a}{b}}) = \sum_{m \geq 0} \frac{1}{m^2 b + a}$$

to obtain

$$a_s^* = \frac{\pi}{it} \cot \frac{\pi s}{t} \;, \quad (s \neq 0),$$

which, upon removal of the common factor π/t, recovers the pair of para. 4. More importantly, this technique can yield pairs which cannot be so easily obtained otherwise and a good example is the pair

$$a_r = w^{r^2} \;, \quad a_s^* = \tfrac{1}{2} \, \sqrt{t}(t+i) \, [\, 1 + i^{2\,s-t} \,] \, w^{-\frac{1}{4} s^2} \;.$$

a_0^* is the Gauss sum which occurs in the theory of quadratic residues. Indeed the DFT seems to have as many intriguing links with Number Theory as it does with digital harmonic analysis.

REFERENCES

1.   I.J. Good, <u>Analogues of Poisson's summation formula</u>, American Mathematical Monthly, <u>69</u>, 259-266, 1962.
2.   I.J. Good, <u>A new finite series for Legendre polynomials,</u> Proc. Camb. Phil. Soc. <u>51</u>, 385-388, 1955.
3.   M. Abramowitz, et al, <u>Handbook of Mathematical Functions,</u> Dover Publications Inc., New York, 1965.

<u>DISCUSSION</u>

In discussion B.W. Conolly recalled that computational savings were inherently available in calculating multidimensional DFTs (if one went about it the right way) and indeed that this was the key to the effectiveness of FFT algorithms applied to one-dimensional DFTS. The one method required the factors of the modulus to be mutually coprime; the other method made no such demand. Professor Conolly suggested that additional savings in the calculation of multidimensional transforms might be achieved by applying one dimensional FFT methods in each stage of the multidimensional calculation.

AN ADAPTIVE BEAMFORMER WHICH IMPLEMENTS CONSTRAINTS USING AN
AUXILIARY ARRAY PREPROCESSOR

LLoyd J. Griffiths

Electrical Engineering Department, University of
Colorado, Boulder, Colorado 80302

ABSTRACT: This paper presents a method of adaptive beamforming
which makes use of a fixed, conventionally-weighted array and an
auxiliary, adaptive array. The beamformed output is taken as the
difference between the conventional and adaptive outputs. It is
assumed that the sensor outputs, which are common to both arrays
have been time-delay steered to the direction of interest. Similar
structures have been described previously in which the auxiliary
array is adapted using linearly-constrained adaptive algorithms.
In the present work, it is shown that linear constraints can also
be implemented by preceding the auxiliary adaptive array with a
fixed preprocessor W_s. The adaptive algorithm then reduces to
the simple; unconstrained LMS procedure. Thus, the proposed implementation separates out, in a natural manner, the three basic
components of a linearly constrained adaptive beamformer: the
conventional beamformer, the linear constraints, and the adaptive
algorithm. This separation has significant advantages in practical implementations and suggests extensions to other beamformer
structures.

1. DEFINITION OF TERMS

The K-dimensional vector of received array outputs at the kth
sample is denoted as $\underline{X}(k)$. It is presumed that these outputs
have been presteered in the direction of interest such that

$$\underline{X}(k) = s(k)\underline{1} + \underline{N}(k) \tag{1}$$

where:

    $s(k)$ = scaler signal of interest

$\underline{1}$ = K-dimensional vector of ones

$\underline{N}(k)$ = vector of noise samples.

In conventional beamforming, an output $y_C(k)$ is derived as the weighted sum of the elements of $\underline{X}(k)$,

$$y_C(k) = \underline{W}_C^T \underline{X}(k) \tag{2}$$

The elements of $\underline{W}_C$ are the fixed conventional weighting coefficients which are generally selected to provide a prespecified main-lobe width and sidelobe level. Tchekychev polynomial or Dolph [1] weights are a common choice. Figure 1 depicts a schematic representation of this beamformer.

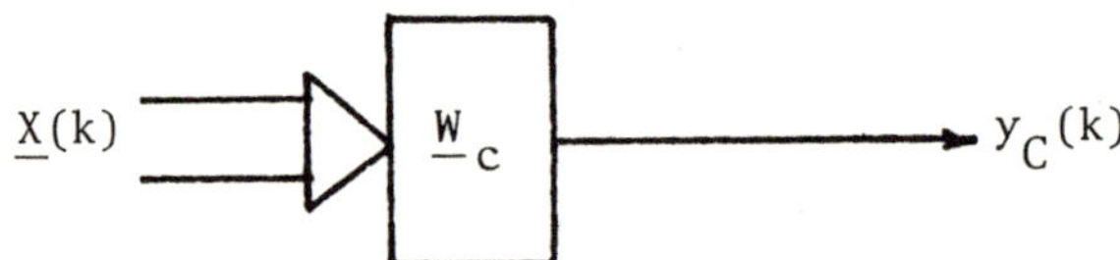

Fig. 1   Conventional beamformer structure

In linearly-constrained adaptive beamforming [2], an output $y_A(k)$ is formed using continuously time-varying coefficients as follows:

$$y_A(k) = \sum_{\ell=0}^{L-1} \underline{W}_A^T(\ell,k)\underline{X}(k-\ell) \tag{3}$$

The columns of adaptive weights $\underline{W}_A(\ell,k)$ are adapted so as to minimize $E[y_A^2(k)]$ subject to a set of linear constraints. Those constraints may be expressed as

$$\underline{C}^T(\ell)\underline{W}_A(\ell,k) = f(\ell) \quad , \quad \ell=0,1,..L-1 \tag{4}$$

where the elements of $\underline{C}(\ell)$ and the scalars $f(\ell)$ determine the constraint. One common constraint which ensures that the desired signal appears undistorted at the output, is to set $\underline{C}(\ell) = \underline{1}$ and $f(\ell) = 0$, except for $f(L/2) = 1$. For purposes of simplicity, we will restrict out attention to this constraint in the following discussion. Extensions to the more general case proceed in an obvious manner.

Figure 2 shows a schematic of the linearly-constrained adaptive beamformer structure as originally proposed by Frost [2]. The

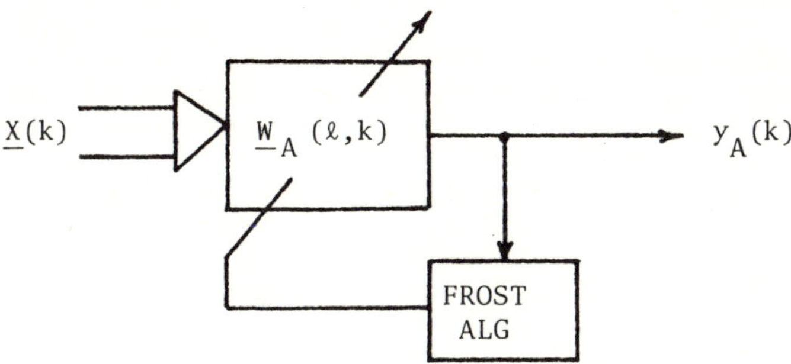

Fig. 2   Linearly-constrained adaptive beamformer structure

$W_A(\ell,k)$ coefficients are updated using a procedure which incorporates the constraints in Eq(4) as an integral part of the algorithm.

2.   AUXILIARY ARRAY BEAMFORMERS

It has been shown [3,4] that the beamformer in Fig. 2 can also be implemented using an auxiliary, adaptive array as depicted in Fig. 3.   In this form, the conventional beamformer has been

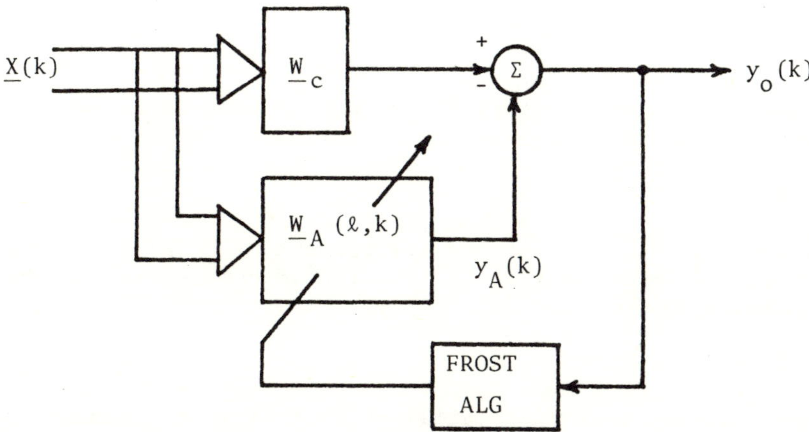

Fig. 3 Auxiliary array implementation of the linearly-constrained adaptive beamformer.

separated out and Frost's algorithm is used to update the $\underline{W}_A(\ell,k)$
'subject to the constraint that all $f(\ell)=0$. This ensures that the
desired signal $s(k)$ cannot appear in $y_A(k)$ and the adapted output
$y_A(k)$ works toward eliminating the noise and interference terms
from the conventional beamformer output. The structures in
Fig's. 2 and 3 are identical for all values of k, regardless of
$\underline{W}_C$, provided that they are initialized to the same overall beamformer at k=0.

The structure of interest in this paper is depicted in Fig. 4.

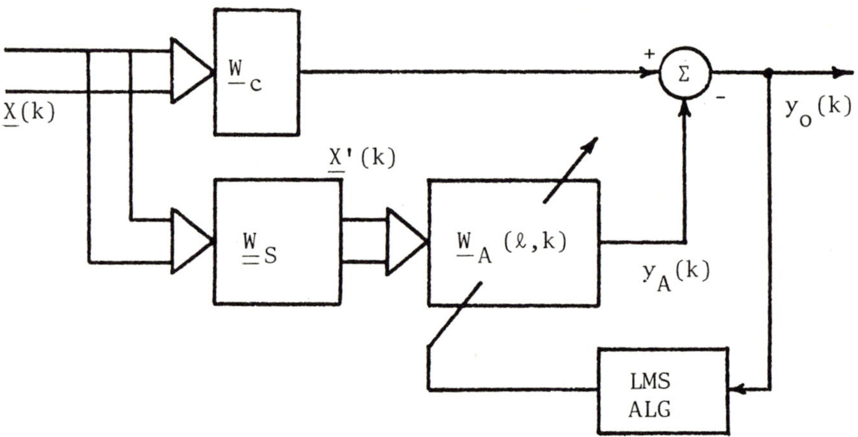

Fig. 4   Hardware-constrained auxiliary array adaptive beam-
         former structure.

A fixed preprocessor matrix filter $\underline{\underline{W}}_S$ is used to derive a vector
of modified array signals $\underline{X}'(k)$ as follows:

$$\underline{X}'(k) = \underline{\underline{W}}_S^T \underline{X}(k) \qquad (5)$$

Equivalently, if $x_i'(k)$ is the ith element of $\underline{X}'(k)$ and $\underline{V}_i$ is the
ith column of $\underline{\underline{W}}_S$, then $x_i'(k)$ is a linear combination of the
elements of $\underline{X}(k)$,

$$x_i'(k) = \underline{V}_i^T \underline{X}(k) \qquad (6)$$

In general, $\underline{\underline{W}}_S$ is a KxK' matrix where K'<K is the number of
elements in $\underline{X}'(k)$. The purpose of this preprocessor is to
block the desired signal $s(k)$ from all $x_i'(k)$. This will be the
case provided that

$$\underline{V}_i^T \underline{1} = 0 \text{ for } i = 1, 2, \ldots K' \tag{7}$$

Since $s(k)$ is absent from $\underline{X}'(k)$, the adaptive weights $\underline{W}_A(\ell,k)$ can be iterated using the simple, unconstrained LMS algorithm due to Widrow [5].

3. PROPERTIES OF THE HARDWARE-CONSTRAINED BEAMFORMER

The structure shown in Fig. 4 has the following properties, provided that the conditions in Eq. (7) are satisfied:

   a) If the matrix $\underline{\underline{W}}_S$ has dimension K'=K-1 and if the columns of $\underline{\underline{W}}_S$ are linearly independent (equivalently, the matrix has rank K-1), then the steady-state mean value of the beamformer shown in Fig. 4 converges to the same vector as the linearly-constrained processors depicted in Figs. 2 and 3.
   b) If in addition to the conditions in a), the columns of $\underline{\underline{W}}_S$ are mutually orthogonal ($\underline{V}_i^T \underline{V}_j = 0$, $i \neq j$), then the structures in Figs. 2-4 are all identical, for all values of k, provided that they are initialized to the same overall beamformer at k=0.

The implication of these properties is that if a) is satisfied and b) is not, then the sequence of adapted beamformers obtained with the structure in Fig. 4 will be different from those obtained using either Fig. 2 or 3. For stationary statistics, however, they will all converge to the same steady-state mean value. In effect, the hardware-constrained case will take a different path to that solution. Since the $\underline{\underline{W}}_S$ which satisfy b) are a subset of those satisfying a), the beamformer in Fig. 4 is the more general structure and contains those in Figs. 2 and 3 as special cases.

Two specific examples of appropriate $\underline{\underline{W}}_S$ for K=4 may be formed using only adders and subtractors (± 1) as follows:

$$\underline{\underline{W}}_S = \begin{bmatrix} 1 & 0 & 0 \\ -1 & 1 & 0 \\ 0 & -1 & 1 \\ 0 & 0 & -1 \end{bmatrix}, \text{ satisfies a)} \tag{8}$$

$$\underline{\underline{W}}_S = \begin{bmatrix} 1 & 1 & 1 \\ 1 & -1 & -1 \\ -1 & -1 & 1 \\ -1 & 1 & -1 \end{bmatrix}, \text{ satisfies a) and b)} \tag{9}$$

Many other forms, not restricted to ± 1 elements are possible.

4. DISCUSSION

The sequence of beamformers shown in Fig. 2-4 represents a progression of structures which may all be used to implement a linearly-constrained adaptive beamformer. The last of these has at least two advantages. First, it separates out in a natural manner, both a conventional beamformer and the constraints as fixed, hardware elements. As a result, adaptation is reduced to its simplest possible form, the unconstrained LMS algorithm. This separation provides a vehicle for computing the sensitivity of the beamformer to element failure or deviations from plane-wave geometry and, in addition, simplifies the practical hardware required to construct the beamformer.

The second advantage lies in the generality of the structure. It is, in effect, the array equivalent to the noise-cancelling system described by Widrow, et al [6] and many of the results in this area carry over directly. One example of interest lies in a study of preprocessors $\underline{\underline{W}}_S$ which are of rank less than K-1. The columns of this matrix may be viewed as the weighting elements of conventional beamformers. Satisfying Eq. (7) is equivalent to having a null in each of these beams at $\theta=0^{\circ}$. Extensions to cases involving multiple, closely-spaced nulls or to beams which also have a zero derivative in the pattern at $\theta=0^{\circ}$ are easily obtained. These, and other interpretations of this beamformer are presently under investigation.

REFERENCES

1. C.L. Dolph, "A current distribution for broadside arrays which optimizes the relationship between beam width and sidelobe level," Proc. IRE, vol. 34, pp. 335-348, June, 1946.
2. O.L. Frost, III, "An algorithm for linearly-constrained adaptive array processing," Proc. IEEE, vol. 60, pp. 926-935, August, 1972.
3. N.W. Owsley, "A recent trend in adaptive spatial processing for sensor arrays: Constrained adaptation," Signal Processing, Pro. NATO ASI, J.W.R. Griffiths, P.L. Stocklin, and C. Van Schooneveld, (Eds.) Academic Press, 1973.
4. L.J. Griffiths, "Adaptive monopulse beamforming," Proc. IEEE (letters), vol. 64, pp. 1260-1261, August, 1976.
5. B. Widrow, et al, "Stationary and non-stationary learning characteristics of the LMS adaptive filter," Proc. IEEE, vol. 64, August, 1976.
6. B. Widrow, et al, "Adaptive noise cancelling: Principles and applications," Proc IEEE, vol. 63, pp. 1692-1716, Dec., 1975.

S U B J E C T 4

DISPLAYS, PATTERN RECOGNITION, HUMAN DECISION

ERROR AND REJECT TRADEOFF FOR NEAREST NEIGHBOR DECISION RULES

Pierre A. Devijver

MBLE Research Laboratory
Brussels, Belgium.

1. INTRODUCTION

In a highly influencial paper, [1], Cover and Hart pointed
out that, in the classification problem, there are two extremes
of knowledge which the statistician may possess. In the first
place, he may know the underlying joint distribution of the
observation x and its class θ, in which case a Bayes procedure
is an optimal decision rule, in the sense that it achieves minimum probability of error of classification. In the other extreme,
the decision procedure is allowed to make use only of the knowledge that can be inferred from a collection of m correctly iden-
tified sample pairs, (x^1, θ^i), $1 \leqslant i \leqslant m$. In the latter situation, Cover and Hart were able to show that, in the large sample
case, and under fairly general conditions -that is without assuming any additional a priori knowledge than that conveyed by the
samples- such a simple strategy as the single nearest neighbor
(1-NN) decision rule has a probability of error which is no greater than twice the Bayes probability of error. They also showed
that the error probability of the k-nearest neighbor (k-NN) rule
is monotonically decreasing in k to the Bayes error probability.
The family of k-NN rules is therefore quite versatile, extending
as it is from the simple, but still efficient 1-NN rule to the
optimal Bayes rule.

In another series of influencial papers, [2], [3], Chow
investigated the application of the Bayes rule to the pattern
classification problem. In particular, he showed that modifying
the Bayes rule by making provision for a reject option is a convenient way to safeguard against excessive misclassification :
it basically consists in converting potential misclassification

into rejection. The modified Bayes rule makes the same decision as the original one whenever the conditional Bayes error probability $e^*(x)$ is no greater than some predetermined rejection threshold λ_r, and rejects the sample pattern otherwise. Quite remarkable is the fact that the (unconditional) probability of error, or error rate, $E^*(\lambda_r)$, is related to the unconditional probability of reject, or reject rate, $R^*(\lambda_r)$, according to the simple relationship, [3]

$$E^*(\lambda_r) = - \int_0^{\lambda_r} \lambda \, d \, R^* (\lambda). \qquad (1)$$

This appeared to be a far reaching result. Some of its implications will be discussed in a subsequent section.

A very similar modification, aimed at allowing rejects with the k-NN rules, has been investigated by Hellman [4]. The modified rule, which is referred to hereafter as the (k,k')-NN rule, makes the same decision as the k-NN rule whenever one, or more than one class scores at least k' among the votes of the k nearest neighbors, otherwise the reject option is exercised. Hellman was able to relate the error rate $E_{k,k'}$ and the reject rate $R_{k,k'}$ of the (k,k')-NN rule to $E^*(\lambda_r)$ and $R^*(\lambda_r)$. For instance, taking k'=k, and letting λ_r be adjusted such that $R^*(\lambda_r) = R_{k,k}$ he showed that $E^*(\lambda_r) \leqslant E_{k,k} \leqslant (1+k/2) \, E^*(\lambda_r)$.

Our interest in this paper with focus on Hellman's (k,k')-NN rule, with the aim of establishing, in terms of this rule, a result that is strictly equivalent to that obtained by Chow in terms of the Bayes rule. More specifically we shall exhibit a general tradeoff relationship connecting the error rate $E_{k,k'}$ to the reject rate $R_{k,k'}$ and we shall show that the error rate can be directly evaluated from the reject function $R_{k+1,j}$ versus j, k' $\leqslant$ j $\leqslant$ k+1. Like for the Bayes rule, our results provide a basis for calculating the (k,k')-NN error rate from the empirical rejection curve, without actually identifying the errors. The relationship between the Bayes and k-NN rules will be further tightened by showing that our results include Chow's as a limiting case. An illustrative example will be discussed in detail. Eventually, the discussion section will raise a number of puzzling questions regarding error rate and reject rate estimation.

In order to enhance the readability of the paper, we shall limit ourselves to the discussion of the two-class case. This is little harm since the formulation of the results is not sensitive to the number of classes. The interested readers shoud have no difficulty in devising their own proofs in the multiclass case, and those who are short on time are advised to consider only the case k'=k.

2. PRELIMINARIES

Let the observation x belong to one of two possible classes, $\theta=\theta_1$ and $\theta=\theta_2$, with a priori probabilities π_1 and π_2 respectively. Suppose that x has the conditional probability density function $p_i(x)$ under $\theta=\theta_i$, i=1,2. We assume that $p_1(x)$ and $p_2(x)$ are both decomposable into a continuous component plus a series of mass points. We let

$$\eta_i(x) = \frac{\pi_i\, p_i(x)}{\pi_1 p_1(x) + \pi_2\, p_2(x)} \qquad (2)$$

designate the a posteriori probability of class θ_i, i=1,2.

Suppose that $\{x^i,\, \theta^i\}_1^m$ is a sequence of independent identically distributed sample pairs with the distribution of (x,θ). If d is some metric defined on the set of observations X and if x' is the nearest neighbor (NN) to x from the sequence $\{x^i\}_1^m$, according to d, then, by virtue of Cover and Hart's convergence lemma, [1], $\lim_m d(x,x') = 0$, with probability 1, and $\lim_m \eta_i(x')=\eta_i(x)$, with probability 1. These convergence properties extend to the kth NN to x, with k fixed. There follows that, in the large sample case, any of the k NNs to x belongs to class 1 with probability $\eta_1(x)$, and to class 2 with probability $\eta_2(x)=1-\eta_1(x)$: We are thus dealing with the outcomes of k Bernouilli trials. This observation leads us to make the following definitions and to state a number of lemmas which will be useful in the result section of the paper.

Let $b(r;n,p)$ be the probability that n Bernouilli trials with probabilities p for success and q=1-p for failure result in r successes and n-r failures. Then, [5],

$$b(r;n,p) = \binom{n}{r} p^r\, q^{n-r}. \qquad (3)$$

Let $S(r;n,p) = \Sigma_{j=r}^{n} b(j;n,p)$ designate the probability of having at least r successes in n trials, and $B(r;n,p) = \Sigma_{j=0}^{r} b(j;n,p)$ designate the probability of having at most r successes in n trials.

Lemma 1.

$$(1-p)\, b(r;n,p) = (1 - \frac{r}{n+1})\, b\,(r;n+1,p) \qquad (4)$$

$$p\, b(n-r;\, n,p) = (1 - \frac{r}{n+1})\, b\,(n-r+1;\, n+1,p) \qquad (5)$$

Lemma 2. (Cf. [5], p. 173, Eqs. (10.7))

$$B\,(r;n+1,p) = B\,(r;n,p) - p\, b(r;n,p) \qquad (6)$$

$$B\,(r+1;\, n+1,\, p) = B\,(r;\, n,p) + q\, b\,(r+1;\, n,p) \qquad (7)$$

Lemma 3.

$$S\,(r;n,p) = p\, S\,(r-1;\, n-1,\, p) + q\, S\,(r;\, n-1,\, p) \qquad (8)$$

$$B\,(r;\, n,\, p) = p\, B\,(r-1;\, n-1,\, p) + q\, B\,(r;\, n-1,\, p) \qquad (9)$$

The proofs of Lemmas 1 and 3 are relegated to Appendix I.

3. MAIN RESULTS

3.1. Error and reject tradeoff for general (k,k')-NN rules

Let $\eta=\eta_1(x)$ be the probability of occurrence of a class-1
NN to x, and $1-\eta=\eta_2(x)$ be the probability of occurrence of a
class-2 NN to x. The probability of occurrence of <u>exactly</u> k_1
class-1 NNs among the first k NNs to x is $b(k_1; k, \eta)$. The probability of occurrence of <u>at least</u> k' class-1 NNs among the first
k-NNs to x is $S(k'; k, \eta)$, whereas $B(k-k'; k, \eta)$ is the probability of occurrence of <u>at most</u> k-k' class-1 NNs -equivalently, at
least k' class-2 NNs- among the first k NNs to x.

For the (k,k')-NN rule, the conditional acceptance probability is

$$A_{k,k'}(x) = Pr \{k_1 \geqslant k' \ U \ k_1 \leqslant k-k' \mid k, x\}.$$

For $k' \geqslant (k+1)/2$, to avoid ties, this becomes,

$$A_{k,k'}(x)=Pr \{k_1 \geqslant k' \mid k,x\} + Pr \{k_1 \leqslant k-k' \mid k, x\}$$

$$=S (k'; k, \eta) + B (k-k'; k, \eta). \tag{10}$$

Assuming class conditional independence, the probability of
correctly classifying x is

$$C_{k,k'}(x) = Pr\{k_1 \geqslant k'\mid k, x\}. \ \ Pr\{x \text{ being from class 1}\}$$

$$+ Pr\{k_1 \leqslant k-k'\mid k, x\}. \ \ Pr\{x \text{ being from class 2}\}$$

$$= \eta S (k'; k, \eta) + (1-\eta) B (k-k'; k, \eta). \tag{11}$$

Likewise, for the conditional probability of error we have

$$E_{k,k'}(x) = (1-\eta) S(k'; k, \eta) + \eta B(k-k'; k, \eta). \tag{12}$$

Let $E\{-\}$ stand for the expectation operator with respect to
the unconditional distribution $p(x) = \pi_1 p_1(x) + \pi_2 p_2(x)$, and let
$A_{k,k'} = E\{A_{k,k'}(x)\}$ be the unconditional acceptance probability.
Similarly, let $C_{k,k'} = E\{C_{k,k'}(x)\}$ and $E_{k,k'} = E\{E_{k,k'}(x)\}$. We
may now state the following theorem :
<u>Theorem 1.</u>   For $k' \geqslant 1 + k/2$

$$A_{k,k'} = C_{k-1,k'-1} + E_{k-1,k'}. \tag{13}$$

<u>Proof.</u>   From Eq. (10) and Lemma 3 we can write

$$A_{k,k'}(x) = \boxed{\begin{aligned} &\eta S(k'-1;k-1,\eta) \\ &+ (1-\eta) B(k-k';k-1,\eta) \end{aligned}} \boxed{\begin{aligned} &+ (1-\eta) S(k';k-1,\eta) \\ &+ \eta B(k-1-k';k-1,\eta) \end{aligned}} \quad (14)$$

Grouping terms as shown by the dashed lines, and comparing the
resulting expressions to those appearing in Eqs. (11) and (12),
one readily sees that Eq. (14) can also be written

$$A_{k,k'}(x) = C_{k-1,k'-1}(x) + E_{k-1,k'}(x)$$

and the theorem follows by taking expectations. It is noted
that the condition $k' \geq 1 + k/2$ is introduced here in order to
avoid ties with the $(k-1, k'-1)$-NN rule.

Let $R_{k,k'}(x)$ and $R_{k,k'} = E\{R_{k,k'}(x)\}$ be the conditional and
unconditional probabilities of rejection respectively. The
first formulation for the error and reject tradeoff for (k,k')-
NN rule is now given by Corollary 1 that follows directly from
Theorem 1 and the obvious relation

$$R_{k,k'} + C_{k,k'} + E_{k,k'} = 1. \quad (15)$$

<u>Corollary 1.</u>   For $k' \geq 1 + k/2$,

$$R_{k,k'} - R_{k-1,k'-1} = E_{k-1,k'-1} - E_{k-1,k'}. \quad (16)$$

From noting the boundary conditions

$$E_{k,k'} = 0, \ R_{k,k'} = 1 \quad \text{if } k' > k; \quad (17)$$

and $R_{k,k'} = 0 \quad \text{if } k' < (k+1)/2, \ k \text{ odd}, \quad (18)$

$$k' \leq k/2, \ k \text{ even};$$

and using the recursivity of Eq. (16), there follows
<u>Corollary 2.</u>   For $k' \geq (k+1)/2$,

$$E_{k,k'} = \sum_{j=k'}^{k} [R_{k+1,j+1} - R_{k,j}]. \quad (19)$$

From Eq. (19) it is seen that for any given k and k' the
error rate $E_{k,k'}$ can be determined from the knowledge of the
reject functions $R_{k+1,j+1}$ and $R_{k,j}$ versus j, $j=k'$, $k'+1,\ldots,k$.
Though, as will be seen hereafter there is still a considerable
amount of redundancy in Eq. (19). In fact, we shall presently
show that the knowledge of $R_{k,j}$, $j=k',\ldots,k$ is not required for
determining $E_{k,k'}$. This somewhat unexpected observation results

from the fact that reject rates are connected by the relation
given in the following theorem :

<u>Theorem 2.</u>   For $k' > 1 + k/2$,

$$R_{k+1,k'+1} - R_{k,k'} = (1 - \frac{k'}{k+1})[\, R_{k+1,k'+1} - R_{k+1,k'}\,]. \tag{20}$$

Equivalently,

$$R_{k+1,k'+1} = R_{k,k'} + \frac{k-k'+1}{k}\,(R_{k,k'} - R_{k+1,k'}). \tag{21}$$

The proof of this theorem is deferred to Appendix II.

By combining Corollary 1 and Theorem 2, we obtain a second
formulation for the error and reject tradeoff :

<u>Corollary 3.</u>   For $k' > \frac{k}{2} + 1$,

$$E_{k-1,k'-1} - E_{k-1,k'} = (1 - \frac{k'-1}{k})\,[\,R_{k,k'} - R_{k,k'-1}\,] \tag{22}$$

$$= (\frac{k}{k'-1} - 1)\,[\,R_{k-1,k'-1} - R_{k,k'-1}]. \tag{23}$$

Likewise, from Corollary 2 and Theorem 2, we have

<u>Corollary 4.</u>   For $k' > (k+1)/2$,

$$E_{k,k'} = \sum_{j=k'}^{k} (1 - \frac{j}{k+1})\,[\,R_{k+1,j+1} - R_{k+1,j}\,] \tag{24}$$

$$= \frac{1}{k+1}\sum_{j=k'}^{k} R_{k+1,j+1} - (1 - \frac{k'}{k+1})\,R_{k+1,k'}. \tag{25}$$

Equation (22) admits of the following interpretation. The
decrease in error rate due to a unit increment in the rejection
threshold in the $(k-1,k'-1)$-NN rule is equal to a fraction
$(k-k'+1)/k$ of the decrease in reject rate due to a unit decrement
in the rejection threshold in the (k,k')-NN rule. Eq. (23)
relates the same decrease in error rate to the decrease in reject
rate subsequent to a unit increment in the number of NNs searched
for a fixed rejection threshold. Eqs. (24) and (25) indicate
that, as was previously asserted, the error rate of the (k,k')-NN
rule can be determined from the knowledge of the reject function
$R_{k+1,j}$, $j = k',k'+1,\ldots,k+1$.

<u>Remark.</u>

In the derivation of the above results, we have been consistently
excluding the case when ties may occur. There are two reasons
for this. First, for the two-class problem, ties can only occur
when $k = 2k'$, in which case $R_{k,k'} = 0$ (as reflected in the boun-

dary condition of Eq. (18)). This is clearly a situation that
is not relevant to our main preoccupation. Second, taking ties
into account would have forced us to reflect in Eqs. (10)-(12)
the necessity of randomizing the decision rule when $k_1=k_2=k/2$.
While there is no essential difficulty to incorporate the necessary refinements, this could only have been done at the price
of a significant loss in the clarity of the derivation.

3.2. Error and reject tradeoff for (k,k)-NN rules

As should be clear from the above, the (k,k)-NN rules are
those requiring unanimous votes from the k NNs for making a
decision; the pattern is rejected whenever the k NNs do not
belong to the same class : A very conservative class of rules
indeed.

Taking k'=k brings about a number of simplifications in the
results of the previous section. In particular, Theorem 1 becomes

$$A_{k,k} = C_{k-1,k-1},\tag{26}$$

whereas Corollaries 1 and 2 yield

$$E_{k-1,k-1} = R_{k,k} - R_{k-1,k-1}\tag{27}$$

for all $k \geq 2$. From Corollaries 3 and 4, we also have

$$E_{k-1,k-1} = \frac{1}{k}[R_{k,k} - R_{k,k-1}],\ k > 2.\tag{28}$$

The simplest form for the error and reject tradeoff is
obtained by taking k = 2 in Eqs. (26) or (27) :

$$E_1 = R_{22}\tag{29}$$

where E_1 stands for E_{11}, i.e. the error rate of the single nearest neighbor decision rule. It is noted that Eq. (29) was first
obtained by Hellman [4], who derived it via a quite different
route. The interpretation of the results in this section is
straightforward and will not be explicited.

3.3. Error and reject tradeoff for the Bayes decision rule

It is known that as $k \to \infty$, with $k'/k \to 1-\lambda_r$, the (k,k')-NN
rule approaches the Bayes rule with threshold λ_r, [4]. This
follows from the fact that, by the strong law of large numbers,
the frequency of occurrence of class i NNs to x, i.e. k_i/k converges to the a posteriori probability $\eta_i(x)$, i = 1,2. Consequently, if the Bayes conditional error probability $e^*(x) = \min$
$(\eta_1(x), \eta_2(x))$ satisfy the condition $e^*(x) \leq \lambda_r$, then $\lim_k \min$

$(\frac{k_1}{k} , \frac{k_2}{k}) \leqslant \lambda_r$, $R_{k,k'}(x) \to 0$ and $E_{k,k'}(x) \to e^*(x)$, whereas if $e^*(x) > \lambda_r$, then $R_{k,k'}(x) \to 1$ and $E_{k,k'}(x) \to 0$. On the basis of this observation, one could suspect that, in the limit, the error and reject rates of the (k,k')-NN rules are bound to obey the law relating the Bayes error and reject rates (Cf. Eq. (1)). We shall presently show that this is indeed the case.

Let $E_{k,k'}$ and $R_{k,k'}$ be designated by $E_k(\lambda)$ and $R_k(\lambda)$ with $\lambda = 1 - k'/k$, and let $\Delta\lambda = 1/k$. Thus, for $R_{k,k'-1} = R_k(\lambda')$, we have $\lambda' = 1 - \frac{k'-1}{k} = \lambda + \Delta\lambda$ and $R_k(\lambda') = R_k(\lambda + \Delta\lambda)$. Next, let $\Delta R_k(\lambda)$ denote the (negative) increment in R_k subsequent to a change from λ to $\lambda + \Delta\lambda$. There follows,

$$(1 - \frac{k'-1}{k}) [R_{k,k'} - R_{k,k'-1}] = (\lambda+\Delta\lambda) [R_k(\lambda) - R_k(\lambda+\Delta\lambda)]$$

$$= (\lambda+\Delta\lambda)(-\Delta R_k(\lambda))$$

$$= - \lambda\Delta R_k(\lambda) - \Delta\lambda\Delta R_k(\lambda). \quad (30)$$

Likewise, let $E_{k-1,k'-1} = E_{k-1}(\lambda'')$ with

$$\lambda'' = 1 - \frac{k'-1}{k-1} = \lambda + \lambda\Delta\lambda + 0^2(\Delta\lambda),$$

and $E_{k-1,k'} = E_{k-1}(\lambda''')$ with

$$\lambda''' = 1 - \frac{k'}{k-1} = \lambda + (\lambda-1) \Delta\lambda + 0^2(\Delta\lambda).$$

It is readily seen that $E_{k,k'}(x)$ is a continuous function of $\lambda(\lambda \geqslant 0)$, thus so is $E_{k,k'} = E\{E_{k,k'}(x)\}$. Making a linearity assumption we obtain

$$E_{k-1,k'-1} - E_{k-1,k'} = E_{k-1}(\lambda+\lambda\Delta\lambda+0^2\Delta\lambda) - E_{k-1}(\lambda+(\lambda-1)\Delta\lambda+0^2\Delta\lambda)$$

$$= \Delta E_{k-1}(\lambda) + 0^2\Delta\lambda, \quad (31)$$

where $\Delta E_{k-1}(\lambda) = \Delta E_{k-1}(\lambda+\Delta\lambda) - \Delta E_{k-1}(\lambda)$. Substitution of Eqs. (30) and (31) into Eq. (22) yields

$$\Delta E_{k-1}(\lambda) + 0^2\Delta\lambda = - \lambda\Delta R_k(\lambda) - \Delta\lambda\Delta R_k(\lambda). \quad (32)$$

We may now sum Eq. (32) over the range of interest, that is for $\lambda = \lambda_r = 1 - k'/k$ to $\lambda = 0$ to obtain

$$E_{k-1}(\lambda_r) + 0^2\Delta\lambda = - \sum_{\lambda=0}^{\lambda=\lambda_r} \lambda\Delta R_k(\lambda) - \sum_{\lambda=0}^{\lambda=\lambda_r} \Delta\lambda\Delta R_k(\lambda). \tag{33}$$

Now letting $k \rightarrow \infty$, $k' \rightarrow \infty$ with $k'/k = 1 - \lambda, \Delta\lambda \rightarrow 0$, the higher order terms vanish and we are left with the following Stieltjes integral

$$E_\infty(\lambda_r) = - \int_o^{\lambda_r} \lambda dR_\infty(\lambda) \tag{34}$$

which one readily recognizes as the error and reject tradeoff for the Bayes decision rule [3] .

Before engaging into a discussion of these results, we shall examine the behaviour of the various tradeoffs with the help of an illustrative example.

4. EXAMPLE

We consider Cover and Hart's example where the real valued random variable x has the triangular densities $p_1(x) = 2x$ and $p_2(x) = 2-2x$, $0 \leqslant x \leqslant 1$, with a priori probabilities $\pi_1 = \pi_2 = 1/2$. The unconditional density is uniform on [0,1] and the a posteriori probabilities are $\eta_1 = x$ and $\eta_2 = 1-x$ respectively. A simple calculation shows [6] that for the Bayes decision rule with rejection threshold λ_r, we have $E^*(\lambda_r) = \lambda_r^2$, $R^*(\lambda_r) = 1-2\lambda_r$ and the tradeoff curve is

$$E^*(\lambda_r) = (1/4) [1-R^*(\lambda)]^2. \tag{35}$$

For the (k,k')-NN rules, we have

$$R_{k,k'} = 1 - A_{k,k'}$$

$$= 1 - \sum_{j=k'}^{k} \binom{k}{j} \int_0^1 [x^j(1-x)^{k-j} + x^{k-j}(1-x)^j] dx$$

$$= 1 - 2 \frac{k-k'+1}{k+1} = \frac{2k'}{k+1} - 1. \tag{36}$$

Introducing this into the expression of $E_{k,k'}$ we obtain

$$E_{k,k'} = \frac{(k-k'+1)(k-k'+2)}{(k+1)(k+2)} . \tag{37}$$

For the probability of correct classification $C_{k,k'}$ one finds

$$C_{k,k'} = \frac{(k-k'+1)(k+k'+2)}{(k+1)(k+2)} . \tag{38}$$

For a given k, $R_{k,k'}$ and $E_{k,k'}$ are uniquely functions of k', $(k+1)/2 \leqslant k' \leqslant k+1$. Elimination of k' yields the tradeoff equation

$$E_{k,k'} = \frac{1 - R_{k,k'}}{4(k+2)} [2 + (k+1)(1 - R_{k,k'})] . \tag{39}$$

Letting $k \to \infty$ and using the notation of Section 3.3 Eq. (39) becomes

$$E_\infty(\lambda) = \frac{1}{4} [1 - R_\infty(\lambda)]^2 \tag{40}$$

as previously asserted in terms of the Bayes rule.

With the distributions in this example, the (k,k)-NN rules are also giving rise to interesting observations. One alternative form for Eq. (26) is

$$C_{k-1,k-1} = 1 - R_{k,k}. \tag{41}$$

On the other hand, from $R_{k,k} = (k-1)/(k+1)$ and $E_{k-1,k-1} = 2/k(k+1)$ we derive the tradeoff relations

$$E_{k-1,k-1} = (1 - R_{k,k})^2/(1 + R_{k,k}), \tag{42}$$

and $E_{k,k} = (1 - R_{k,k})^2/(3 - R_{k,k}).$ $\tag{43}$

Now, from Eqs. (41) and (42) we see that the performance of the (k-1,k-1)-NN rule can be completely determined once we know the reject rate of the (k,k)-NN rule. It is also easily seen, on this example, that the tradeoff is most effective when k is small and becomes more difficult as k increases : For large k, many "would be correct" decisions are converted into rejects.

5. DISCUSSION

The various tradeoff relations presented in the previous sections relate error rate to reject rate. These are two definitely different concepts based on two different bodies of knowledge. Indeed, the very notion of error rate involves comparing the decision being made to that that should have been made, viz., the true class identification of test patterns. At the contrary, a test pattern is accepted or rejected on the sole basis of the votes of its k NNs. The class information for test patterns is

playing no role in this business. Thus, the tradeoff relationships offer the possibility of calculating the error rate without
actually identifying the recognition errors. This observation,
which was first made by Chow in terms of the Bayes rule [3],
and which applies equally-well to the (k,k')-NN rules has led
Fukunaga and Kessel to investigate error estimates based on
unclassified test samples [7]. One chief characteristic of the
error estimate they obtained was a substantial reduction in the
variance of the estimate as compared to that of an estimate based
on error-count. This is a quite paradoxical result since they
were getting a better estimate by ignoring the class identification of test patterns [8]. We would like to uncover another
aspect of the paradox by exhibiting a simple example where a
reject-count yields a much more reliable result than an error-
count.

Let $\{x^i, \theta^i\}_1^m$ be a sequence of iid. random variables with
the distribution of (x,θ), (here we restrict ourselves to the
case where the class conditional densities are continuous).
Suppose we want to estimate the quantities involved in Eq. (29),
viz.,

$$E_1 = R_{22} \tag{29}$$

from the available data, by using the resubstitution scheme [9].
A little tought reveals that this scheme consistently yields an
overly optimistically biased estimate of zero for E_1. At the
contrary, the resubstitution estimate of R_{22} based on $\{x^i, \theta^i\}_1^m$,
is easily seen to be identical to the unbiased deleted estimate
of E_1. Hence, in this example, the reject-counting technique
definitely outweighs the error-counting technique.

Some of our result raise other, interesting questions regarding error and reject estimates. Consider, for example, determining the deleted estimates of the quantities intervening in
Eq. (27), viz.,

$$E_{k-1,k-1} = R_{k,k} - R_{k-1,k-1}. \tag{27}$$

The deleted estimate of $E_{k-1,k-1}$ is the empirical frequency of
occurrence, in $\{x^i, \theta^i\}_1^m$, of samples from class 1, (2) surrounded by k-1 NNs from class 2, (1), whereas the deleted estimate
of $R_{k,k} - R_{k-1,k-1}$ is the empirical frequency of occurrence of
one sample from either class surrounded by k-1 NNs from class 1,
(2) and whose kth NN is from class 2, (1). While both counts
are based on k votes, they expect the sequence of votes to unfold
in a quite different manner. Whether such estimates based on
reject-counts will exhibit some peculiar properties still remains
to be investigated.

REFERENCES

1. T.M. Cover and P.E. Hart, "Nearest Neighbor pattern classifi-
    cation", IEEE Trans. Inform. Theory, Vol. IT-13, pp. 21-27,
    Jan. 1967.
2. C.K. Chow, "An optimum character recognition system using
    decision functions", IRE Trans. Electronic Computers, Vol.
    EC-6, pp. 247-254, Dec. 1957.
3. C.K. Chow, "On optimum recognition error and reject tradeoff",
    IEEE Trans. Inform. Theory, Vol. IT-16, pp. 41-46, Jan. 1970.
4. M. Hellman, "The Nearest Neighbor classification rule with a
    reject option", IEEE Trans. on Syst. Sci. Cybern., Vol. SSC-6,
    pp. 179-185, July 1970.
5. W. Feller, An Introduction to Probability Theory and its
    Applications, Vol. I. 3rd Ed. New York : Wiley, 1968.
6. P.A. Devijver, "Decision theoretic and related approaches to
    pattern classification", MBLE Res. Rept. R 308, Nov. 1975.
7. K. Fukunaga and D.L. Kessel, "Application of optimum error-
    reject functions", IEEE Trans. Inform. Theory, Vol. IT-18,
    pp. 814-817, Nov. 1972.
8. L.N. Kanal, "Patterns in Pattern Recognition : 1968-1974",
    IEEE Trans. Inform. Theory, Vol. IT-20, pp. 697-722, Nov.
    1974.
9. G.T. Toussaint, "Bibliography on estimation of misclassifica-
    tions" IEEE Trans. Inform. Theory, Vol. IT-20, pp. 472-479,
    July 1974.

APPENDIX I.

Proof of lemma 1.

$$b(r; n+1, p) = \binom{n+1}{r} p^r (1-p)^{n+1-r}$$

$$= \frac{n+1}{n+1-r} (1-p) \binom{n}{r} p^r (1-p)^{n-r}$$

$$= \left(1 - \frac{r}{n+1}\right)^{-1} (1-p) \, b(r; n, p)$$

The proof of Eq. (5) is similar.

Proof of lemma 3.

From the combinatorial identity $\binom{n}{j} = \binom{n-1}{j} + \binom{n-1}{j-1}$ and the definition of $S(r; n, p)$, we have

$$S(r; n, p) = \sum_{j=r}^{n} \binom{n-1}{j} p^j q^{n-j} + \sum_{j=r}^{n} \binom{n-1}{j-1} p^j q^{n-j} \qquad (I.1)$$

From noting $\binom{n-1}{n} = 0$, we can write

$$\sum_{j=r}^{n} \binom{n-1}{j} p^j q^{n-j} = q \sum_{j=r}^{n-1} \binom{n-1}{j} p^j q^{n-1-j}$$

$$= q\, S(r;\, n-1,\, p) \qquad\qquad (\text{I.2})$$

Making the change of variable $i = j-1$ in the last term of Eq. (I.1) yields

$$\sum_{j=r}^{n} \binom{n-1}{r-1} p^j q^{n-j} = \sum_{i=r-1}^{n-1} \binom{n-1}{i} p^{i+1} q^{n-i-1}$$

$$= p \sum_{i=r-1}^{n-1} \binom{n-1}{i} p^{i} q^{n-1-i}$$

$$= p\, S(r-1;\, n-1,\, p) \qquad\qquad (\text{I.3})$$

Substitution of Eqs. (I.2) and (I.3) in (I.1) yields Eq. (8). The proof of Eq. (9) is similar.

APPENDIX II

<u>Proof of theorem 2.</u>
It is to be shown that

$$R_{k+1,k'+1} - R_{k,k'} = \left(1 - \frac{k'}{k+1}\right) \left[R_{k+1,k'+1} - R_{k+1,k'} \right]. \qquad (\text{II.1})$$

From Eq. (10) and $R_{k,k'} = 1 - A_{k,k'}$, we have

$$R_{k+1,k'+1}(x) - R_{k,k'}(x) = \begin{bmatrix} B(k';k+1,\eta) \\ -B(k'-1;k,\eta) \end{bmatrix} \begin{bmatrix} - B(k-k'\ ;k+1,\eta) \\ + B(k-k';k,\eta) \end{bmatrix}$$

Grouping terms as indicated and using lemma 2 yields

$$R_{k+1,k'+1}(x) - R_{k,k'}(x) = (1-\eta)\, b(k';k,\eta) + \eta b(k-k';k,\eta) \qquad (\text{II.2})$$

$$= \left(1 - \frac{k'}{k+1}\right)\left[b(k';k+1,\eta) + b(k-k'+1;k+1,\eta) \right] \qquad (\text{II.3})$$

In going from Eq. (II.2) to Eq. (II.3) we have been using Lemma 1. On the other hand,

$$R_{k+1,k'+1}(x)-R_{k+1,k'}(x) = \boxed{\begin{array}{c} B(k';k+1,\eta) \\ -B(k'-1;k+1,\eta) \end{array}} \boxed{\begin{array}{c} -B(k-k';k+1,\eta) \\ +B(k-k'+1;k+1,\eta) \end{array}}$$

$$= \; b(k';k+1,\eta)+b(k-k'+1;k+1,\eta). \quad (II.4)$$

Substitution of Eq. (II.4) in Eq. (II.3) yields

$$R_{k+1,k'+1}(x)-R_{k,k'}(x) = (1-\frac{k'}{k+1})[\,R_{k+1,k'+1}(x)-R_{k+1,k'}(x)\,]\,,$$

and the theorem follows by taking expectations.

TWO-DIMENSIONAL SPECTRAL ANALYSIS FOR IMAGE CODING, PROCESSING
AND RECOGNITION: AN APPROACH BASED ON SOME PROPERTIES OF THE
HUMAN VISUAL SYSTEM *

C. Braccini, G. Gambardella

Istituto di Elettrotecnica, Università di Genova
Genova, Italy

INTRODUCTION

The main purpose of this work is showing the use in the field
of digital image processing (such as coding, enhancement and recognition) of what has been learned so far about the functioning of
the peripheral visual system in man.

Being such a system essentially modeled as a spectral analyzer, emphasis will be put on those areas and techniques of image
processing which are closely related to two-dimensional spectral
analysis. However, owing to the tutorial nature of the paper, a
short account of some general problems and techniques in digital
image processing is also given (section 1).

In section 2 a mathematical model is presented of the early
processing stages in the visual system with reference to still
monochrome images.

Section 3 is devoted to the theory of constant Q filtering,
which is part of the model, and presents, anyhow, a quite general
interest.

Among the numerous techniques of image coding and enhancement,
emphasis is put on those taking advantage of the above mentioned
model. Sections 4 and 5 are devoted, in this sense, respectively
to image coding and image enhancement techniques.

---

* This work has been partly supported by the Consiglio Nazionale
  delle Ricerche (C.N.R.)

In section 6 the problem of texture analysis is dealt with, where a correspondence exists, as we shall see, with some further properties of the peripheral visual system in man.

Finally, in section 7, mention will be given to the object detection problem through two-dimensional matched filtering. This is a fundamental technique in the field of two-dimensional signal processing and constitutes an extension of a well known result in statistical communication theory.

1. AN INTRODUCTION TO DIGITAL IMAGE PROCESSING TECHNIQUES

This section is intended to give a brief account of some general problems and techniques in digital image processing. Our approach does not relate to the whole field being mainly oriented toward signal processing rather than artificial intelligence (say "scene analysis") techniques. Even in the restricted area of signal processing, we will not cover some relevant applications, such as the reconstruction of objects from their projections, which have already led to special purpose machines implementation. The specific subjects that we shall briefly discuss are the following:

a) representation and filtering of images [1, 2]
b) improvement of the quality of images: enhancement and restoration [1, 2, 3]
c) redundancy reduction (coding) for transmission and storage purposes [1, 3, 4]
d) pattern classification [5, 6].

a) The digital representation of images usually implies such problems as sampling, quantizing and displaying. The theoretical basis for dealing with sampling is the Nyquist theorem. Sample quantization, on the other hand, does not follow a uniform rule: the noise effects and the properties of the human observer should be taken into account in order to find, for each class of images, the best quantization rule.

The problem of image displaying strongly relates to the field of "computer graphics", which deals with displaying devices as well as with the software techniques for handling pictures as inputs and outputs from the computer.

As far as digital filtering of images is concerned, it is possible to use and sometimes generalize techniques that have already proved effective in one-dimensional filtering (such as the fast convolution algorithms based on the Fast Fourier Transform). On the other hand the design of two-dimensional filters requires the solution of some specific and new problems [2].

b) The aim of image enhancement is improving the quality of an image for either pre-processing purposes or allowing a more satisfactory subjective viewing. In the latter case the enhancement techniques may take full advantage of a proper modeling of the visual system.

This does not generally apply to the case of the restoration techniques which are aimed to counteract the degradation effect due to an imaging system. Modeling such a degradation is not an easy task since many effects may be involved [1], such as diffraction within the optical systems, sensor or film non-linearities, aberrations, atmospheric turbulence, image motion blur, geometric distortions, noise due to the electronic circuitry or to the film granularity. In the general case the degradation process is non-linear, although the linearity assumption is usually made. In its general form the restoration problem is that of estimating an unknown image from a measured one, under some assumptions on the degradation system.

In its simplest form the restoration problem is that of recovering the image f from the measured image g= Hf where H is a linear operator. Inverting the system g= Hf is a typical problem of numerical analysis which, however, is severely complicated by the dimensionality and the ill conditioning of the system. The solutions via the inverse filter are faced, indeed, with serious problems of realizability. The restoration becomes even more difficult when the degradation is space-variant and affected by noise. In some circumstances, however, satisfactory solutions have been found [7, 8]. Some of the techniques which appear to be useful in these cases are the "singular value decomposition" [9] and the projection iterative method [10].

When the degradation, besides being linear and shift invariant, is affected by additive indipendent noise:

$$g = Hf + n$$

the optimum filter (in the sense of the MMSE) can be found and the solution implemented in several ways (not necessarily in the Fourier domain [11]). This (Wiener) approach requires the a priori knowledge of the point spread function (PSF) of the degradation system together with the image and noise covariances. Knowing the

latters is not required, however, if the constrained least squares
estimation is used [12]. Further mention of this technique will
be given in section 5. Related to this approach are the techniques
adding linear equality and inequality constraints [13] and imply-
ing the solution of quadratic programming problems. A peculiar
and efficient way of constraining the estimate is that based on
the Maximum Entropy Method [2, 14].

In the case of an unknown PSF, this can be estimated by
averaging image degraded segments in the log-spectral domain.Such
a technique is an application of the homomorphic filtering theory
[15, 16], and is referred to as "blind deconvolution".

Finally, owing to their general interest , we want to mention
the techniques of bandwidth extrapolation (of band-limited images)
aimed to improve the spatial resolution. This approach,usually
referred to as "superresolution" [2, 17], may be severely constrained, however, by noise and stability problems.

c) A few words will be spent now on the redundancy reduction techniques (see also section 4). These are essentially aimed to take
advantage (for bit saving purposes) of the strong correlation
existing among the samples of most images. A classical approach
is that of "transform coding" [3, 18]. Whenever the receiver of
the coded-decoded image is the human eye, then the coding techniques may take advantage also from the known performance, that is
the model, of such a receiver. In other words the redundancy can
be reduced so that the degradation effect is hardly perceived by,
or is less disturbing to, the observer.

d) The image processing techniques for pattern recognition tasks
are difficult to classify or summarize being generally task and
image dependent.

It is possible to state, however, that their theoretical
base essentially lies in decision theory, parameter estimation
theory and clustering.

Sometimes patterns are classified according to their textures
and therefore texture discrimination techniques become relevant
and have been already implemented in special purpose machines
(see, e.g.[19]).

Some of these techniques will be described in section 6.

Among the most widely used techniques for pattern classification and recognition, template matching and matched filtering are
worth mentioning.

A brief account of the latter will be given in section 7.

2. MODEL OF THE PERIPHERAL VISUAL SYSTEM

As far as perception of still monochrome images is concerned, there is general agreement about the existence of three very early processing stages within the human visual system. These stages are characterized respectively by:
a) a non linear convex transformation of the stimulus intensities;
b) a band-pass linear filtering;
c) a spectral analysis performed through almost constant-Q band-pass filters.

a) The response of the system to intensity changes of the stimulus is indeed of a logarithmic type for most of the intensity range [1, 20, 21, 22]. An equivalent formulation of this property is that the just noticeable intensity difference ΔI is an almost constant fraction ($\sim 2\%$) of the intensity I, regardless to the form of the stimulus.

Such a logarithmic transformation is essentially attributed to the primary receptors (rods and cones) and therefore it is localized at the beginning of the processing within the visual system.

b) The next processing is a linear filtering of the stimulus. Linearity holds at least for a moderate range of the intensities. This filtering can be modeled in terms of a convolution of the stimulus with a shift-invariant point-spread function, or, equivelently, in terms of filtering (say product) in the domain of spatial frequencies [1, 20, 21, 22] .

The amplitude of the characteristic function of the filter, usually called the Modulation Transfer Function (MTF), has been evaluated at spatial frequencies of different orientations. Most measurements have been performed at threshold level, that is looking for the minimum contrast allowing the detection of a given sinusoidal grating.

The so measured M.T.F.'s show up a band-pass shape with a maximum value placed at a spatial frequency of a few cycles per degree.

A mathematical isotropic model of this function is, e.g., the one suggested by Mannos and Sakrison [22]:

$$A(\omega) = \left[c + (\omega/\omega_0)^K \right] \, \exp \left[- (\omega/\omega_0)^h \right]$$

where $\omega = \omega_0$ corresponds to the maximum of $A(\omega)$ and the meaningful range of ω is about 0.2 ÷ 40 cycles/degree.

This band-pass filtering can be looked at as the combination of a low-pass and a high-pass effect. The latter is usually attributed to the so called "lateral inhibition" phenomenon taking place already at the retinal level.

On the other hand the low-pass effect is attributed to the finite size and spacing of the photoreceptors. It should be noticed that the stimulus reaching the retina has already undergone a low-pass filtering through the lenses of the eye. This is, however, a second order effect (usually) and can be ignored with respect to the processing taking place at the receptor level.

c) A further processing of the stimulus taking place within the peripheral visual system is a particular type of spectral analysis. There is, infact, enough evidence about the existence "within the nervous system of linearly operating indipendent mechanisms selectively sensitive to limited ranges of spatial frequencies"[23]. Such an evidence has been reached on the base of experiments where the stimuli are "gratings", either simple or complex (that is made up of several sinusoidal components). It has been shown that the sinusoidal components of a complex grating are detected indipendently at threshold, provided that the components are separated by at least one octave. Further support to this hypothesis of spatial frequency channels acting independently comes from the experiments on adaptation to gratings and noise-masking.

The latter effect has been studied over a variety of stimuli, i.e. not limited to gratings. It has been shown that not only gratings, but also entire scenes, can be recognized in "visual noise" if the spatial frequencies making up the noise are, in any given direction, sufficiently different from the spatial frequency components of the grating or the scene. From a quantitative point of view, it has been shown that, given a sinusoidal input and a band of noise centered at its frequency, the masking effect increases with the band of noise up to a "critical" value of $\pm$ one octave, and does not increase further when the band is widened beyond this range [24].

On the base of the above mentioned evidence it is possible to assume the existence, for any given direction, of a bank of parallel band-pass filters and subsequent processors associated to them. As far as the bandwidths of these filters are concerned, the evidence is that they are almost proportional to the corresponding center frequencies [22, 24, 25].

By putting together the three processing stages mentioned

above (a, b and c) one may end up with the following crude functional schema of the processing performed by the peripheral visual system:

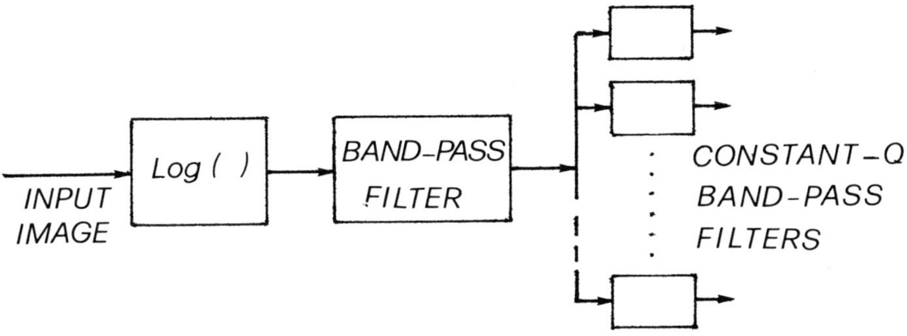

It should be stressed that this schema represents only a part of the processing taking place within the peripheral visual system. This is particularly true since the physical location, along the visual pathway, of the constant-Q band-pass filters is still very much uncertain.

As we shall see, the utility of the above schema is twofold: first of all it allows a synthetic representation of many properties or "effects" characterizing the visual perception; second, it gives good hints in designing efficient methods of image processing, for enhancement and coding purposes.

The next section is devoted to the theory of spectral analysis performed through banks of constant - Q band-pass filters, which are an essential part of the above schema.

3. PROPERTIES OF CONSTANT-Q BAND-PASS FILTERS

Let us call $f(x,y)$ the stimulus (say image) undergoing a spatial frequency analysis along the direction θ , through a parallel bank of band-pass filters. Under the assumption that the filters are all the same (except that for their center frequencies), we maintain that the spectral analysis undergone by $f(x,y)$ can be mathematically modeled through the following Fourier integral:

$$\iint_{-\infty}^{+\infty} f(\xi - x, \eta - y)\, w(x,y) e^{-j\omega(x \cos\theta + y \sin\theta)} dx\, dy \qquad (1)$$

where (ξ, η) are the coordinates of the "running point" (analogous
to the "running time" in the case of the time waveforms) and
$w(x,y)$ is a low-pass weighting function originating all the
point spread functions of the band-pass filters.

In fact, integral (1) can also be written as:

$$\iint_{-\infty}^{+\infty} f(\xi - x, \eta - y)w(x,y)\,\cos\omega(x\,\cos\theta + y\,\sin\theta)\,d\,x\,d\,y \; -$$

$$-j \iint_{-\infty}^{+\infty} f(\xi - x, \eta - y)w(x,y)\sin\omega(x\,\cos\theta + y\,\sin\theta)\,d\,x\,d\,y$$

where the two integrals can be interpreted as the output (namely
a function of (ξ, η)) of two Fourier-complementary band-pass filters
whose point spread functions an respectively given by
$w(x,y)\cos\omega(x\,\cos\theta + y\,\sin\theta)$ and $w(x,y)\sin\omega(x\,\cos\theta + y\,\sin\theta)$.
The parameter ω clearly plays the role of center frequency of
these two filters.

The envelope of the output of any of these two filters can
be taken as a measure of the short space amplitude spectrum of
$f(x,y)$ and it is essentially coincident with the magnitude of
integral (1). Therefore, in order to compute such a magnitude,
or its square (which is the short-space energy spectrum of $f(x,y)$),
it is sufficient, from the physical point of view, only one of
the two Fourier-complementary filters. Of course, as many filters
are necessary as the ω values at which an estimate of the energy
spectrum of $f(x,y)$ is wanted.

We have then shown how integral (1) is an adequate mathematical representation of the spectral analysis performed through
a bank of band-pass filters all having the same equivalent low-
pass point spread function $w(x,y)$. The frequency resolution of
such a spectral analysis is given by the bandwidth of the filters
and therefore it is inversely proportional to the spatial extent
of $w(x,y)$ along the direction θ . Consequently the resolution
of the analysis does not depend on ω . Furthermore, in case
$w(x,y)$ is a radially symmetric function, the resolution is also
θ -invariant.

Let us now face the problem of finding a mathematical representation of the short space spectral analysis performed by means
of band-pass filters whose bandwidth depend on their respective
center frequencies ω [26]. We maintain that such a representation is given by the integral:

$$\iint_{-\infty}^{+\infty} f(\xi - x, \eta - y)w(x,y,\omega)e^{-j\omega(x\cos\theta + y\sin\theta)}\,dxdy \qquad (2)$$

In fact, by means of a few formal steps, strictly parallel to those performed with respect to (1), one can show that, at any given ω, the magnitude of integral (2) is essentially coincident with the envelope of the output of a band-pass filter whose point spread function is

$$h(x,y,\omega_c) = w(x,y,\omega_c)\cos\omega_c(x\cos\theta + y\sin\theta)$$

where the parameter ω has been substituted by ω_c in order to emphasize its meaning of center frequency of the filter. The function $w(x,y,\omega_c)$, on the other hand, has the meaning of the low-pass equivalent function (practically the envelope) of $h(x,y,\omega_c)$.

Since, this time, such a low-pass equivalent function depends on ω_c, it follows that the bandwidths of the filters, and, hence, the frequency resolution of the spectral analysis represented by (2), also depend on ω_c. If, e.g., the width of $w(x,y,\omega_c)$ along the direction θ is inversely proportional to ω_c, then the band-width of the filters grows proportionally to ω_c, so that integral (2) ends up to represent just the spectral analysis performed through a bank of constant-Q filters.

Obviously (2) is not a Fourier integral, although it may be considered a generalization of it [27]. The relationship between Fourier transforms and (2) is better understood if the latter is written in the form:

$$\left[\iint_{-\infty}^{+\infty} f(\xi - x, \eta - y)w(x,y,\omega_c)e^{-j\omega(x\cos\theta + y\sin\theta)}\,dxdy\right]_{\omega=\omega_c}$$

The above expression shows that (2) is the value at $\omega = \omega_c$ of the Fourier transform, along the direction θ, of the function $f(\xi - x, \eta - y)w(x,y,\omega_c)$.

Integral (2) has several interesting mathematical properties. We shall show two of them.

a) The first one refers to the square magnitude of integral (2). It is possible to show that if this square magnitude is ideally averaged over all the values (ξ, η), one obtains:

$$\iint_{-\infty}^{+\infty} \Phi_\theta(\nu,\mu)\,\left|W_\theta(\nu-\omega_c,\mu,\omega_c)\right|^2\,d\nu\,d\mu \qquad\qquad (3)$$

where $\Phi_\theta(\omega,\mu)$ is the energy density spectrum of $f(x,y)$ along
the directions θ and $\theta + \pi/2$, while $W_\theta(\omega,\mu,\omega_c)$ is the
Fourier transform of $w(x,y,\omega_c)$ computed along the same directions θ and $\theta + \pi/2$. The frequencies ω and μ are respectively associated to the directions θ and $\theta + \pi/2$.

Integral (3) shows that the result of the above averaging
(which can be considered as simply one global result of the
processing performed through the bank of band-pass filters) is
a "smoothing" of the energy density spectrum of the image $f(x,y)$.
Such a smoothing is performed along the frequency axis associated
to the direction θ and is frequency dependent.

If, e.g., the function $w(x,y,\omega_c)$ represents a family of
constant-Q filters acting along the direction θ, then the smoothing is larger at the higher spatial frequencies ω_c, since the
bandwidth of $W_\theta(\omega,\mu,\omega_c)$ along the ω axis grows proportionally to ω_c.

b) The second formal property of integral (2), that we want to
show, relates to the transformation that it undergoes when the
signal $f(x,y)$ is linearly scaled in (x,y). This scaling operation
is, in the case of images, highly significant from the physical
point of view, since it corresponds to a change in the distance
between an object and its observer.

If the hypothesis is made that the output of the band-pass
filter bank might be used (by the visual system or an artificial
processor) for image recognition, then it would be highly desirable that the form of such an output remains unchanged after
$f(x,y)$ has been scaled to $f(kx, ky)$. This form-invariance can be
imposed in mathematical terms by writing that integral (2) applied
to $f(kx, ky)$ is, on its turn, a linearly scaled version (in ξ, η
and ω), of integral (2) applied to $f(x,y)$.

It is possible to show that the class of weighting functions
$w(x,y,\omega)$ allowing this form invariance is:

$$\left\{\omega^p \cdot \gamma(x\omega, y\omega)\right\} \qquad\qquad (4)$$

where $\gamma(\ )$ is any real function of the two products $x\omega$ and
$y\omega$, and p is any real constant. Of course some constraints
exist for $\gamma(\ )$ and p because of the physical meaning of

$w(x,y,\omega)$ which should be preserved.

It is not difficult to recognize that class (4) is coincident with the class of weighting functions corresponding to constant-Q band-pass filters.

It can be noticed, in particular, that the above form-invariance property can not be satisfied by means of any bank of constant bandwidth filters.

It remains to be seen how, in practice, the above shown property of constant-Q filters might be exploited.

A logarithmic representation of the ω axis (internal to the processor) would imply, for instance, that a linearly scaled pattern in the frequency domain (consequent to a linear scaling of the image) would result affected by a simple shift along the logarithmic axis. Such a shift would make its recognition easier (through, e.g., a matched filter) together with an estimate of its range. From this point of view the very interesting works of Altes [28] should be mentioned. Following a different path, that is through the study of sonars (both animal and artificial) he has pointed out the same nice property of the constant Q filters that we have found and described above.

4. SPATIAL FREQUENCY DOMAIN IMAGE CODING

One of the main applications of digital image processing is the design and implementation of efficient techniques for image representation through a reduced number of bits. Of course, the coding must be performed according to some quality criterion which may depend on the particular application (visual fidelity for subjective viewing, decision error rate in an automatic classification system).

Almost all the proposed coding methods are designed in such a way as to take advantage of the strong correlation existing, in most natural images, between neighboring pixels. For this reason a linear transformation is usually applied to the image, in order to obtain, in the transformed domain, a set of elements less correlated than the original ones and unevenly contributing to the information content and to the image quality. A "select and save" scheme can thus be used (e.g. based on the energy content of the elements) to achieve bit reduction within the quality constraints.

Some classical coding techniques, like DPCM [4], widely used for both one and two-dimensional signals, perform the above mentioned partial decorrelation. The optimal transformation, for both decorrelation and energy compaction, is the two-dimensional Karhunen-Loeve transform. The shortcomings of the K-L transform are the lack of a fast algorithm and the difficulties of estimating the image covariance matrix (whose eigenvectors are the transform basis).

Other transforms, like the Discrete Cosine, the Fourier and the Hadamand-Walsh, provide a partial decorrelation and energy compaction. For example, the energy concentration in the Fourier domain of a large class of pictorial images can be as high as 95% in about 1% of the samples around the origin. Therefore, classical transform coding techniques [3, 18] tend to quantize only part of the transform coefficients according to some suitable quantization rules (like the logarithmic one or the maximum entropy rule).

The above mentioned approaches to image coding are essentially based on objective image properties such as the correlation among pixels. On the other hand, other approaches exist, which properly take into account the properties of the visual system (see section 2).

An example in this class is Sakrison's approach [22], that has been developed in the framework of the rate-distortion theory. Sakrison, in fact, looks for a distortion measure $d(u, \tilde{u})$, between the "transmitted" image u and the received (or decoded) image $\tilde{u}$, which is in good agreement with the subjective evaluation of image quality. This distortion function $d(u, \tilde{u})$ is chosen within a class for which it is possible to compute the rate-distortion function $R(\bar{d})$ and simulate the optimal coding.

This allows, for a fixed image and a fixed transmission rate, evaluating the coded image quality corresponding to each function $d(u, \tilde{u})$ chosen within the class.

By so doing it becomes at least possible to select the most appropriate $d(u, \tilde{u})$ in the class.

In designing its class of distorsion measures, Sakrison refers to what presumably is the detection mechanism in the visual system for stimuli consisting of sinusoidal luminance patterns (gratings). This mechanism seems to be based on the computation of the energy (rms) at the output of the bank of band-pass filters shown in the schema of section 2. Sakrison furtherly suggests modeling the distorsion (between two images) at the threshold level, that is when the distorsion is such as to produce a just noticeable difference between the two images.

On the base of these hypothesis and by means of simulations and experiments he ends up with a distorsion measure which essentially consists in the maximum energy difference exhibited by the two images at the output of the various band-pass filters.

Another approach to coding, also based on the properties of the peripheral visual system (see section 2), is the one suggested by Braccini and Gambardella [29]. Their approach tries to take advantage of the observation that, as far as the amplitude spectrum of an image is concerned, the visual system has a lower resolution at the high spatial frequencies with respect to the low ones.

This property is exactly what the psychophysical experiments have shown (see section 2) and, as we have seen in section 3, is embodied in the processing schema made of parallel constant-Q band-pass filters.

Based on this observation the following coding-decoding processing steps have been suggested:

a) given the image u , compute $v = \log(u)$;

b) compute the magnitude $|V|$ and the phase φ_v of the Fourier transform of v;

c) perform weighting and frequency-dependent smoothing of $|V|$, so that $S(|V|)$ is computed, according to the model of the peripheral visual system;

d) code $S(|V|)$ with a fixed bit rate, much lower than the bit rate associated to φ_v (which is simply quantized);

e) decode and reconstruct $|V|$ through an estimate $|\hat{V}|$;

f) associate $|\hat{V}|$ and φ_v in order to reconstruct v and u through the estimates $\hat{v}$ and $\hat{u}$.

The preliminary experiments performed for this type of coding are encouraging, although step (a) has been ignored and the decoding algorithm has been a very crude one.

One implication of the above coding technique is that the amplitude spectrum of a pictorial image carries, for the human observer, less information than the phase spectrum. It is still uncertain whether this property has also an objective foundation, i.e. be also an intrinsic attribute of images.

Anyhow this property has been already pointed out by other researchers and used for coding purposes. For instance Tescher [30] has implemented an adaptive technique essentially based on the choice of a variable number of bits for quantizing each sample of the Fourier transform of an image. More bits are used for the

phase samples than for the amplitude samples. Furthermore the
number of quantization levels for the amplitude samples is set
proportional to their variance,being the latter estimated, at each
spatial frequency, through the amplitude levels previously quantized. This coding technique corresponds, practically, to a
smoothing of the amplitude spectrum which is larger at the high
spatial frequencies compared to the low ones. The total bit saving
obtained by Tescher is impressive, since the quality of the reconstructed image is still good with a bit rate as low as 0.5 bits
per pixel.

5. ENHANCEMENT AND RESTORATION

The purpose of the image enhancement techniques is to improve
the image quality according to some criterion which might be the
satisfaction of the human viewer or the efficiency and reliability
of some subsequent machine processing. Differently from the case
of restoration, in the case of enhancement no degrading system
is taken into account and no reference is made to an original
(ideal) image to be estimated.
In what follows the enhancement techniques aimed to improve
the image quality for the human viewer will be shortly presented.
Emphasis will be put however on those methods taking advantage
of a simplified version of the visual system model presented in
section 2.
First of all we mention four enhancement techniques which
do not make direct use of the above model.

a) Contrast enhancement: each pixel amplitude is rescaled (linearly or not) in order to better exploit the available dynamic range
of the display device.

b) Histogram modification: it is used to enhance images whose
gray level histogram is biased toward a region of the gray scale
(this happens for many natural images, in particular if they are
linearly quantized). In this technique the image is scaled so
that the resulting histogram has the desired shape: good results
have been obtained by approximating a uniform histogram (histogram equalization), with the purpose of presenting as many gray
shades as possible, or, equivalently, maximizing the first order
entropy of the output image.

c) Noise cleaning, by means of linear filtering or non-linear

smoothing operations aimed to eliminate those pixels presenting
too large variations with respect to the surrounding elements.

d) Edge sharpening.Psychophysical experiments show,indeed, that
an image where the edges have been somehow emphasized is subjec-
tively more pleasing. This effect is obtained by means of some
form of high-pass filtering or differentiation.

Similar results are achieved through non linear operations
in the spatial frequency domain, such as the "coefficient rooting"
[31] which consists in taking a power (with exponent less than
one) of the image transform amplitude, and the "cepstrum" where
instead, the logarithm of the image spectrum is taken [15]. Both
techniques yield a sort of energy equalization in the spatial
frequency domain, which can be thought as equivalent to some
sort of high-pass filtering since natural images spectra present,
in general, high amplitudes at low frequencies and small amplitudes at high frequencies.

Let us consider now some enhancement techniques exploiting
a simplified model of the peripheral visual system, such as the
one discussed in section 2, limited to the first two blocks, that
is to the receptors logarithmic transformation followed by the
spatial band-pass filter. The output of this filter is, in the
simplified model, the signal to be processed at higher levels
for classification and recognition purposes. Therefore, it seems
worthwhile to operate in such a way as to let this signal carry
the maximum amount of information.

A first method toward this goal is to equalize the photoreceptors output histogram instead of the original image histogram (the band-pass filter is neglected). In other words, the
original intensity I must be replaced by a monotone increasing
function $f(I)$ so that the probability density function of the
log-transformed variable is constant. The resulting amplitude
density function of f turns out to be hyperbolic and the method
is usually referred to as histogram hyperbolization [32]. Subjective viewing experiments have shown that in many cases the result
is actually better when hyperbolization instead of equalization
is applied.

A second method of improving the output signal from the
photoreceptors, consists of high-pass filtering the logarithm
of the original image, in agreement with the model.

The overall enhancement system becomes a particular form
of the homomorphic filter [15], namely the cascade of three blocks.
The first block computes the image logarithm, the second one is

a high-pass filter and the last block performs the inverse-log
operation, that is the exponential. This kind of processing can
be furtherly motivated by noticing that an image intensity, I,
can be modeled as the product of illuminance, i, and reflectance,
r, and that the log operation acts upon the two components in
such a way that

$$\log I = \log i + \log r.$$

Since approximately the energy content of i tends to be
concentrated at low frequencies, while the reflectance components
are more spread toward high frequencies, the high-pass homomorphic
filtering has the effect of enhancing the image details (related
to the reflectance component) and leaving unaffected (or attenua-
ting, if desired) the illuminance component. In this way, a tra-
de-off can be achieved between sharpness enhancement and dynamic
range constraints.

Of course, the two above described methods can be combined
together: in this case, the exponential block in the homomorphic
system is replaced by the histogram hiperbolization operation,
acting upon the high-pass filter output.

We describe now an image restoration technique where an
important role is played by the eye frequency response (i.e. the
second block of the model illustrated in section 2).

The method, that belongs to the class of constrained least
squares restoration of images degraded by a linear system and
indipendent additive noise [33], leads to an extension of the
Wiener filter. The problem is formulated as the minimization of
a linear operation c on the image f, subject to some constraint.
The latter may be a bound on the estimation error, while the objective function may be related, e.g., to the solution oscilla-
tions. In the Fourier domain the optimum filter transfer function
is

$$H_c = H^*/(|H|^2 + \lambda |C|^2) \tag{5}$$

where H is the frequency response of the degrading system, C
is the Fourier transform of c and * denotes conjugation; λ
can be determined iteratively.

If the operator c is suitably chosen in relationship with
the image autocovariance matrix, the transformation cf yields a
whitened representation of f: the method then minimizes the length

of the whitened object vector f subject to the side constraint.
The whitening procedure, though being of basic importance in the
statistical theory of least squares estimation, can hardly be
implemented due to the lack of a priori information about f. In
[34] the "maximum ignorance assumption" has been proposed, i.e.
f is assumed to be a realization of a white random process. In
this case the covariance of the perceived image is the one induced by the visual system, of which only the band-pass transfer
function H_e is taken into account: in (5) C is substituted by
$1/H_e$. The results of the restoration of blurred and noisy images,
based on this approach and shown in 34 , are definitely superior
to the results achieved through the Wiener filter.

6. TEXTURE ANALYSIS

Texture can be defined as the repetitive arrangement,over an
area, of an elementary pattern, that possibly presents statistical variations in size, shape and spatial shift. The ability to
perform automatic texture analysis (by measuring, e.g., the edges
of a texture area and its coarsness) is an essential tool in picture segmentation and image classification for various applications (e.g. earth resources mapping).
The most commonly used textural measurements are taken
either in the Fourier or in the spatial domain.
A first measure of an area texture coarsness can be extracted
from the power spectrum of that area [35]. In fact, the power
spectrum radial distribution depends on the texture coarsness in
the sense that a coarse texture exhibits high values in the
spectrum near the origin, while the values are more spread out
in the case of a fine texture. A first set of useful features
aimed to texture discrimination are therefore the averages of
the spectrum taken over rings centered at the origin. Similarly,
the angular energy distribution can be used to discriminate the
texture directionality (if many edges in the θ direction are
present, high spectrum values are concentrated around the direction $\theta + \pi/2$). A second set of features are thus the spectrum
averages taken over wedge-shaped regions. If the averaging regions
are chosen to be intersections of rings and wedges, the resulting
features are sensitive to both size and orientation of textures.
Another class of textural measures, taken in the spatial
domain, is based on the assumption that the textural information

lies in the relationship existing, in average, between the gray
levels [36]. A set of gray level dependence (or co-occurrence)
matrices $P(r,\theta)$ can be built, whose elements p_{ij} represent the
co-occurrence probability of levels i and j, r radial units apart
in direction θ (r and θ fixed for any matrix). Intuitively, if
a region is characterized by a relatively flat gray scale (no
texture), the high values of P are clustered around the main
diagonal. The same holds true if the texture is coarse and r
small with respect to the element size, while for small texture
elements (with respect to r) the matrix is spread out.

On the basis of the above remarks, a variety of texture
measures have been proposed and used, that are extracted from
matrices $P(r,\theta)$ by essentially measuring the clustering of their
elements.

It is worth noticing that all the texture measures above
mentioned, i.e. power spectrum and co-occurrence (or joint probability) matrices, are based on second order statistics of the
image. This approach, usually based on objective texture properties, can also be supported by psychophysical experiments [37].
These experiments, in fact, have shown that texture discrimination in human perception is essentially based on the first and/
or second order statistical differences in the textures. In the
texture discrimination model proposed by Julesz the stimulus
"is first analyzed by local features extractors that can detect
only simple features such as dots and edges of given sizes and
orientations. Then the outputs of these simple extractors are
evaluated by a global processor that can compute only second-or
first- order statistics (that is can compare at most two such
outputs)" [37].

7. MATCHED FILTERING IN IMAGE PROCESSING

One of the most commonly used techniques to detect objects
in an image is the two-dimensional matched filtering [3], which
represents an extension of the one-dimensional technique so
widely applied in communications.

The 2-D matched filter is a spatial filter whose output is
not an image but a measure of correlation between the input image
and the reference object. This correlation measure, obtained by
maximizing the signal to noise (indipendent and additive) energy
ratio at some output point, can then be used to decide whether

an object (or which one among many) is present in the image.

The natural implementation domain of the filter is the Fourier plane, where it is invariant to object translations in either direction (so that the object is "detected" anywhere in the image), but not to rotations. The filter transfer function is

$$H_M(\omega_1, \omega_2) = F^*(\omega_1, \omega_2)\exp\left[-j(\omega_1 x + \omega_2 y)\right]\Big/ N(\omega_1, \omega_2)$$

where F is the Fourier transform of the object, N is the noise power spectrum density and (x,y) is the object offset. At these coordinates in the correlation plane a maximum occurs if the object is in the scene.

The main drawback of the matched filter is the fact that its output (correlation) is essentially sensitive to the image energy and not to its spatial structure, although it is often desired to detect objects on the basis of edge or gradient information rather than energies. To improve the discriminating power between objects of different shape but similar size and energy, the derivative matched filter has been proposed [3]. The general form of the pth order derivative matched filter is

$$H_p = (\omega_1^2 + \omega_2^2)^P H_M$$

which includes, for p = 0, the normal matched filter H_M , and for p = 1 the so-called Laplacian matched filter.

The derivative matched filter can be implemented as the cascade of two blocks: the first one is a generalized derivative with edge enhancing effects, and the second block is the normal matched filter.

REFERENCES

[1] W.K. Pratt, Digital Image Processing, I.P.I., Univ. of South
    Cal., 1975.
[2] T.S. Huang, Ed., Picture Processing and Digital Filtering,
    Springer-Verlag, New York, 1975.
[3] H.C. Andrews, Computer Techniques in Image Processing, Acade-
    mic Press, New York, 1970.
[4] A. Habibi and G.S. Robinson, Computer 7, 22 (1974).

[5]   R.O. Duda and P.E. Hart, <u>Pattern Classification and Scene Analysis</u>, Wiley, New York, 1973.

[6]   B.S. Lipkin and A. Rosenfeld, Eds., <u>Picture Processing and Psychopictorics</u>, Academic Press, New York, 1970.

[7]   A.S. Sawchuk, Proc. IEEE 60, 854 (1972).

[8]   G.M. Robbins and T.S. Huang, Proc. IEEE 60, 862 (1972).

[9]   H.C. Andrews and C.L. Patterson, IEEE Trans. ASSP-24, 26 (1976).

[10] T.S. Huang, D.S. Barker and S.P. Berger, Appl. Optics 14, 1165 (1975).

[11] W.K. Pratt, IEEE Trans. C-21, 636 (1972).

[12] B.R. Hunt, IEEE Trans. C-22, 805 (1973).

[13] N.D.A. Mascarenhas and W.K. Pratt, IEEE Trans. CAS-22, 252 (1975).

[14] B.R. Frieden, J. Opt. Soc. Am. 62, 511 (1972).

[15] A.V. Oppenheim, R.W. Schafer and T.G. Stockham, Jr., Proc. IEEE 56, 1264 (1968).

[16] T.G. Stockham, Jr., T.M. Cannon and R.B. Ingebretsen, Proc. IEEE 63, 678 (1975).

[17] B.R. Frieden, Appl. Optics 9, 2489 (1970).

[18] P.A. Wintz, Proc. IEEE 60, 809 (1972).

[19] Leitz T.A.S. (Texture Analyzer System).

[20] T.G. Stockham, Jr., Proc. IEEE 60, 828 (1972).

[21] T.N. Cornsweet, <u>Visual Perception</u>, Academic Press, New York, 1970.

[22] J.L. Mannos and D.J. Sakrison, IEEE Trans. IT-20, 525 (1974).

[23] F.W. Campbell and J.G. Robson, J. Physiol. 197, 551 (1968).

[24] C.F. Stromeyer and B. Julesz, J. Opt. Soc. Am. 62, 1221 (1972).

[25] B. Julesz and J.E. Miller, Perception 4, 125 (1975).

[26] G. Gambardella, J. Opt. Soc. Am. 65, 99 (1975).

[27] G. Gambardella, IEEE Trans. CT-18, 455 (1971).

[28] R.A. Altes, Progr. Rep. PR 144, Electromagnetic Systems Laboratories, Sunnyvale, Cal.

[29] C. Braccini and G. Gambardella, Proc. of the 22nd International Comm. Conf., Genova, Italy, 383 (1975).

[30] A.G. Tescher, USCIPI Rep. 510, Univ. of South Cal. (1973).

[31] H.C. Andrews, A.G. Tescher and R.P. Kruger, IEEE Spectrum 9, 20 (1972).

[32] W. Frei, USCIPI Rep. 560, Univ. of South Cal., 72 (1975).

[33] B.R. Hunt, IEEE Trans. C-22, 805 (1973).

[34] B.R. Hunt, Proc. IEEE 63, 693 (1975).

[35] J.S. Weszka, C.R. Dyer and A. Rosenfeld, IEEE Trans. SMC-6, 269 (1976).

[36] R.M. Haralick, K. Shanmugam and I. Dinstein, IEEE Trans. SMC-3, 611 (1973).

[37] B. Julesz, E.N. Gilbert, L.A. Shepp and H.L. Frisch, Perception 2, 391 (1973).

SPECTRUM ANALYSIS

C. v. Schooneveld, P.J.A. Prinsen

Physics Laboratory NDRO, The Hague, The Netherlands

A. INTRODUCTION *

Let $u(t)$ be a continuous, real, scalar, stationary random
process with mean value zero, covariance-function $R(\tau)$ and power
density spectrum $S(f)$:

$$E\{u(t)\}=0 \quad E\{u(t)u(t-\tau)\}=R(\tau) \quad S(f)=\int_{-\infty}^{\infty}R(\tau)\exp-2\pi jf\tau \; d\tau \quad (1)$$

Spectrum analysis is concerned with the estimation of $S(f)$ from a
finite length record $x(t)$ [0,T] of one realisation (fig. 1):

$$\begin{aligned} x(t) &= u(t) \quad ; \; o<t<T \\ &= 0 \quad\quad ; \; t<0, \; t>T \end{aligned} \quad (2)$$

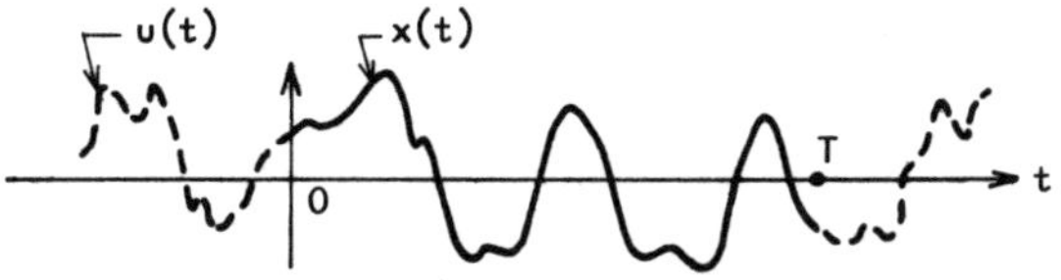

Fig. 1. Finite length record $x(t)$, [0,T]

For convenience, the independent variable t is referred to as
"time" (sec.) and the conjugate variable f as "frequency" (cycles/
sec. or Hz), although other physical quantities may of course appear in specific applications.

Let $Q(f)$ be the estimate of $S(f)$. In general, $Q(f)$ will differ from $S(f)$ for two reasons (fig. 2):

---

* This paper is the first part of a tutorial presentation on "spectrum analysis and imaging systems". The second part, related to
spatial imaging systems, is not included because of page limitation.

G. Tacconi (ed.), Aspects of Signal Processing, Part 2, 563-590. All Rights Reserved.
Copyright © 1977 by D. Reidel Publishing Company, Dordrecht-Holland.

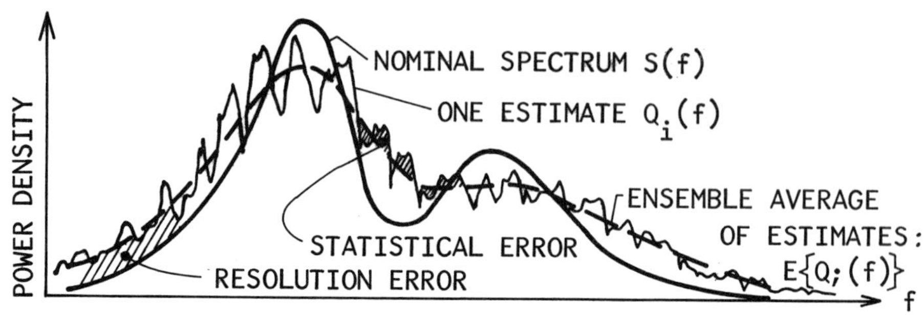

Fig. 2. <u>Spectral estimation errors.</u>

1. <u>Resolution errors</u>. If our estimation scheme is successively
applied to a large number of records $x_1(t), x_2(t), \ldots$, the ensemble average $E\{Q(f)\}$ of the estimates $Q_1(f), Q_2(f), \ldots$ differs
from $S(f)$:

$$\text{resolution error} = E\{Q(f)\} - S(f) \tag{3}$$

Fundamentally, the resolution error is related to the finite
recordlength T, although in practice the resolved frequency
interval Δf is often made large in comparison to the truncation
limit $\frac{1}{T}$.

2. <u>Random errors</u>. The random deviations of the successive estimates $Q_1(f), Q_2(f), \ldots$ from their ensemble average $E\{Q(f)\}$ constitute the random error in the spectral estimate:

$$\text{random error} = Q(f) - E\{Q(f)\} \tag{4}$$

Random errors have the same origin as the resolution error:
truncation of $u(t)$. The finite record length prevents an effective suppression of accidental fluctuations which are more
characteristic for the particular record rather than for the
process as a whole.

In general, resolution errors increase when random errors
are made to decrease, and conversely. Usually, a reasonable
compromise is achieved when both errors are of approximately
equal magnitudes.

Spectrum analysis is based on one of two definitions of $S(f)$ (lit. Papoulis):

1. $S(f) = \int_{-\infty}^{\infty} R(\tau) \exp{-2\pi j f \tau}\, d\tau$ (Wiener – Khintchin)   (a)

    where $R(\tau) = E\{u(t)u(t-\tau)\}$
            $=$ covariance function of $u(t)$ (b)

$$\left. \right\} \quad (5)$$

2. $S(f) = \lim_{T \to \infty} E\{P(f)\}$ (a)

    where $P(f) = \frac{1}{T}\left| \int_0^T x(t)\, \exp{-2\pi j f t}\, dt \right|^2$

            $=$ periodogram of $x(t)$ (b)

$$\left. \right\} \quad (6)$$

The equations suggest 3 basic approaches to spectrum analysis (Lit. Blackman and Tukey, Bingham et al, Parzen, Welch, Jenkins and Watts, Oppenheim and Schafer):

a. Correlogram method. Estimate covariance function from finite length record $x(t)$, thereby replacing the ensemble average in (5-b) by a time average. Compute power spectrum as fouriertransform of estimated covariance function.
b. Periodogram method. Calculate periodogram of finite length record $x(t)$. Replace ensemble average in eq. 6 -a by an average over adjacent frequencies in the periodogram.
c. Averaged short-time periodograms (filter bank method)
Divide recordlength T in (M+1) sections, each of length L. Calculate periodogram of each short-time section. Replace ensemble average in eq. 6-a by an average over the (M+1) successive periodograms. This method is equivalent to a parallel bank of bandpassfilters, squarers and integrators.
    Fig. 3 summarizes the 3 approaches and emphasizes the two essential steps: frequency decomposition to achieve spectral resolution and averaging to obtain statistical stability of the estimate $Q(f)$ (i.e. small random errors.) More detailed block diagrams are presented in figs. 4 and 5.
Correlogram and periodogram method (fig. 4). Both methods are fourier transforms of each other and both lead from the same record $x(t)$ [0,T] to the same spectral estimate $Q(f)$. They are specifically suited for discrete time spectrum analysis by means of a computer. The record $x(t)$ is sampled and stored, prior to the actual spectrum analysis. Essentially, the correlogram approach follows the path $1 \to 2 \to 3 \to 4 \to 8$ and the periodogram method runs along $1 \to 5 \to 6 \to 7 \to 8$. Also, mixed routes are often used. For example, the path $1 \to 2 \to 6 \to 7 \to 3 \to 4 \to 8$ requires no convolutions but only fouriertransforms and multiplications. An additional advantage of this path is the possibility to obtain plots of the measured covariance function $p(\tau)$, as well as of the periodogram $P(f)$, as intermediate results of high diagnostic value. (For many applications, a plot of $P(f)$ is already sufficient and the step(s) from $P(f)$

to $Q(f)$ can often be omitted). A disadvantage of the mixed path
is that the signal $x(t)$ must be augmented with zeroes to double
length (2T) to avoid a circular correlation at point 3, in stead
of the desired linear correlation. This results in double length
fouriertransforms and in oversampling of the periodogram $P(f)$
and the estimate $Q(f)$ by a factor 2.

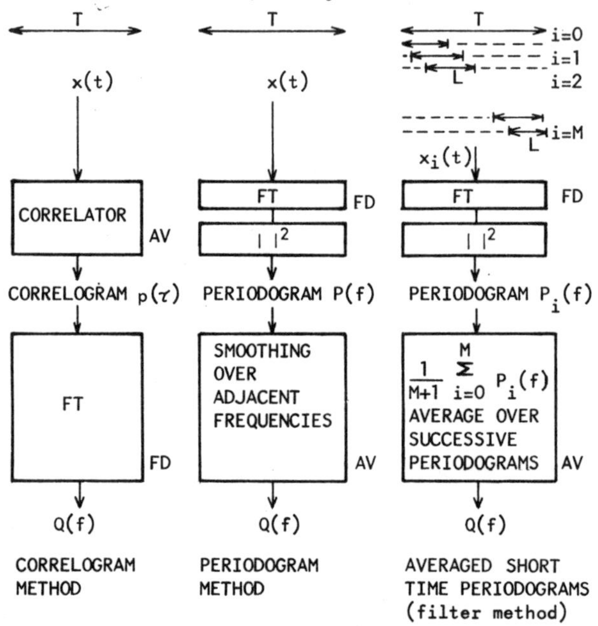

Fig. 3. <u>Basic approaches spectrum analysis;</u> FD = "frequency
decomposition, AV = averaging

In some applications, the bandwidth of $x(t)$ may be so large
that sampling becomes impossible or that storage and subsequent
datahandling becomes prohibitive. In that case, a parallel bank
of correlators (either analog or digital) can be used to calculate the covariance function $p(\tau)$ in real time for a limited num-
ber of lagvalues: $\tau = 0, \Delta, 2\Delta, \ldots, (N-1)\Delta$. Values of $p(n\Delta)$ are stored
and spectrum analysis is achieved via the path 3→4→8. However,
this method invites the danger of spectral leakage because of the
truncation of $p(\tau)$. It is therefore preferred to store $x(t)$ over
its entire length whenever possible and to calculate $p(\tau)$ over its
entire length. The limits to resolution are then set by the record length T, rather than by the length $(N-1)\Delta$ of the truncated
covariance function $p(\tau)$. See section C for this method.
<u>Averaged short-time periodograms</u> (fig. 5). This method is ana-
lyzed in section B. The blockdiagram presents the method in the
same format as fig. 4: frequency to the right and time to the
left. The method is the digital computer analogon of the well-
known (bandpass filters + wattmeters) method. Usually, the path

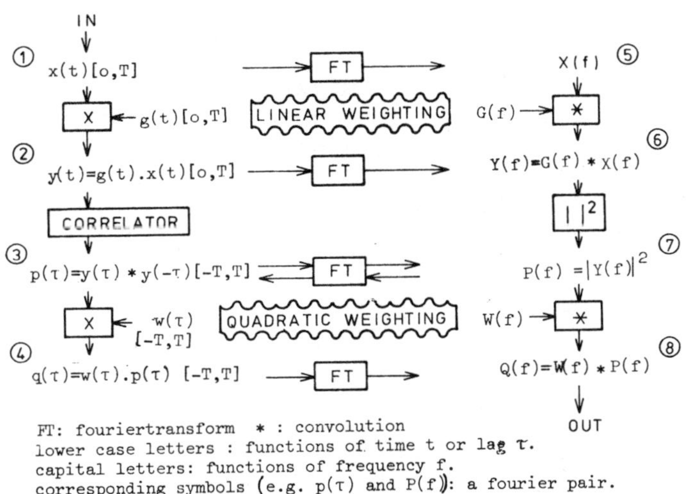

FT: fouriertransform * : convolution
lower case letters : functions of time t or lag τ.
capital letters: functions of frequency f.
corresponding symbols (e.g. p(τ) and P(f)): a fourier pair.

Fig. 4. Correlogram and periodogram method

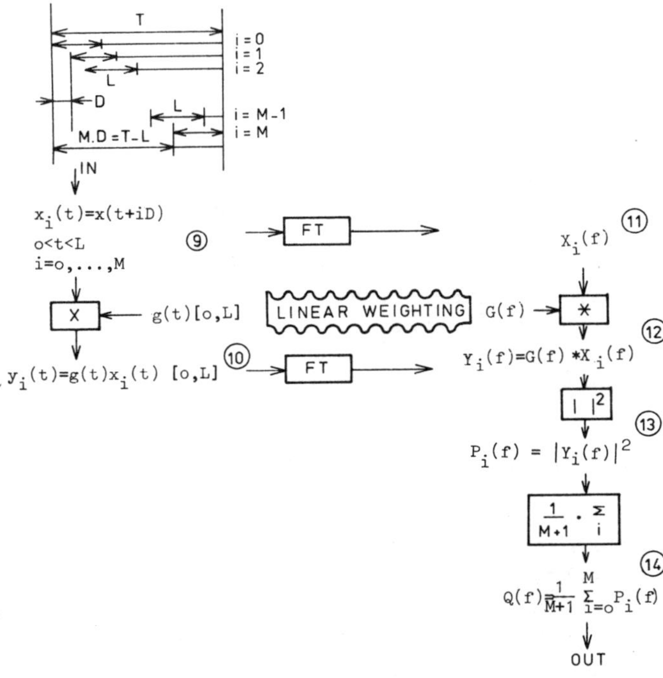

Fig. 5. Averaged short time periodograms (filterbank method)

9→10→12→13→14 is followed.

Compared to fig. 3, the operations of linear and quadratic weighting have been introduced in figs. 4 and 5.

Quadratic weighting is applied only in the periodogram and correlogram methods. It is applied at the "powerlevel" or "quadratic level". Purpose: to control the balance between resolution errors and random errors in the spectral estimate $Q(f)$. Quadractic weighting is effected by either using a weighting function $w(\tau)$ [-T,T] in step 3→4 or a convolution kernel $W(f)$ in step 7→8 (both functions being each other's fouriertransform). Usually, $w(\tau)$ is given an effective length T_{quad}<<T. It is a symmetric function, smoothly tapered towards the ends of the interval [-T,T] (fig. 8).

Linear weighting is applied in all 3 methods. It is applied before the correlator or squarer, or at the "linear level". In the correlogram and periodogram methods, its aim is to obtain a further control over spectral resolution, in particular to suppress leakage in spectra with a large dynamic range. In the method of averaged short-time periodograms, linear weighting is the only means to control resolution. The weighting function $g(t)$ exists on the interval [0,T] (correlogram and periodogram method)or [0,L] (averaged short time periodograms). Like the quadratic weighting function $w(\tau)$, it is normally a symmetric and smoothly tapered function (fig. 8).

Parametric Methods (Section D). When the observed signal $x(t)$ is obtained as the output of a system with rational power transfer function of finite order, fed by white noise, then its power spectrum can be completely specified by a finite number of parameters (e.g. poles and zeroes, or coefficients of a difference equation). This suggests a different method of spectrum analysis, viz. the direct estimation of these parameters from the observed signal. Once these parameters are estimated the power spectrum estimate can easily be computed. When no generator model is given, this method should be generalized to include an estimation procedure for the number and type of the parameters. This method is treated in section D.

B. FILTERBANK AND AVERAGED SHORT TIME PERIODOGRAMS

One of the oldest methods of spectrum analysis is the parallel bank of bandpassfilters, each of them followed by a wattmeter (squarer-integrator); fig. 6. (Lit. Blackman, Tukey). The filters have frequency transmission functions $H_k(f)$ of appropriate * bandwidth B, centered around tuning frequencies f_k (k= 0,1,2,...). The filters can be either analog or digital. For simplicity, we assume in this particular case that the input

* We assume that all filterbandwidths are equal to B and that the spacing of centerfrequencies f_k is in the order of B, or less. Generalization to unequal bandwidths, e.g. a bank of third octave filters, is straightforward.

signal has been present on the filters for a sufficient time to
guarantee that transients have died out. The estimate of the power
density spectrum $S(f)$ at $f=f_k$ is given by $Q(f_k) =$
$\frac{1}{T} \int_0^T y^2(t)dt$, where $y(t)$ is the output of the kth filter. If the
experiment is repeated a number of times for different sections
of the input signal, we find that $Q(f_k)$ fluctuates around its
ensemble average: $Q(f_k) = E\{Q(f_k)\}+ Q'(f_k)$, where

$$E\{Q(f_k)\}=E\{y^2(t)\}=\int_{-\infty}^{\infty}|H_k(f)|^2 S(f)df \qquad (7)$$

$$var\{Q'(f_k)\}=E\{[Q-E\{Q\}]^2\} = \frac{1}{T^2}\int\int_0^T E\{y^2(t)y^2(\Theta)\}\, dt\, d\Theta - E^2\{y^2(t)\}$$
$$(8)$$

Eg. 7 is essentially a convolution: the estimate $Q(f_k)$ equals the
original spectrum $S(f)$, integrated over a spectral window
$|H_k(f)|^2$. In order to obtain a correct estimate for a spectrum $S(f)$
which does not change appreciably over the bandwidth B, the transmission function must be normalized as indicated below:

$$\int_{-\infty}^{\infty}|H_k(f)|^2 df=1 \qquad (9)$$

(For digital filters with sampling distance Δ, the transmission
function is periodic: $H_k(f)=H_k(f+\frac{m}{\Delta})$. The integration limits in
eq. 9 are then $-1/2\Delta$ and $+ 1/2\Delta$).

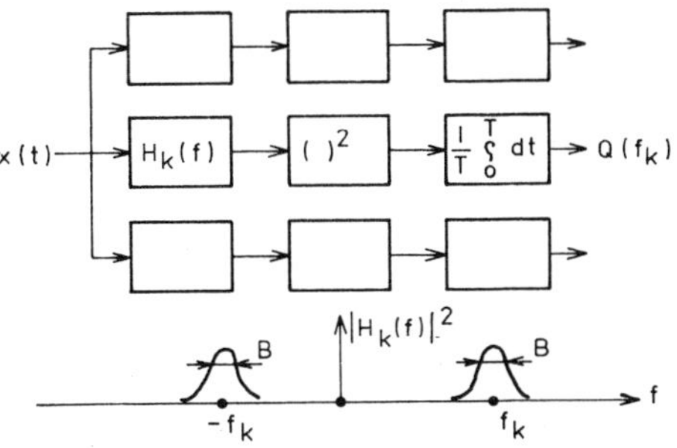

Fig. 6. __Parallel bank of bandpassfilters__

Resolution. Resolution errors were defined in section A as the
difference between the average estimate, $E\{Q(f_k)\}$, and the actual spectrum $S(f)$. Two aspects are usually distinguished, although there is no qualitative difference between the two:
spectral smoothing and spectral leakage.
Spectral smoothing.
Imagining a closely spaced raster of frequencies f_k, eq.7 will result in a smoothed image, $E\{Q(f_k)\}$, of the original spectrum $S(f)$.
(fig. 7-b). The smoothing bandwidth is given by the peakwidth of
the spectral window $|H_k(f)|^2$.

Spectral leakage

Leakage occurs when $S(f)$ exhibits differences in magnitude which
are large compared with the attenuation of $|H_k(f)|^2$ on the slopes
of the passband (or large in comparison with the main peak vs
sidelobe ratio of $|H_k(f)|^2$ in those cases where $H_k(f)$ is the
fouriertransform of a finite length function). Leakage can then
set a limit to the dynamic range of the spectrum analysis procedure. The powerspectrum $S(f)$ in fig. 7-c has a strong peak at
frequency f_o, surrounded by weak structure at nearby frequencies.
At frequency f_k, the output of the convolution (7) will be controlled by power, leaking from the strong peak of $S(f)$ at $f=f_o$
via the skirts (or sidelobes) of the window $|H_k(f)|^2$, rather than
by the weak power of $S(f)$ at $f=f_k$ which is transmitted via the
main peak of $|H_k(f)|^2$. Consequently, the weak structure around
the peak of $S(f)$ is overwhelmed by spectral leakage. Suppression
of spectral leakage requires steep slopes, or low sidelobes, in
the window $|H_k(f)|^2$.

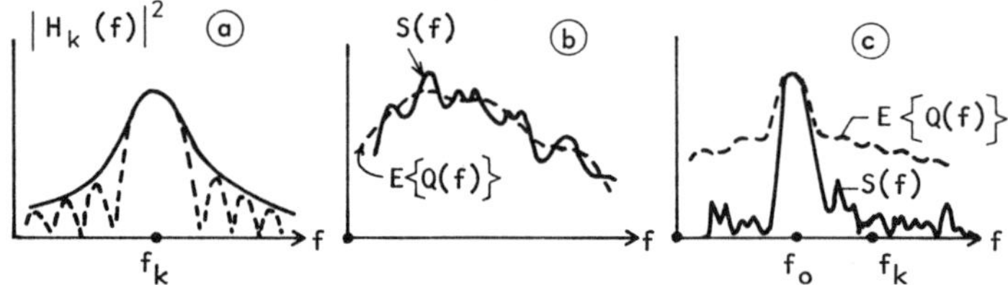

Fig. 7. Spectral window (a), spectral smoothing (b),
         spectral leakage (c).

Statistical stability. Eq. 8 gives the variance of the random
component $Q'(f_k)$ of the spectral estimate. For a gaussian * process, eq. 8 reduces to

$$\text{var}\{Q'(f_k)\} = \frac{2}{T} \int_{-T}^{T} \left(1 - \frac{|\varepsilon|}{T}\right) R_y^2(\varepsilon)\, d\varepsilon,$$

where $R_y(\varepsilon)$ is the covariance function of $y(t)$. Under the assumption of a long averaging time (more precisely, $T \gg$ effective duration of $R_y^2(\varepsilon)$), the expression simplifies to

$$\text{var}\{Q'(f_k)\} \approx \frac{2}{T} \int_{-\infty}^{\infty} |H_k(f)|^4 S^2(f)\, df \tag{10}$$

Assuming finally that $S(f)$ does not change appreciably over the
width of the spectral window $|H_k(f)|^2$, we obtain from eqs. 7 and
10:

---

* If a,b,c,d are gaussian with mean values zero:

$E\{abcd\} = E\{ab\}.E\{cd\} + E\{ac\}.E\{bd\} + E\{ad\}.E\{bc\}.$

(Lit. Papoulis)

$$E\{Q(f_k)\} \approx S(f_k) \cdot \int_{-\infty}^{\infty} |H_k(f)|^2 \, df$$

$$\text{var}\{Q'(f_k)\} \approx \frac{2}{T} S^2(f_k) \cdot \int_{-\infty}^{\infty} |H_k(f)|^4 \, df \qquad (11)$$

We now define the effective bandwidth of the spectral window as

$$B = \frac{1}{2} \frac{[\int_{-\infty}^{\infty} |H_k(f)|^2 \, df]^2}{\int_{-\infty}^{\infty} |H_k(f)|^4 \, df} \qquad (12)$$

The factor $\frac{1}{2}$ in eq. 12 is related to the "double-sided-ness" of $H_k(f)$ (fig. 6). Combining eqs 11 and 12 we obtain for the relative variance of $Q(f)$:

$$\text{relvar}\left\{Q(f_k)\right\} \overset{\text{def}}{=} \frac{\text{var}\{Q'(f_k)\}}{E^2\{Q(f_k)\}} = \frac{1}{BT} \qquad (13)$$

Relvar$\{Q(f_k)\}$ can be regarded as a noise-to-signal ratio of the spectral estimate $Q(f_k)$. Thus, a large product of window width B and observation time T results in small statistical errors and, conversely, in a high statistical stability of $Q(f_k)$.

Observe that resolution requires a small value of B, and statistical stability a large value. The problem of finding a suitable compromise represents the classical dilemma of stochastic spectrum analysis (as well as of many other areas in the signal processing field).

Eq. 13 indicates that the rms value of the statistical error in $Q(f)$ is proportional to the average value, $E\{Q(f)\}$ This implies that the statistical error has a constant r.m.s. value on a logarithmic scale (decibels). Actually, it can be shown that the random component in $Z(f) = 10 \log Q(f)$ has an r.m.s. value of 5.5 dB.

The method of averaged short-time periodograms is an analogon of the filterbank method; it is specifically designed for discrete spectrum analysis (Lit. <u>Welch</u>). Fig. 5 gives the blockdiagram.

The recordlength T is <u>divided</u> in (M+1) pieces of length L with mutual time shifts D<L. The (M+1) short-time recordings are defined as

$$x_i(t) = x(t+iD) \; ; \; 0<t<L \; ; \; i=0,\ldots,M \qquad (14)$$

Each signal is multiplied with a weighting function g(t) of duration L. Usually, g(t) is a real function, symmetric with respect to $t=\frac{1}{2} L$ and smoothly tapered towards the ends of the interval

[O,L]. The short time periodograms are defined* as:

$$P_i(f) = \left| \int_0^L g(t)x_i(t)\exp-2\pi jft\,dt \right|^2$$

$$= E\{P_i(f)\} + P_i'(f) \tag{15}$$

$E\{P_i(f)\}$ represents the ensemble average and $P_i'(f)$ the random fluctuations, in the same way as in eqs. 7 and 8. They can be derived via straightforward but somewhat tedious computation from the definition of $P_i(f)$. For a gaussian process $x(t)$ we find:

$$E\{P_i(f)\} = |G(f)|^2 * S(f) \qquad (a)$$

$$\mathrm{var}\{P_i'(f)\} = E^2\{P_i(f)\} \qquad (b) \tag{16}$$

where $*$ denotes convolution and where $G(f)$ is the fourier transform of $g(t)$:

$$G(f) = \begin{cases} \int_0^L g(t)\exp - 2\pi jft\ dt & \text{(continuous time)} \\[2mm] \Delta \sum_{n=o}^{N-1} g(n\Delta)\exp - 2\pi jf\ n\Delta & \text{(discrete time)} \end{cases} \tag{17}$$

(Actually, eq 16-b gives a value which is too low by a factor of 2 in exceptional cases, i.e. at frequency f=o and/or when important leakage occurs. Since we are mainly interested in the order of magnitude of variances, we shall omit this complication).

The spectral estimate $Q(f)$ is obtained by averaging the (M+1) periodograms:

$$Q(f) = \frac{1}{M+1} \sum_{i=o}^{M} P_i(f) = E\{Q(f)\} + Q'(f) \tag{18}$$

Consequently,

$$E\{Q(f)\} = |G(f)|^2 * S(f) \quad \text{and} \quad \mathrm{var}\{Q'(f)\} = \frac{E^2\{Q(f)\}}{M+1} \tag{19}$$

---

* For discrete spectrum analysis:

$$P_i(f) = \left| \Delta. \sum_{n=o}^{N-1} g(n\Delta)x_i(n\Delta)\exp-2\pi jfn\Delta \right|^2,$$

where Δ=sampling distance and L=NΔ. Usually, $P_i(f)$ is computed via an FFT and $P_i(f)$ is obtained only at the frequencies $f_k = \frac{k}{N\Delta}$; k=o,..., N-1.

The variance expression in eq. 19 assumes that the random components $P'_i(f)$ of the short-time periodograms are statistically independent. The smallest timeshift D which gives (approximately) uncorrelated periodograms equals the inverse of the effective bandwidth B of the time series $P_i(f)$; ...,i-1,i,i+1,.... Under the condition that $S(f)$ does not vary appreciably over the width of the spectral window $|G(f)|^2$, we get

$$D = \frac{1}{B}, \text{ where } B = \frac{[\int_{-\infty}^{\infty} |G(f)|^2 df]^2}{\int_{-\infty}^{\infty} |G(f)|^4 df} \tag{20}$$

(For discrete time analysis, the integration limits in eq. 20 are $-1/2\Delta$ and $+1/2\Delta$). The number of signal pieces (fig. 5) is given by $(M+1) = \frac{T-L}{D} + 1$. Using eq. 20, and under the assumption of a long integration time $(T \gg D = \frac{1}{B})$ we obtain:

$$(M+1) \approx BT \tag{21}$$

Combination of eqs 21 and 19 leads to:

$$\text{relvar } \{Q(f)\} \overset{\text{def}}{=\!=} \frac{\text{var}\{Q'(f)\}}{E^2\{Q(f)\}} \approx \frac{1}{BT} \tag{22}$$

The analogy with the filterbank will be clear. The fouriertransform (15) represents a bank of bandpassfilters with power transfer functions $|G(f)|^2$. The filter outputs are squared (eq. 15) and averaged over time (eq. 18). The expression for $E\{Q(f)\}$ in eq. 19 is comparable to eq. 7 and eqs 20,22 correspond to eqs. 12 and 13, respectively.

C. CORRELOGRAM AND PERIODOGRAM METHODS

Fig. 4 gives the blockdiagram of the correlogram and of the periodogram method. (Lit. <u>Blackman and Tukey</u>, <u>Bingham et al</u>, <u>Parzen</u>). Both approaches are treated as a unity because of their mathematical equivalence; starting from a given record x(t) they both lead to the same estimate $Q(f)$.

Many of the quantities in fig. 4 are related via fouriertransforms. In principle, these can be continuous time transforms or discrete time transforms, based on samples of x(t) at distance Δ. In the latter case, the DFT is used (discrete fourier transform) and use is made of the FFT algorithms (fast fourier transform; lit <u>Bingham</u>). Although digital computers are virtually the only means to do the required calculations, we shall present the results of the present section in the format of continuous time transforms. The basic ideas remain unchanged but a simpler and more compact notation is available. If required, the discrete time counterparts of the expressions given below can be easily

found. The main difference is in the periodicity of spectral window functions, which introduces the danger of aliasing components
in the spectral estimate $Q(f)$, unless suitable anti-alias filters
are used before sampling the signal $x(t)$. In addition, it may be
necessary to add zeroes to the signals when a discrete convolution or correlation is computed via the DFT, in order to avoid
an undesired   circular convolution. Finally, it should be kept
in mind that fouriertransforms of discrete weighting functions
differ from their continuous counterparts.

The   continuous fouriertransform is defined as:

$$A(f) = \int_{-\infty}^{\infty} a(t)\exp\!-2\pi jft\,dt \quad;\quad a(t) = \int_{-\infty}^{\infty} A(f)\exp 2\pi jft\,df \qquad (23)$$

and convolutions in the time (lag) domain and frequency domain
assume the form:

$$a(t) * b(t) = \int_{-\infty}^{\infty} a(\Theta)b(t-\Theta)d\Theta \quad;\quad A(f)*B(f) = \int_{-\infty}^{\infty} A(\varphi)B(f-\varphi)d\varphi \qquad (24)$$

Frequency functions extend generally from $f = -\infty$ to $f = \infty$
Functions of time or lag often have a finite extent; in this case
the time integrals above have finite limits. Further, frequency
functions are given by a capital letter and time (lag) functions
by a lower case letter, with the exception of the fourierpair
consisting of $R(\tau)$ (covariance function) and $S(f)$ (power density
spectrum).

If the spectrum analysis is repeated for many records $x(t)$,
taken from the same ensemble, the functions $p(\tau)$, $q(\tau)$, $P(f)$ and
$Q(f)$ of fig. 4 will fluctuate around their ensemble averages.
Denoting the random departures from the average values by a prime:

$$p(\tau) = E\{p(\tau)\} + p'(\tau) \quad (a) \quad \xleftarrow{\;FT\;} \quad P(f) = E\{P(f)\} + P'(f) \quad (c)$$
$$(26)$$
$$q(\tau) = E\{q(\tau)\} + q'(\tau) \quad (b) \quad \xleftarrow{\;FT\;} \quad Q(f) = E\{Q(f)\} + Q'(f) \quad (d)$$

The ensemble averages and the (co)variances of the random components can be found from the definitions of $p(\tau)$, $q(\tau)$, $P(f)$,
$Q(f)$ (fig. 4 and eqs. 23, 24). The derivations are straightforward but somewhat tedious. The results are listed below.

$$E\{p(\tau)\} = gg(\tau).R(\tau) \qquad (a) \quad \xleftarrow{\;FT\;} \quad E\{P(f)\} = |G(f)|^2 * S(f) \quad (c)$$
$$(27)$$
$$E\{q(\tau)\} = w(\tau).gg(\tau).R(\tau) \quad (b) \quad \xleftarrow{\;FT\;} \quad E\{Q(f)\} = W(f) * |G(f)|^2 * S(f) \quad (d)$$

where $gg(\tau) = g(\tau) * g(-\tau) = \int g(t)g(t-\tau)dt \qquad (28)$

Variances and covariances are defined as central moments. Taking
$P(f)$ as an example:

$$\left.\begin{array}{l} \text{cov}\{P'(f_1),P'(f_2)\} \overset{\text{def}}{=\!=\!=} E\{P'(f_1)P'(f_2)\} \\[3mm] \text{var}\{P'(f_1)\} \overset{\text{def}}{=\!=\!=} E\{P'^2(f_1)\} \end{array}\right\} \tag{29}$$

We find:

$$\text{cov}\{p'(\tau_1),p'(\tau_2)\} = k \int_0^T\!\!\int g(t)g(t-\tau_1)g(\Theta)g(\Theta-\tau_2)\} \times \\ \times R\{t-\Theta\}R\{(t-\Theta)-(\tau_1-\tau_2)\}dtd\Theta \tag{30}$$

$$\text{cov}\{P'(f_1),P'(f_2)\} = k.\left|\int_{-\infty}^{\infty} G(\bar\varphi-f_1)G^*(\varphi-f_2)S(\varphi)d\varphi\right|^2 \tag{31}$$

$$\text{var}\{P'(f)\} = k.[|G(f)|^2 * S(f)]^2 = k..E^2\{P(f)\} \tag{32}$$

$$\text{var}\{Q'(f)\} = k\int_{-\infty}^{\infty}\!\!\int S(\varphi_1)S(\varphi_2)\left|\int_{-\infty}^{\infty} W(a)G[a-(f-\varphi_1)] \times \right. \\ \left. \times G^*[a-(f+\varphi_2)]da\right|^2 d\varphi_1 d\varphi_2 \tag{33}$$

<u>Conditions</u>. Eqs 30,31,32,33 were derived under 3 assumptions:

$$\left.\begin{array}{l} 1)\ x(t)\ \text{is a gaussian process.} \\ 2)\ w(\tau)\ \text{real and symmetric : } w(\tau)=w(-\tau) \\ 3)\ g(t)\ \text{real and symmetric : } g(\tfrac{1}{2}T+t)=g(\tfrac{1}{2}T-t) \end{array}\right\} \tag{34}$$

<u>Factor k</u>. The factor k in eqs. 30,31,32,33 is related to the omission, for reasons of simplicity, of a second part which ought to appear in each of the expressions, but which is usually negligible. In exceptional cases, the second part assumes a maximum value equal to the first part. (This happens, for example, in eqs. 32,33 when f=o or when important leakage occurs). Consequently, the value of k is normally equal to 1 and can at most be equal to 2.

<u>Normalisation</u>. Eqs. 27-c and d show that the linear and quadratic weighting functions must be normalised as indicated below, in order to get the correct values for $E\{P(f)\}$ and $E\{Q(f)\}$ in the case of a constant power density spectrum $S(f)=S$:

$$\left.\begin{array}{ll} \displaystyle\int_{-\infty}^{\infty}|G(f)|^2 df= \int_0^T g^2(t)dt=1 & \text{(a)} \\[4mm] \displaystyle\int_{\infty}^{\infty} W(f)df = w(o) = 1 & \text{(b)} \end{array}\right\} \tag{35}$$

C.1 <u>Periodogram P(f)</u>

Eq. 27-c shows that the ensemble average of $P(f)$ equals the power density spectrum $S(f)$, viewed through a spectral window $|G(f)|^2$. Since $|G(f)|^2$ is related to the linear weighting function $g(t)$, we shall refer to it as the "linear spectral window" (fig.8). It is the fouriertransform of a finite length function, $gg(\tau)= g(\tau) * g(-\tau)$, and it will therefore consist of a main lobe and sidelobes (fig. 8) The main peak bandwidth of the linear spectral window, B_{lin}, is in the order of $\frac{1}{T_{lin}}$, where T_{lin} is the effective duration of $gg(\tau)$. (eq.44, gives a definition of T_{lin}). Thus, a certain degree of smoothing occurs in the average periodogram; any fine structure of $S(f)$ on a scale smaller than B_{lin} is lost in $E\{P(f)\}$. Without linear weighting (i.e. $g(t)= T^{-\frac{1}{2}} =$ constant), $T_{lin}=T$ and $B_{lin} \approx \frac{1}{T}$. For a tapered function $g(t)$ the value of T_{lin} will be somewhat less, $T_{lin} \lesssim T$ and $B_{lin} \gtrsim \frac{1}{T}$.

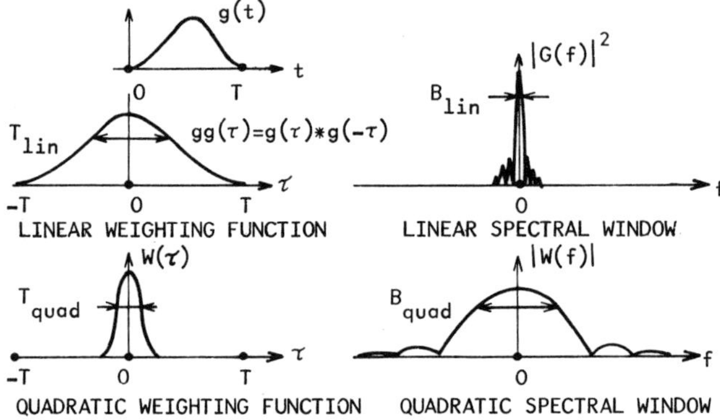

Fig. 8. <u>Linear and quadratic weighting</u>

Eq. 32 gives the variance of the random component $P'(f)$ of the periodogram. Apparently, the r.m.s. value of $P'(f)$ equals the average value of $P(f)$:

$$\text{r.m.s.}\{P'(f)\} =E\{P(f)\} \text{ (putting } k=1 \text{ in eq. 32)} \tag{36}$$

Eq. 36 reflects the fact that for a gaussian process $x(t)$ the quantity $P(f) = \left| \int_{0}^{T} g(t)x(t)\exp-2\pi jft \, dt \right|^2$ has an exponential distribution, for which the mean and r.m.s. values are indeed equal.

Fig. 9-b shows the periodogram of a record $x(t)$, taken from a random process with the power spectrum $S(f)$ of fig. 9-a. Horizontal axis: $f=o$ to $f=.5$ kHz. Recordlength: $T=1.024$ sec. Linear weighting function: raised cosine, $g(t)=c. \{1+\cos2\pi(\frac{t}{T} - \frac{1}{2})\}$, where c satisfies eq. 35-a. Fig. 9-c shows the corresponding linear window $|G(f)|^2$, centered arbitrarily at $f_0= .45$ kHz. Figures 9 have vertical scales in decibels. Because of the proportionality of $\text{var}\{P'(f)\}$ and $E^2\{P(f)\}$ (eq. 36), the random fluctuations in $P(f)$ assume a constant r.m.s. value (5,5 dB) on a logarithmic scale; ref. section B. The pictures were obtained via a discrete

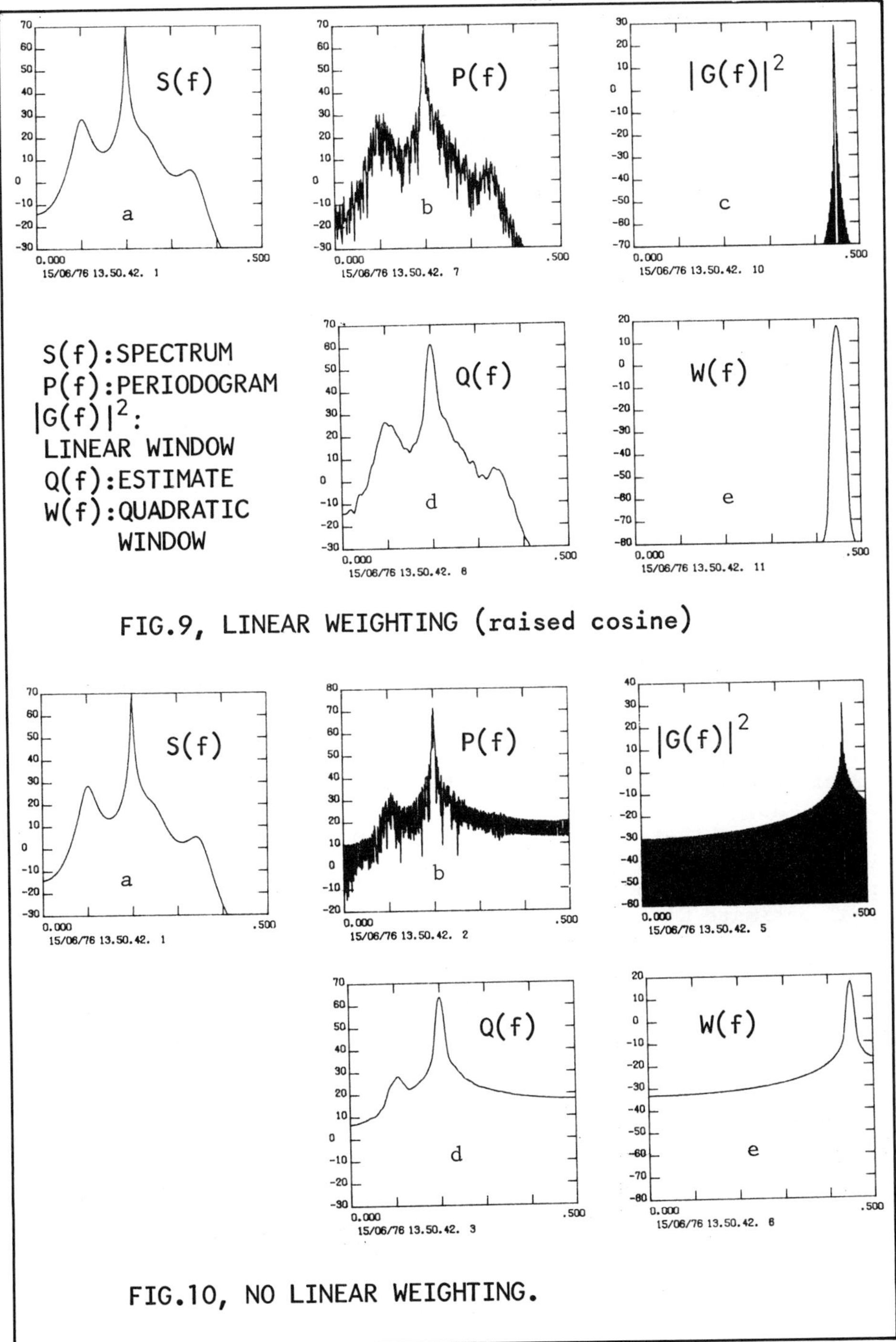

FIG.9, LINEAR WEIGHTING (raised cosine)

FIG.10, NO LINEAR WEIGHTING.

time spectrum analysis, following the path $1\to2\to6\to7$ in fig. 4.
(Fig. 9-c was obtained by using a complex harmonic $\exp 2\pi j f_0 t$ as
input in fig. 4).

Eq. 31 indicates that the rate of fluctuation of $P'(f)$ depends on $G(f)$. The random process $P'(f)$ becomes uncorrelated for a frequency shift Δf equal to the main peak bandwidth B_{lin} of $|G(f)|^2$. Thus, the correlation distance in $P'(f)$ is in the order of $\Delta f = 1/T_{lin}$.

C.2 Spectral estimate Q(f)

Fig. 4 shows that the estimate $Q(f)$ is a smoothed version of the periodogram, $Q(f)=W(f) * P(f)$, irrespective of the way in which $Q(f)$ is actually computed (i.e. via path $1\to2\to3\to4\to8$, fig.4, or via $1\to5\to6\to7\to8$, or via any other path). The smoothing function $W(f)$, called the "quadratic spectral window"(fig. 8), is the fouriertransform of a finite length function $w(\tau)$. Thus, it exhibits a main lobe and sidelobes, in the same way as the linear spectral window.

The smoothing is introduced intentionally to reduce the random fluctuations in $Q(f)$, relative to those in $P(f)$. The variance is approximately reduced by a factor

$$F \approx B_{quad} \cdot T_{lin},\tag{37}$$

where B_{quad} is the effective bandwidth of the window $W(f)$ and T_{lin} is the inverse of the correlation distance in the periodogram. (B_{quad} is defined by eq. 44). (We show below that the simple result of eq. 37 is equivalent to the complicated expression of eq. 33).

Observe that the step from $P(f)$ to $Q(f)$ does not destroy the proportionality of random fluctuations and average values;proportionality of the random error is maintained in the final spectral estimate $Q(f)$ (unless serious leakage occurs in the periodogram as discussed below).

Statistical stability of $Q(f)$ clearly requires a fairly wide window $W(f)$. For example, a variance reduction by a factor $F=100$ (still leaving a 10% random fluctuation in $Q(f)$)calls for a bandwidth*$B_{quad} = \dfrac{100}{T_{lin}} \approx 100.B_{lin}$. On the other hand, a large value of B_{quad} reduces the resolution in $Q(f)$ and finestructure of $S(f)$ which is still present in the periodogram may get lost in $Q(f)$. Here we face again the dilemma between resolution and statistical stability.

* As a consequence, the effective duration T_{quad} of the quadratic weighting function $w(\tau)$ will be relatively short: $T_{quad} \approx 1/B_{quad} = .01 \cdot T_{lin}$. When using the correlogram branch of fig. 4 ($1\to2\to3\to4\to8$), it is therefore often sufficient to calculate $p(\tau)$ only over a limited range of lag values.

Fig. 9 d shows an example of a spectral estimate $Q(f)$, obtained by convolving the periodogram of fig. 9-b with the quadratic window $W(f)$ of fig. 9-e, arbitrarily centered at $f = .45$ kHz. (Actually, fig. 9-e shows the "analysis profile" $V(f)$, defined below, but in the case of fig. 9 the difference is negligible).

In some cases it may be desired to force the estimate $Q(f)$ to be non-negative for all frequencies. Since $Q(f) = P(f) * W(f)$, and since $P(f) \geqslant 0$ by definition, this can always be achieved by using a non-negative quadratic window $W(f)$ or, alternatively, by using a quadratic weighting function $w(\tau)$ which can be written as the auto-correlation of some other function: $w(\tau) = a(\tau) * a(-\tau)$.

Analysis profile and resolution errors. Resolution errors were defined in section A as the differences between the estimate's ensemble average, $E\{Q(f)\}$ and the actual spectrum $S(f)$. Eq. 27-d shows that $E\{Q(f)\}$ equals $S(f)$, viewed through a spectral window $V(f)$:

$$E\{Q(f)\} = V(f) * S(f); \quad \text{(a)}$$
$$V(f) = |G(f)|^2 * W(f) \quad \text{(b)}$$

$$(38)$$

$V(f)$ is called the "analysis profile" or "instrument profile". It is the impulse response of the linear transformation 38-a which maps $S(f)$ into $E\{Q(f)\}$. In general, the analysis profile will consist of a main lobe and sidelobes, in the same way as its constituents $|G(f)|^2$ and $W(f)$. The main peak width of $V(f)$ will be very nearly equal to B_{quad} because $B_{quad} >> B_{lin}$. On the other hand, the sidelobes of $V(f)$ depend in general on the sidelobe properties of both $|G(f)|^2$ and $W(f)$.

As discussed in section B, resolution errors are divided in spectral smoothing and spectral leakage.

Reduction of the smoothing error requires an analysis profile with a narrow peak, i.e. a small value of B_{quad}, but statistical stability sets a lower limit to B_{quad}. When the two requirements are in conflict, the only way out is to increase the recordlength T, thereby increasing the bandwidth-time product of eq. 37 without increasing the value of B_{quad}.

Suppression of spectral leakage requires an analysis profile with a low and/or rapidly decreasing sidelobe level. This is achieved by properly tapering the fouriertransform $v(\tau)$ of $V(f)$:

$$v(\tau) = gg(\tau) . w(\tau), \quad [-T,T] \quad \text{(a)}$$
$$\text{where } gg(\tau) = g(\tau) * g(-\tau) \quad \text{(b)}$$

$$(39)$$

On the necessity of linear weighting in addition to quadratic weighting. For spectra $S(f)$ with large dynamic differences at nearby frequencies, leakage reduction requires a tapering of

both $g(t)$ and $w(\tau)$ (lit. <u>Bingham</u> et al; <u>Sloane</u>). This is not
immediately clear from eq. 39. Without a linear taper, $gg(\tau)$ assumes a triangular shape on $[-T,T]$. Consequently, any desired
function $v(\tau)$ within the class of functions limited to $[-T,T]$,
and any corresponding analysis profile $V(f)$, can be obtained
by substitution of a properly selected quadratic weight $w(\tau)$ in
eq. 39. The obvious advantage would be a simplification of the
spectrum analysis procedure. Further, the value of T_{lin} in eq. 39
reaches a maximum when linear weighting is omitted.

However, the argument above neglects the presence of the
random component $P'(f)$ in the periodogram. The expressions for
$P(f)$ and $Q(f)$ are:

$$P(f) = E\{P(f)\} + P'(f) = [|G(f)|^2 * S(f)] + P'(f)$$

$$Q(f) = E\{Q(f)\} + Q'(f) = [W(f) * |G(f)|^2 * S(f)] + [W(f) * P'(f)] \tag{40}$$

Without a linear taper, the window $|G(f)|^2$ becomes

$$|G(f)|^2 = T \cdot \left(\frac{\sin \pi f T}{\pi f T}\right)^2 \tag{41}$$

The strong sidelobes of $|G(f)|^2$ may lead to leakage in the average
periodogram, $E\{P(f)\} = |G(f)|^2 * S(f)$. Fig. 11 gives an illustra-
tion for a spectrum $S(f)$ with a steep slope at $f=f_s$. Appreciable
values of $E\{P(f)\}$ occur in the region $f>f_s$ and they are accompa-
nied by a random component $P'(f)$ of comparable r.m.s. value
(eq. 36). The application of a suitable quadratic weight $w(\tau)$ to
improve the analysis profile, as discussed above, is equivalent
in the frequency domain to a deconvolution of $E\{P(f)\}$. In this
way, leakage in the average estimate $E\{Q(f)\} = W(f) * |G(f)|^2 * S(f)$
can indeed be perfectly suppressed. Unfortunately, this mechanism
has no effect on the random component $P'(f)$. The result is that
the estimate's random component $Q'(f) = W(f) * P'(f)$ maintains
appreciable values in those areas of the spectrum where leakage
in $E\{P(f)\}$ was **successfully removed** (frequencies $f>f_s$ in fig. 11),
and nothing has been gained. The only way to avoid this situation is to use a well-tapered function $g(t)$ to prevent leakage
in the periodogram.

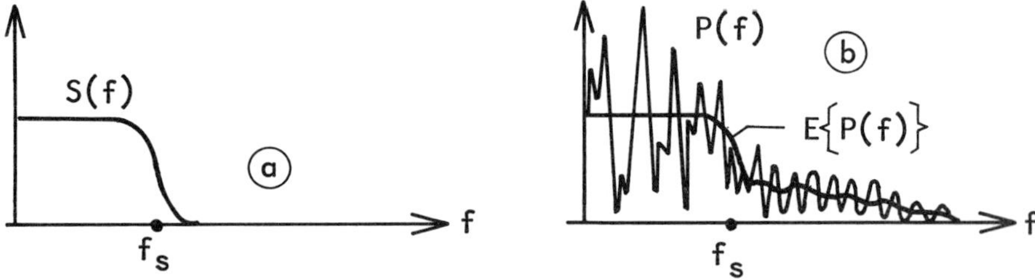

Fig. 11 <u>Power spectrum $S(f)$ (a) and periodogram without linear</u>
<u>weighting (b)</u>

Summarizing, the following policy emerges. A quadratic weight $w(\tau)$ or $W(f)$ is applied to avoid leakage which might otherwise arise in the step from $P(f)$ to $Q(f)$ (or, equivalently, from $p(\tau)$ via $q(\tau)$ to $Q(f)$; fig. 4). This measure is insufficient if leakage occurs already in the periodogram. In that case, a linear weight $g(t)$ must be introduced.

Fig. 10 shows the periodogram $P(f)$ and the spectral estimate $Q(f)$ for the same record as used in fig. 9, but without linear weighting. The leakage in the periodogram is evident, as well as the disastrous effect in $Q(f)$.

<u>Statistical stability.</u> Eq. 33 gives a general expression for the variance of $Q'(f)$. It can be simplified when $|G(f)|^2$ behaves like a deltafunction with respect to $W(f)$ and when $V(f) = |G(f)|^2 * W(f)$ does the same with respect to $S(f)$. The 1st condition implies $T_{lin} \gg T_{quad}$ (footnote page 16) and the 2nd condition implies the absence of leakage. Putting $k=1$ in eq. 33, we obtain:

$$\text{var } Q'(f) \approx S^2(f) . \int_{-\infty}^{\infty} W^2(f) df . \int_{0}^{T} g^4(t) dt \tag{42}$$

Under the same conditions, eq. 27-d is approximated by

$$E\{Q(f)\} \approx S(f) . \int_{-\infty}^{\infty} W(f) df . \int_{0}^{T} g^2(t) dt \tag{43}$$

(Actually, the integrals in eq. 43 are normalized to 1). We now define a relative variance of the estimate in the same way as in eqs. 13 and 22 and we find

$$\text{relvar } \{Q(f)\} \stackrel{\text{def}}{=\!=} \frac{\text{var}\{Q'(f)\}}{E^2\{Q(f)\}} \approx \frac{1}{B_{quad} T_{lin}} = \frac{1}{F}$$

$$\text{where } B_{quad} = \frac{[\int_{-\infty}^{\infty} W(f) df]^2}{\int_{-\infty}^{\infty} W^2(f) df} \quad \text{and} \quad T_{lin} = \frac{[\int_{0}^{T} g^2(t) dt]^2}{\int_{0}^{T} g^4(t) dt} \tag{44}$$

Eq. 44 corresponds to eq 37. The expressions for B_{quad} and T_{lin} define the effective quadratic bandwidth and linear duration. Both quantities have already been used before in a qualitative sense. Relvar $\{Q(f)\}$ can be regarded as a noise-to-signal ratio of the spectral estimate. Thus, statistical stability is proportional to the bandwidth-time product $F = B_{quad} . T_{lin}$.

The value of T_{lin} reaches a maximum, equal to the record-length T, in the absence of linear tapering. Thus, the penalty for introducing a linear weight $g(t)$ consists of a decrease of T_{lin} (usually by a factor in the order of 2).

Example Fig. 12 compares 9 spectral estimates, all belonging
to the power spectrum S(f) of fig. 9. The estimates were made
from 3 different recordlengths T, as tabulated below. Raised
cosines were used as linear weighting functions. The corresponding effective durations, T_{lin}, are indicated in the table.
The 3 periodograms thus obtained are shown in fig. 12, together
with the linear spectral windows $|G(f)|^2$. Subsequently, each of
the 3 periodograms was convolved with 3 quadratic windows W(f)
of gaussian shape. The effective bandwidths are tabulated below.
Fig. 12 presents the estimates Q(f), together with the correspon-
ding analysis profiles V (f) = W(f) $*$ $|G(f)|^2$. The estimates were
obtained via a discrete time spectrum analysis, following the
path $1 \to 2 \to 6 \to 7 \to 3 \to 4 \to 8$ in fig. 4. Horizontal scales in fig. 12 are
from 0 to .5 kHz; vertical scales are in decibels (note the shifts
in the vertical scales). The quantity F in the table below corresponds to eqs 37 and 44.

	T = .256 sec T_{lin} = .13 sec	T = 1.024 sec T_{lin} = .52 sec	T = 4.096 sec T_{lin} = 2.1 sec
B_{quad} = 56 Hz	F = 7.4	F = 30	F = 120
B_{quad} = 14 Hz	F = 1.9	F = 7.4	F = 30
B_{quad} = 3.5 Hz		F = 1.9	F = 7.4

C.3 On the proportionality of the random errors in the spectral estimate

One of the remarkable results of the previous spectrum analysis procedures is the proportionality of the average estimate,
E{Q(f)}, and the r.m.s. value of the random component Q'(f).
Cf. eqs. 13,22 and 44. This property is often very useful because for spectra S(f) with large dynamic differences, we are
indeed interested in constant relative errors rather then in an
error of constant absolute magnitude.

Proportionality of random errors and average values is almost
trivial for spectrum analysis methods which rely directly or in-
directly on bandpassfilter systems (this includes periodogram
methods), but is less trivial for the correlogram method. When
following the path $1 \to 2 \to 3 \to 4 \to 8$ in fig. 4, the proportionality of
E{Q(f)} and r.m.s. {Q'(f)} can be traced back to the correlation
which exists between the random errors $p'(\tau_1)$ and $p'(\tau_2)$ in the
measured covariance function $p(\tau)$ (eq. 30). Without the precise
correlation structure of eq. 30, proportionality of error would not
exist.

One example is when successive values of $p(\tau)$ (τ=o,Δ,...,
$(N-1)\Delta$) are not determined from one and the same record x(t)
[o,T], as in fig. 4 or fig. 13-a, but from N successive non-
overlapping pieces $x_o(t),........,x_{N-1}(t)$, each of length T (fig.
14-a). The random errors $p'(n\Delta)$; n=o,..., N-1, will then be un-

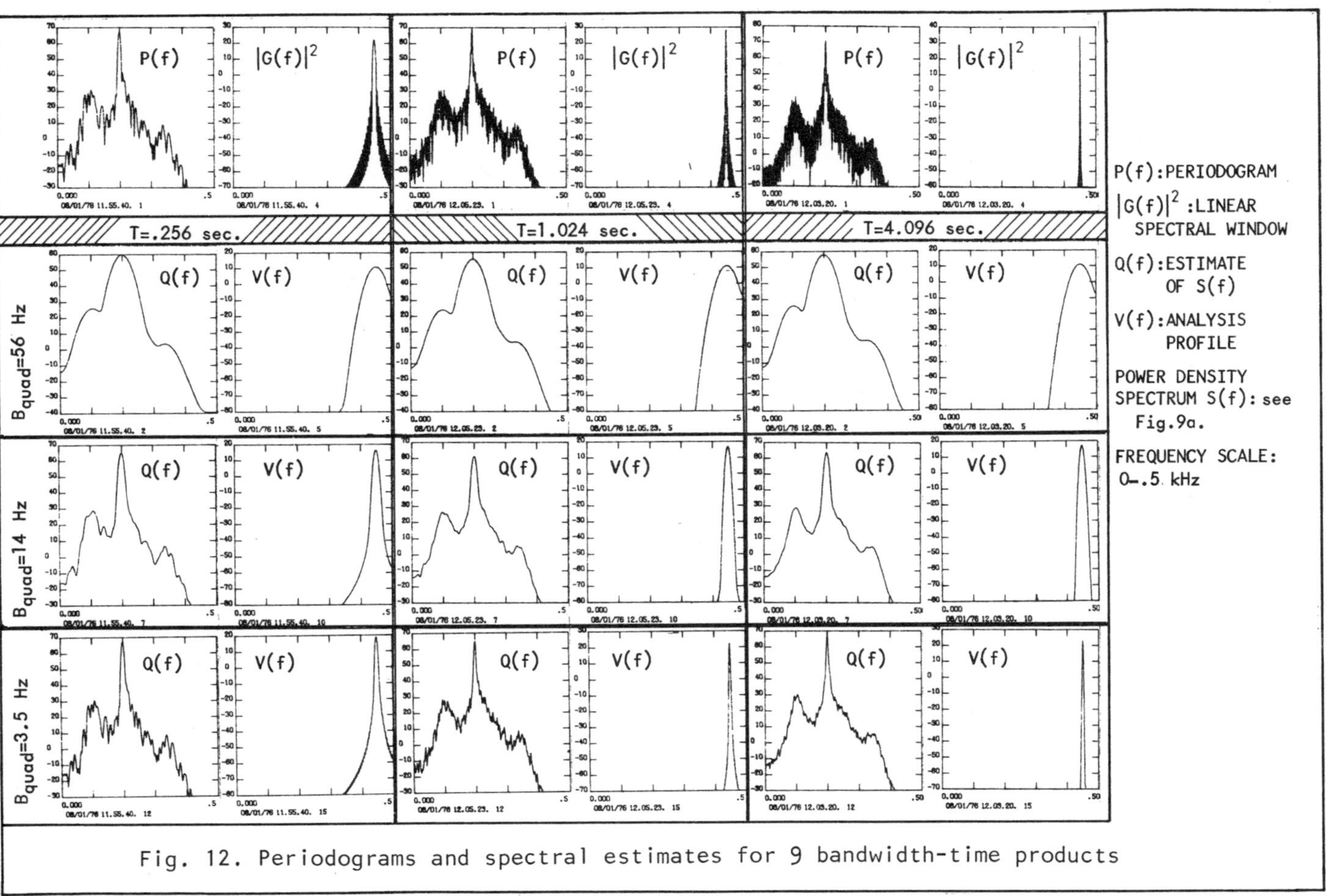

Fig. 12. Periodograms and spectral estimates for 9 bandwidth-time products

correlated. Consequently, the variance of the random errors in $Q(f)$ will be independent of frequency ("white noise") and error proportionality is lost. However, the average error variance remains unaltered, as shown below.

Let $S(f)$ have a maximum frequency f_m (fig. 15) and let $p(\tau)$ be determined for $\tau = o, \Delta, \ldots, (N-1)\Delta$, where $\Delta = 1/2.f_m$. No linear weighting is used (i.e. $g(t) = \frac{1}{\sqrt{T}}$). The measured covariances are quadratically weighted:

$$q(n\Delta) = w(n\Delta).\ p(n\Delta), \tag{45}$$

and are fouriertransformed to obtain the estimate $Q(f)$:

$$Q(f) = \Delta \sum_{n=-(N-1)}^{(N-1)} q(n\Delta).\exp{-2\pi jfn\Delta}, \quad \text{where } q(-n\Delta) = q(n\Delta) \tag{46}$$

First we apply this program to covariance estimates $p(n\Delta)$, obtained from one record $x(t)$ $[o,T]$. (fig. 13-a). Eqs. 42,43,44 are applicable and we find

$$E\{Q(f)\} \approx S(f) \qquad \text{(a)}$$
$$\text{var}\{Q'(f)\} \approx \frac{1}{B_{quad}T}.S^2(f) \quad \text{(b)} \tag{47}$$

Secondly, we repeat the procedure for independent covariance estimates, obtained from N successive records of length T (fig.14-a). The average value of $p(n\Delta)$ is $R(n\Delta)$ and the random errors $p'(n\Delta)$ are now uncorrelated: $E\{p'(n\Delta)p'(m\Delta)\} = \text{var}\{p'(n)\}.\delta_{mn}$. We find from eq. 46:

$$E\{Q(f)\} = \Delta \sum_n w(n\Delta)R(n\Delta)\exp{-2\pi jfn\Delta} = \int_{-f_m}^{f_m} S(a)W(f-a)da$$
$$\approx S(f).\int_{-f_m}^{f_m} W(f)df \tag{48}$$

$$\text{var}\{Q'(f)\} = \Delta^2 \sum_{nm} w(n\Delta)w(m\Delta)E\{p'(n\Delta)p'(m\Delta)\}\exp{-2\pi jf\,(n-m)\Delta} =$$
$$= \Delta^2.\sum_n w^2(n\Delta).\text{var}\{p'(n\Delta)\}$$

and, assuming that $\text{var}\{p'(n\Delta)\}$ is approximately independent of n:

$$\text{var}\{Q'(f)\} \approx \text{var}\{p'(o)\}.\Delta^2 \sum_n w^2(n\Delta) = \Delta.\text{var}\{p'(o)\}.\int_{-f_m}^{f_m} W^2(f)df \tag{49}$$

After normalisation of $W(f)$ (eq.35), we get

$$E\{Q(f)\} \approx S(f)$$
$$\text{var}\{Q'(f)\} \approx \frac{\Delta.\text{var}\{p'(o)\}}{B_{quad}} \tag{50}$$

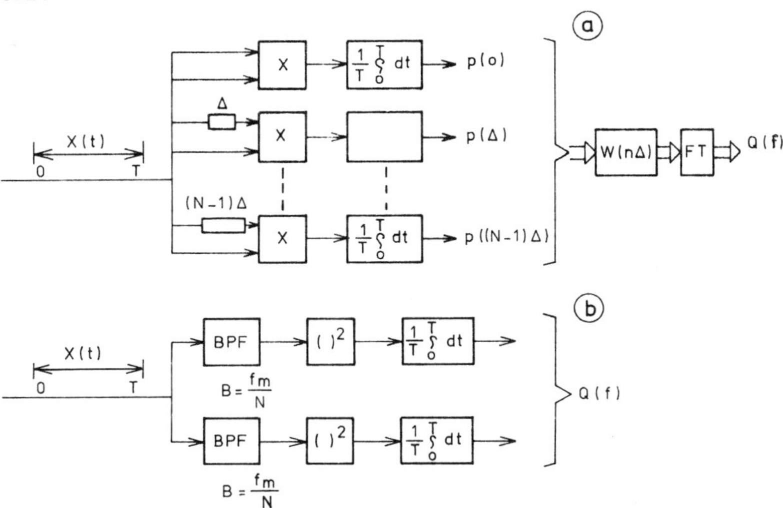

Fig. 13. Parallel correlator bank (a) and parallel filter bank
(b). Both systems use T seconds observation time and
yield proportional random errors.

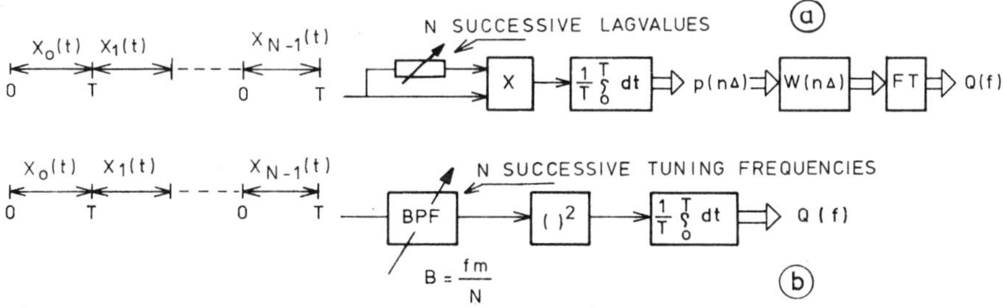

Fig. 14. Stepped lag value (a) and stepped frequencies (b).
Both systems use NT seconds observation time. Fig. a
yields frequency independent random errors and fig. b
proportional errors.

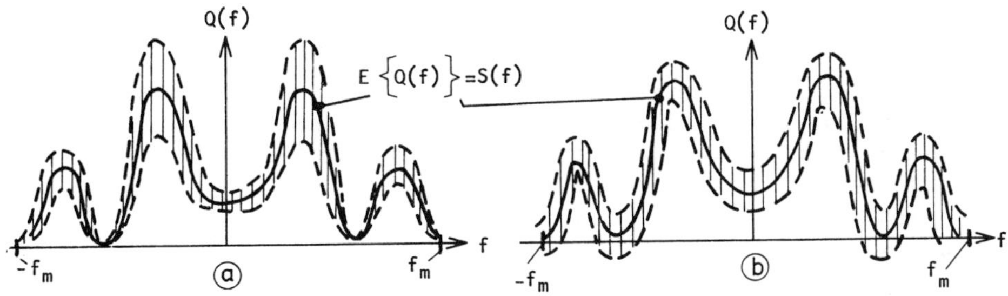

Fig. 15. Proportional errors (a) and frequency independent
errors (b).

Eq. 30 allows the following approximation: $\text{var}\{p'(o)\}\approx$
$\frac{1}{T}\int_{-\infty}^{\infty} R^2(\varepsilon)d\varepsilon = \frac{1}{T}\int_{-f_m}^{f_m}S^2(f)df$. Consequently:

$$E\{Q(f)\} \approx S(f) \qquad\qquad (a)$$
$$\text{var}\{Q'(f)\} \approx \frac{1}{B_{quad}T}\cdot\frac{1}{2f_m}\int_{-f_m}^{f_m}S^2(f)df \qquad (b)$$
$$(51)$$

Eq. 47 (fig. 13-a) corresponds to proportional random errors, whereas eq. 51 (fig. 14-a) corresponds to a frequency independent random error. Fig. 15 illustrates both situations. Observe that the average error variance is the same in both cases:

$$\overline{\text{var}\{Q'(f)\}}\overset{\text{def}}{=}\frac{1}{2f_m}\int_{-f_m}^{f_m}\text{var}\{Q'(f)\}df = \frac{1}{B_{quad}T}\cdot\frac{1}{2f_m}\int_{-f_m}^{f_m}S^2(f)df. \qquad (52)$$

Thus the average error variance is distributed uniformly over all frequencies when the covariances are determined from separate records. When using only one record to determine $p(n\Delta)$, the same average error variance is distributed proportionally to $E^2\{Q(f)\}$.

A second disadvantage of the method of fig. 14-a is that $Q(f)$ may assume slightly negative values in those frequency zones where $S(f)$ is weak.

Obviously, a third disadvantage of fig. 14-a, when compared to fig. 13-a, is the inefficient use of observation time: NT seconds for N lagvalues, rather than T seconds. In this respect, the procedure of fig. 14-a is equivalent to a bandpassfilter system where one filter (bandwidth f_m/N) is tuned to N successive frequencies, the integration time for each frequency being equal to T seconds (fig. 14-b). Observe that the filter system would have the advantage since it yields proportional random errors. (For the sake of completeness, fig. 13-b shows the "parallel filterbank which is equivalent to fig. 13-a).

D. PARAMETRIC METHODS

Thus far spectral analysis was conceived as estimating samples of the power spectrum of a random process of which a realisation of finite duration was observed. As a result, a compromise had to be found between spectral resolution and statistical stability. No attention was paid to the question by how many independent parameters (e.g. poles and zeroes) the power spectrum was determined. In contrast with this approach, the parametric methods first determine the number and type of the parameters (estimated or arbitrarily chosen) of a model, e.g. consisting of a differential or difference equation. Then the model parameters are estimated assuming that the observed signal was generated as the output of

the modelprocess fed by white noise. Once the model parameters are
estimated the powerspectrum estimate $Q(f)$ is uniquely determined.
The structure of the signal generator model determines the type
of the parametric method. In the autoregressive (AR) method a
model is used consisting of an all pole filter (autoregressive
system), resulting in a linear problem. In the ARMA method
(autoregressive moving average) a model is used with
poles and zeroes. In the latter method occasionally the total
number of parameters may be less for a required accuracy, but
a non-linear system of equations results, for which only a numerical solution exists (Lit. Tretter-Streiglitz).
Estimator choice. For estimating the parameters several approaches
can be used. One method is the least mean square estimator, assuming a quadratic error criterion. The results are identical when
the method of maximum likelihood is used, assuming a Gaussian signal.
Estimation scheme. The problem of estimating the parameters can
be formulated either as a Prediction error minimisation problem
(Makhoul) or as a model fitting problem or as a problem of finding a whitening filter for the observed signal, all leading to
identical results. The relationship with Burg's Maximum Entropy
Method is shown by Van den Bos   As an example we consider the
whitening filter approach, using a least mean square estimator
and allowing only an AR generator model while the observed signal
consists of samples x_n of a random process.See fig. 16. The input of the unknown generator is a stationary white noise process
z_n with zero mean. We try to estimate z_n by an linear transformation of past values of x_n.

$$\hat{z}_n = \gamma \sum_{k=0}^{P} b_k x_{n-k}, \quad b=1 \tag{53}$$

In terms of system design this is equivalent with a whitening
filter determined by P zeroes (or feed forward coefficients) and
a gain factor γ. By choosing the parameters $\gamma, b_1, \ldots b_P$ we minimize the error $e_n = \hat{z}_n - z_n$ in l.m.s. sense. Once we have estimated the parameters of the whitening filter we announce the inverse transferfunction of it as the estimate of the power spectrum $Q(f)$:

$$Q(f) = \frac{g}{\left| \sum_{k=0}^{P} b_k \exp j2\pi f k \Delta \right|^2}, \quad b_0=1, \quad g=\frac{1}{\gamma} \tag{54}$$

Thus the signal generator is implicitly modelled by an autoregressive system of order P and gain factor $g=1/\gamma$, fed by white noise
with unit power.

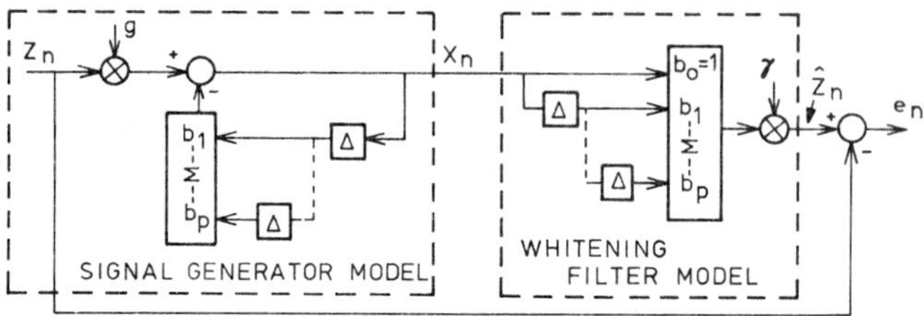

Fig. 16. <u>Whitening filter approach</u>

Using the fact that $E\{z_n x_{n-m}\}$ =o for m>o, and requiring that
$E\{\hat{z}_n^2\}= E\{z_n^2\}$ =1 for the estimation of γ, l.m.s.optimization
leads to a linear system in $g, b_1, \ldots b_P$

$$R_{xx}(o) + R_{xx}(1).b_1 + \ldots + R_{xx}(P).b_P = g^2$$
$$R_{xx}(1) + R_{xx}(o).b_1 + \ldots + R_{xx}(P-1)b_P = 0$$
$$\cdots\cdots\cdots\cdots\cdots\cdots\cdots\cdots\cdots\cdots\cdots\cdots$$
$$R_{xx}(P) + R_{xx}(P_{-1}).b_1 + \ldots + R_{xx}(o)b_P = 0$$

(55)

where $R_{xx}(k) = E\{x_n x_{n-k}\}$
The $R_{xx}(k)$ can be computed as time averages. Since x_n is known
only over a finite interval, a suitable weighting of the signal
is required to avoid spectral leakage in cases of large dynamic
differences, as indicated in section C2. (linear weighting).
Now the system (55) can be solved for the parameters $g, b_1, \ldots b_P$.
By these parameters the power spectrum estimate is completely
determined (see (54)).
<u>Estimating the order.</u> By this method the signal generator is mo-
delled by a P-pole process. If the actual generator contains
zeroes, the whitening filter should contain either poles or an infinite number of zeroes. A very elegant and efficient solution
scheme for the system of (55) exists (Durbin, Levinson). In this
scheme the system is solved recursively. At each step the number
of parameters b_k is augmented, while the error variance is calculated. The recursive process is stopped when the error variance
becomes stationary, thus establishing an estimator for the order
of the process (Lit.:<u>M</u>akhoul, Levinson)
<u>Relative estimation error.</u> Due to estimation errors some correla-
tion is left in $\hat{z}_n$. Therefore, the powerspectrum of $\hat{z}_n$ vz.
$S_{\hat{z}\hat{z}} \overset{def}{=} \gamma^2 . S(f).|B(f)|^2$ is not uniform. (Here B(f) is the Fourier
transform of b_n n=o,$\ldots$P). However, since $E\{\hat{z}_n^2\}=1$, we have
$$\Delta \int_{-1/2\Delta}^{1/2\Delta} S_{\hat{z}\hat{z}}(f)df=1.$$

Now, since $Q(f) = 1/(\gamma.|B(f)|)^2$, the integral yields (Lit.:
Makhoul)

$$\Delta \int_{-\frac{1}{2}\Delta}^{\frac{1}{2}\Delta} \frac{S(f)}{Q(f)}\, df = 1 \tag{56}$$

With the above procedure the powerspectrum was estimated using
the same data record as was used in fig. 9 and fig. 12 (with
T=1.024 sec at a 1 kHz sampling rate.) In this experimental situation the data were generated by a known digital filter having 4 multiple pole pairs in the z-plane to a total of 20 poles. Fig. 17 a
shows the estimated powerspectrum using 21 parameters $(g,b_1,..$
$b_{20})$. The ratio $S(f)/Q(f)$ is plotted in fig. 17-b. The dotted
line in fig. 17-a represents the actual powerspectrum.

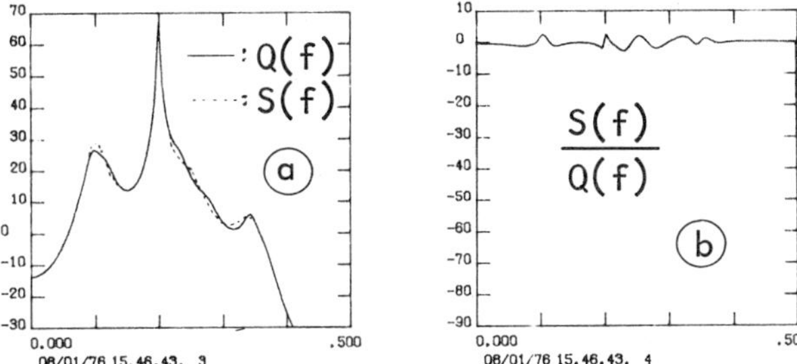

Fig. 17. Spectral estimate (a) and relative error (b)

Literature

Papoulis, A. Probability, random variables and stochastic pro-
cesses. McGraw-Hill, 1965.
Blackman, R.B.; Tukey, J.W. The measurement of power spectra.
Dover Public., 1959.
Bingham, C.;Godfrey, M.D.; Tukey J.W. Modern techniques of
power spectrum estimation. IEEE Trans AU-15, June 1967
Parzen, E. Statistical spectral analysis. Proc. NATO-ASI on sig-
nal processing, Enschede, 1968.
Welch, P.D. The use of the FFT for the estimation of power spec-
tra: time averaging over short, modified periodograms. IEEE
Trans AU-15,No.2 , June 1967.
Jenkins, G.M.; Watts, D.G. Spectral analysis and its applications.
Holden - Day, 1968.
Oppenheim, A.V., Schafer, R.W. Digital signal processing. Pren-
tice - Hall, 1975.
Parzen, E. Time series analysis papers. Holden-Day, 1967
Brigham, E.O. The fast fourier transform. Prentice - Hall, 1974.
Sloane, E.A. Comparison of linearly and quadratically modified
spectral estimates of gaussian signals. IEEE Trans AU-17, No.2,
june 1969.

Makhoul, J. Linear prediction: a tutorial review. Proc. IEEE-63,
Vol. 63 No. 4, april 1975.
Burg, J.P. Maximum entropy spectral analysis, presented at the
37 th Annu. Meeting Soc. of Exploration Geophysicists, Okla-
homa City, Oct. 1967.
Van den Bos, A. Alternative interpretation of maximum entropy
spectral analysis. IEEE Tr. IT-17, no. 4.
Tretter, S.A. - K.Steighlitz, Power spectrum identification in
terms of rational models. IEEE Trans. AC-12, april 1967.
Levinson, N. The Wiener RMS error criterion in filter design and
prediction. Appendix B in "Extrapolation, interpolation and
smoothing of stationary time series" by N. Wiener.

AUTOREGRESSIVE AND MAXIMUM LIKELIHOOD SPECTRAL ANALYSIS METHODS*

R. T. Lacoss

Applied Seismology Group, Lincoln Laboratory,
Massachusetts Institute of Technology,
Cambridge, Massachusetts

1. INTRODUCTION

Three power spectral density estimation problems are considered
in this presentation. These are the estimation of power spectral
density for a single scalar time series (or equivalently a line
array), estimation of power spectral density matrices for vector
valued time series, and the estimation of wavenumber or frequency-
wavenumber spectra for random fields. The paper is primarily a
tutorial coverage of autoregressive (also called maximum entropy)
and maximum likelihood methods of spectral analysis. In addition
to standard material on these subjects a recent generalization of
the Burg algorithm from scalar to vector time series is included.

Section 2 below considers the estimation of power spectral
densities for a single scalar time series. This topic is presented
in the greatest detail since fundamental ideas can be understood
without the notational or conceptual complications resulting from
consideration of vector value processes or fields. Both the
autoregressive and so-called maximum likelihood methods are given
and discussed for this single scalar time series situation.
Section 3 is more limited in scope but deals with the autoregressive
method for the estimation of power spectral density matrices for
vector valued time series. Finally, we briefly present the
maximum likelihood method of frequency-wavenumber analysis using

---

*This work was sponsored by the Advanced Research Projects
 Agency of the Department of Defense.

G. Tacconi (ed.), Aspects of Signal Processing, Part 2, 591-615. *All Rights Reserved.*
Copyright © 1977 by D. Reidel Publishing Company, Dordrecht-Holland.

data from an array of sensors in two or more dimensions. There
are no satisfactory extensions of the autoregressive method to
handle this case.

A very limited but annotated bibliography is included at the
end to supplement the material given here.

2. SPECTRAL ANALYSIS FOR A SINGLE SCALAR TIME SERIES

Given a set of samples of a time series with the samples
taken with uniform spacing in time there are several standard
methods by which we can choose to estimate the power spectral
density of a stationary random process from which the samples are
assumed to be taken. These include:
1. Calculation of some form of sampled autocorrelation function which is then tapered and
   Fourier transformed to obtain the spectrum.
2. Calculation of a periodogram for the entire
   time series and then smoothing over nearby
   frequencies to obtain stability.
3. Breaking the data into shorter segments, obtaining a periodogram for each segment, and
   then averaging the periodograms.
These methods are all roughly equivalent. For our purposes when
there is a need to refer to the standard way to estimate spectra
we will refer to the first of the above. In particular we will
discuss matters relating to bias in the estimation of autocorrelation functions and assumptions about the values of autocorrelation
functions outside of the interval in which they have been estimated.

There is a basic conflict involved in the estimation of
spectra. This is the conflict between resolution, which requires
autocorrelation functions to be known or estimated out to large
lag values, and stability, which requires that autocorrelation
functions be estimated only out to small lags to obtain the
largest number of degrees of freedom in the estimation of spectra.
More formally let R be a sampled autocorrelation function obtained
from a time series of length T. To estimate the spectrum we
multiply R by a taper function, w, and take the Fourier transform
of the product. The expected value of the spectral estimate,
assuming the sampled autocorrelation function is an unbiased
estimate of the true autocorrelation function, is the frequency
domain convolution of the true spectrum and the Fourier transform
of the taper. The spectrum is visible only after it is smeared
out by this window function. The resolution can be improved by
increasing the width of w in lag space. Let L be the half width
of w in lag space. That is, we are effectively using estimates
of R out to lags of plus and minus L. The frequency resolution

will be on the order of 1/L. On the other hand if we define
stability to be the ratio of the square of the expected value of
the estimated spectra to the variance of the estimated spectra
then the stability of a standard spectral estimate is on the
order of T/L. As we originally noted high resolution requires
large L and high stability requires small L.

At least two new methods of estimating spectra have been
developed which are motivated by a desire to circumvent the
conflict. These are the autoregressive method, also known as the
maximum entropy method, and the maximum likelihood method, which
is in fact not a maximum likelihood method for the estimation of
spectra. We begin by presenting the essentials of both these
methods in the case that the autocorrelation function is exactly
known on an interval [-L,L]. The selection of L and other statistical questions are deferred for the present.

2.1 Getting spectra given an interval of the autocorrelation function

In this section, and elsewhere in this presentation, we will
assume that we are dealing with a discrete time process. Time
takes on only integer values. Such a sequence could be obtained
by sampling a continuous time process once per second. We assume
once per second sampling only to simplify notation. Thus for us
the Nyquist frequency will be 1/2 Hz.

Let $R(j)$ be the autocorrelation function of our discrete
process and let $P(f)$ be the power spectral density of the continuous time process from which it is derived by sampling. If the
continuous process is band limited to $|f|<|1/2|$ then $R(j)$ and
$P(f)$ are related by

$$P(f) = \sum_j R(j)e^{-i2\pi f j}, \quad |f| < 1/2 \tag{2.1}$$

and

$$R(j) = \int_{-1/2}^{1/2} P(f)e^{i2\pi f j}df . \tag{2.2}$$

Now suppose that $R(j)$ is known for the interval $j = -L, \ldots,+L$
and we want to know $P(f)$. Can we somehow extend R outside the [-
L,+L] interval and get a best guess at $P(f)$ in that way rather
than just using the known values of R with or without a taper to
adjust the window function? The answer is yes and is the basis
of the autoregressive method of power spectral density estimation.
We now develop this method and will then also note the sense in
which it is also a maximum entropy method.

Imagine that the data available to estimate R is a segment
of a stationary random process and that the process can be thought
of as the result of passing a white noise sequence, $u(k)$, through
an all pole discrete time filter. The resulting process which is
available to us is the autoregressive process which satisfies

$$x(t) = -\alpha(1)x(t-1)...-\alpha(L)x(t-L)+u(t). \tag{2.3}$$

In this representation the order of the filter, L, is assumed
known and it is convenient to assume that it is the same as the
maximum lag out to which the autocorrelation function is known.
The $\alpha(k)$ are the coefficients which define the operation of the
all pole filter. Now suppose that the process under study is
passed through an all zero filter in an attempt to turn it back
into white noise. The output of this all zero filter, again
assuming that L is given, is then

$$v(t) = x(t)+a(1)x(t-1)....+a(L)x(t-L) \tag{2.4}$$

where the $a(k)$ define the filter. If $a(1),...,a(L)$ are selected
to minimize the mean square value of the filter output they will
satisfy the equation

$$[1,a(1)...a(L)] \begin{bmatrix} R(0) & . & . & . & R(L) \\ . & . & & & \\ . & & . & & \\ . & & & . & \\ R(L) & & & & R(L) \end{bmatrix} = [p,0...0] \tag{2.5}$$

Thinking of $1,a(1),...a(L)$ as a convolutional filter it can be
considered as a prediction error filter. That is, it is the
error resulting from the application of a filter $a(1),...a(L)$
designed to best predict the next sample of x as a linear combination of the previous L samples. More specifically we have a
forward prediction error filter since we are thinking of predicting
the future given past values. If the process under study really
is generated by passing white noise through an all pole filter of
order L then it should be clear that the $a(k)$ will be equal to
the $\alpha(k)$. The p on the right side of equation (2.5) is the power
out of the prediction error filter. In an ideal world it is also
equal to the power of the white noise process which is assumed to
pass through an all pole filter to generate the process.

Equations such as (2.5) are very common in discrete filter
theory and there have been developed fast recursive algorithms
for solving such systems of equations. Suppose that such a
system of L simultaneous equations has been solved for $a(1),...,a(L)$

and we want the solution of the corresponding L+1 equations which
are obtained using one more value of lag in the autocorrelation
function. If $a'(1),\ldots,a'(L+1)$ are the solution of the larger
set of equations it is not difficult to show that

$$[1,a'(1),\ldots,a'(L),a'(L+1)] =$$

$$[1,a(1),\ldots,a(L),0]+c[0,a(L),\ldots,a(1),1] \qquad (2.6)$$

where c is a constant with magnitude less than one. When solving
the equations one gets c as a function of the elements of the R
matrix and solution of the system of L equations. This recursive
solution of the equations is generally known as the Levinson
method. The bound on the magnitude of c follows from the Toeplitz
structure of the R matrix and the fact that R must be positive
definite (or at least non-negative definite).

The above recursion imposes the following relationship:

$$R(L+1) = -a(1)R(L)-a(2)R(L-1)\ldots-a(L)R(1)-cp. \qquad (2.7)$$

Thus if we know all the items on the right side we can calculate
the value of $R(L+1)$. If we don't already know $R(L+1)$ and we
don't know anything about c except that it is bounded between
plus and minus one it seems reasonable to set c to zero and use
(2.7) to generate $R(L+1)$. In fact we can recursively generate
all of the larger values of lag. What this does is extend the
autocorrelation function such that each new point is half way
between the limits imposed by the earlier values and by the
requirement that the autocorrelation function be non-negative
definite.

In principal we now know how to extend R to larger lags. We
can do this and Fourier transform the extended function according
to equation (2.1) to get the power spectral density. In fact
what is done is equivalent but fortunately is much easier to do.
Given $R(0),\ldots R(L)$ we can calculate $a(1),\ldots a(L)$. From this we
can obtain the frequency response function of the prediction
error filter. This frequency function is

$$A(f) = (1+a(1)z+\ldots+a(L)z^L)\Big|z=e^{-i2\pi f} . \qquad (2.8)$$

If $P(f)$ is the power spectral density of the process going into
the prediction error filter then $P(f)A(f)A^*(f)$ is the power
spectral density of the output of the filter. If that output is
white with power spectral density p, given by the solution of
(2.5) then

$$P(f) = \frac{p}{|A(f)|^2} . \qquad (2.9)$$

Given R, or estimates of R, out to a lag of L the autoregressive
method of spectral estimation consists of solving equation (2.5)
for the a's and p and simply evaluating 2.9 as a function of
frequency.

We have noted above that the autoregressive method of
spectral estimation is also known as the maximum entropy method.
This is because a completely independent variational approach to
getting P given R over a range of lags results in the same algorithm
as obtained above for the autoregressive approach. In particular
the maximum entropy approach is to find P(f) to maximize

$$J = \int_{-1/2}^{1/2} \ln P(f)df \qquad (2.10)$$

subject to the constraints

$$R(j) = \int_{-1/2}^{1/2} P(f)e^{i2\pi fj}, \; j=0,1,\ldots L. \qquad (2.11)$$

The constraints simply enforce the requirement that the power
spectral density and the autocorrelation function must constitute
a Fourier transform pair. If the process under study is Gaussian
then maximizing (2.10) is maximizing the entropy in the process
as explained by Burg (1975) in his thesis. The only point we
wish to note is that this approach makes no assumptions about the
functional form of P(f) and in particular no assumption is made
that the process is in fact autoregressive.

The maximum likelihood approach is totally different from
the autoregressive approach. It is explained in Lacoss (1971)
which in turn is based on Capon (1969) which treated a more
complicated array processing problem. Imagine that the process
under study has added to it a complex sinusoidal time series of
unknown amplitude and phase and we wish to design a convolutional
filter with L weights which will pass the sinusoid without distortion and reject as much of the noise as possible. The mean
square value of the output of such a filter can be taken as a
measure of the power spectral density of the noise. The rationale
for this is that the filter rejects power at all frequencies
except that of the hypothetical sinusoid and passes power at the
sinusoid frequency without attenuation. Conceptually a different
filter is designed for each frequency of interest. Fortunately
we do not actually need to design each filter to find out what
the output power will be.

Let E be the vector
$$E = \text{col}(1,e^{i2\pi f},\ldots e^{i2\pi f(L-1)}) \qquad (2.12)$$

where f is the frequency of the assumed sinusoid. The requirement
that the filter pass the sinusoid of interest without distortion
simply imposes a constraint that the scalar product of E with a
vector formed by the L filter weights is unity. Let a denote the
filter weight vector and let R again be the autocorrelation
function of the process. The mean square value of the output
of the filter is then $a^T R a$. When this is minimized subject to
the constraint the result is

$$P(f) = \frac{1}{E^T R^{-1} E} \qquad\qquad (2.13)$$

which is the form of the maximum likelihood estimate of the
spectrum. The superscript T denotes transpose conjugate. Of
course for spectral estimation the true autocorrelation matrix R
is not known and some form of sampled estimate is used in expression
(2.13).

Now we have just described an algorithm for estimation of
spectra and called it maximum likelihood although we discussed it
in terms of the power output from linear filters designed to pass
certain sinusoids without distortion. In fact we could have
assumed our noise process to be Gaussian and undertaken to develop
maximum likelihood estimates of the hypothetical sinusoids. The
same filters would have resulted as were obtained from the minimum
variance unbiased approach. That is, the filters we are discussing
can be considered as maximum likelihood estimators of a signal of
known frequency. The mean square output of such a filter, fed
only with noise, as a function of the assumed frequency, is what
we call the maximum likelihood spectrum estimate. It clearly is
not that but the terminology is already well established so it is
too late to change it . It is however important to realize the
rather strange sense in which we can legitimately call this kind
of estimate of a spectrum a maximum likelihood estimate.

The literature is filled with examples of the application of
the autoregressive and, to a lesser extent, the maximum likelihood
algorithms to limited sections of known autocorrelation functions.
Both Lacoss (1971) and Ulrych (1975) contain such exercises and
we will not duplicate such experiments here. We do however
reproduce a table from Lacoss (1971) to emphasize some important
characteristics of the different spectral analysis methods.
Consider the somewhat degenerate case of pure sinewave in white
noise. Let the mean square value of the noise be unity and let α
be the mean square value of the sinewave. We then can look at
the size of the peak in the spectrum due to the sinewave, the
width of that peak, and the power in the peak which can be
approximated by the product of the height and width. The result
is shown in the following table for the direct transform of a

truncated autocorrelation function (Bartlett), for the autoregressive, and for the maximum likelihood methods. L is the filter length used or maximum lag in the autocorrelation fucntion as it has been in all of the above.

Table 1. Approximate bandwidth and peak values of spectral
estimates based on the sample correlation function
$\delta_n + \alpha \cos \lambda_v n$ with $\alpha L \gg 1$ and $L \gg 1$

Estimate	Bandwidth, W(hz)	Peak, P	$W \cdot P$
Bartlett	$\dfrac{\sqrt{6}}{\pi L}$	$\dfrac{\alpha}{2}$	$\dfrac{\sqrt{6}\alpha}{2\pi L}$
Maximum liklihood	$\dfrac{2}{\pi L}\sqrt{\dfrac{6}{\alpha L}}$	$\dfrac{\alpha}{2}$	$\dfrac{1}{\pi L}\sqrt{\dfrac{6\alpha}{L}}$
Autoregressive	$\dfrac{4}{\pi L^2 \alpha}$	$\dfrac{\alpha^2 L^2}{4}$	$\dfrac{\alpha}{\pi}$

Note that the peak value for both the standard method and the maximum likelihood method is proportional to the power α but the peak for the autoregressive method is proportional to the square of the power. However, the true spectrum is an impulse and it is the area of the impluse which represents power. If we look at the area of the peaks for the different methods we see that the autoregressive and standard methods give areas proportional to the power as required but the maximum likelihood method gives an area proportional to the square root of power. This is not an attack on the maximum likelihood method but a caution about the interpretation of spectra for very narrow band processes which cannot be exactly resolved by any of the methods. Other examples in which the process is resolvable will be found in Lacoss (1971).

2.2 Statistical matters and the selection of filter lengths

We want to discuss three different topics related to the statistical aspects of power spectral density estimation. First will be a simple experiment to characterize the performance of traditional and the newer methods in the face of errors in the estimated autocorrelation function. Second we discuss the actual

estimation of the autocorrelation function and present the Burg
algorithm which does it in a very special way which also generates
the prediction error filter coefficients required for the autoregressive spectral estimation method.   Lastly we discuss the selection
of filter lengths, L, to achieve good compromise between resolution
and statistical stability.   In that context the criteria of
Akaike (see Jones, 1974 or Ulrych, 1975) and Parzen (1973) are
given.

A simple experiment was completed and reported in Lacoss
(1971) to study the sensitivity of the various spectral estimation
methods to the exact autocorrelation function.   For that experiment
the autocorrelation function for a sinewave in white noise was
used with the power in the sinewave being about twice that in the
white noise.   Table 2 is a summary of the results obtained in
this numerical experiment.   Note that the pointwise fluctuations
of the autoregressive spectra are much greater than those of the
other methods.   Nevertheless the integral of the power in the
peak of the autoregressive method fluctuates no more than the
peak values of the other two methods.   Recall from above that
this is the proper comparison to make for sinewaves in noise and
sinewaves are in fact not resolved by any method at all.   Before
leaving this topic we note that Baggeroer has done some complementary theoretical work on these points which are reported elsewhere
in this volume.

The next item to discuss is the estimation of autocorrelation
functions.   Given samples $x(0),...x(T)$ of a time series the usual
formuli for estimation of autocorrelation functions are

$$\hat{R}(j) = \frac{1}{T+1} \sum_{k=0}^{T-j} x(k)x(k+j) \qquad (2.14)$$

and

$$\hat{R}(j) = \frac{1}{T+1-j} \sum_{k=0}^{T-j} x(k)x(k+j). \qquad (2.15)$$

The first of these guarantees a positive definite autocorrelation
function estimate at the cost of assuming a zero extension of the
data which causes the estimate to be biased.   The second removes
the bias but can result in a non-positive definite function.
Burg noted these difficulties and proposed the method given below
to avoid the problems.

It is possible to use (2.14) or (2.15) to obtain autocorrelation functions for classical, autoregressive, or maximum likelihood
spectral estimates.   When the maximum lag used, L, is small
relative to data length, T, the differences between them and the
Burg method will be small.   In this paper we do not wish to

Table 2. Summary of the effects of perturbations on peaks of estimated spectra

Units sine wave power	Units white noise power	Standard deviation of perturbations	Percentage deviation from unperturbed spectra			
			Bartlett peak value	Maximum likelihood peak value	Autoregressive peak value	Autoregressive power in peak
2	1	0.1	+1.5	+1.8	-2.8	+1.0
2	1	0.1	-0.8	-0.9	+45.5	+1.0
2	1	0.1	-3.0	-0.9	+11.2	-1.0
2	1	0.1	+0.8	0.0	-30.8	-1.0
2	1	0.1	0.0	0.0	-8.4	0.0
2	1	0.1	+3.8	+3.7	-16.7	+2.9
2	1	0.1	-3.0	-0.9	0.0	-1.0
4	1	0.1	-0.8	-0.9	+1.9	-1.0
1	1	0.1	-2.9	-3.7	0.0	-1.9
2	1	0.2	-3.1	-2.2	+27.0	-1.9
4	1	0.2	-1.6	-1.4	+26.6	-1.5
1	1	0.2	-5.9	-5.4	+28.6	-3.7
2	1	0.282	-3.7	-4.6	+590.0	0.0
4	1	0.282	-2.0	-2.4	-34.0	+9.3
1	1	0.282	-7.0	-8.8	+755.0	0.0

champion the general superiority of any of the methods but simply to give a tutorial description. Since the Burg method is less known it gets correspondingly more attention.

If the time series of interest is an interval from a stationary random process there is no significant statistical distinction between forward time and backward time. Motivated by this fact we define the forward prediction error series

$$e_+(t) = x(t)+a(1)x(t-1)+\ldots+a(L)x(t-L) \tag{2.16}$$

and the backward prediction error series

$$e_-(t) = x(t-L)+a^*(1)x(t-L+1)+\ldots+a^*(L)x(t). \tag{2.17}$$

where the $a(i)$ are the forward prediction error coefficients. The backward error filter coefficients are the complex conjugates of the forward filter coefficients. This is required for complex time series and is included here for completness. We could define a cost function

$$J = \sum_{t=L}^{N} \{|e_+(t)|^2+|e_-(t)|\} \tag{2.18}$$

which equally weights e+ and e_ and find the $a(j)$, j=1...L which minimize J. Given these a's we could then get spectra from them through equations like (2.8) or (2.9). This is a valid approach although the resulting equations for the filter weights are not of the form of equation (2.5), and there is no guarantee that the weights can satisfy an equation like (2.5) with the R matrix non-negative definite. Motivated by this Burg went one more step to develop a recursive method to generate filter weights such that all conditions on them appear to be satisfied.

Given the error series e+,e_ as defined above for a forward prediction error filter [1,a(1)...a(L)] let ε+ and ε_ be the corresponding error sequences for a filter [1,a'(1), ...a'(L+1)] which has one more point. Some study of the previously discussed Levinson recursion shows that

$$\varepsilon_+(t) = e_+(t) + ce_-(t-1) \tag{2.19}$$

and

$$\varepsilon_-(t) = e_-(t-1) + c^*e_+(t) \tag{2.20}$$

where the scalar constant c is the same as appeared in equation (2.6). Burg simply noted that if a(1)...a(L) are known we can get a'(1)...a'(L+1) given c and that if c is selected to minimize

$$J = \sum_{t=L+1}^{T} \{|\varepsilon_+(t)|^2+|\varepsilon_-(t)|^2\} \tag{2.21}$$

it is guaranteed to be less than one in magnitude. This is the
basis for his proposed method for recursively generating longer
and longer filters. The procedure is given more explicitly in
the following pseudo-flowchart. Note that if the forward and
backward error series are thought of as vectors the bounding of
$|c|$ by unity follows from simple geometry.

One might get the impression from the Burg recursion described
above that the number of filter weights to be used for estimation
of spectra can be increased without limit up to the ultimate
limit imposed by the number of waveform data points available for
analysis. This is not true. Just as statistical stability
restricts the number of autocorrelation lags used in traditional
methods it also limits the maximum number of filter weights which
can be used for autoregressive spectral analysis.

The selection of the correct filter length, L, for spectral
analysis and modeling is a complex and important problem. Every
possible method and element of information should be applied in
selecting L. Not the least important are cleverness, intuition,
and experience as well as a priori information about the actual
process if it is believed to be autoregressive in nature.

Some criteria based upon quite sophisticated theoretical
statistical considerations have been developed for the selection
of L. In general they call for the calculation of some function
of L and selecting L which minimizes that function. Three of
these are given here for the case of scalar time series. These
are the Final Prediction Error (FPE) criterion,

$$FPE(L) = \left[\frac{T+1+L}{T-1-L}\right] \hat{p}(L) \ , \tag{2.22}$$

Akaike's Information Criterion (AIC),

$$AIC(L) = \ln \hat{p}(L) + 2 \frac{L}{T} \ , \tag{2.23}$$

and Parzen's CAT criterion,

$$CAT(L) = \frac{1}{T} \left(\sum_{j=1}^{L} \tilde{p}^{-1}(j) \right) - \tilde{p}^{-1}(L) \tag{2.24}$$

where $\hat{p}(L)$ is the mean square prediction error given from equation
2.5 and $\tilde{p}(j)=[T/(T-j)]\hat{p}(j)$. The derivations of these generally
assume equation (2.5) is used with the autocorrelation obtained
by equation (2.15) rather than by the Burg or some other method.
In any case we use the recursion $p(L+1)=(1-|c|^2)p(L)$ for the

DETAILS OF BURG RECURSION

Initialize:

$$p = \frac{1}{T+1} \sum_{t=0}^{T} |x_t|^2$$

$$L = 0$$

Define:

$$c_{++} = \sum_{t=L+1}^{T} |e_+(t)|^2$$

$$c_{--} = \sum_{t=L+1}^{T} |e_-(t-1)|^2$$

$$c_{+-} = \sum_{t=L+1}^{T} e_+(t)e_-^*(t-1)$$

Then

$$c = \frac{-2c_{+-}}{c_{++}+c_{--}}$$

$$p \rightarrow (1-|c|^2)p$$

$$\begin{bmatrix} 1 \\ a_1 \\ \cdot \\ \cdot \\ \cdot \\ a_L \\ a_{L+1} \end{bmatrix} \rightarrow \begin{bmatrix} 1 \\ a_1 \\ \cdot \\ \cdot \\ \cdot \\ a_L \\ 0 \end{bmatrix} + c \begin{bmatrix} 0 \\ a_L \\ \cdot \\ \cdot \\ \cdot \\ a_1 \\ 1 \end{bmatrix}$$

$$L \rightarrow L+1$$

$$e_+(t) \rightarrow e_+(t)+ce_-(t-1)$$

$$e_-(t) \rightarrow e_-(t-1)+c^*e_+(t)$$

error to use at the next stage. The only difference is how c is
selected. It is beyond our scope to go into the derivation of
these criteria for L. Interested readers can get more information
from the Parzen, Akaike and Ulyrich references included at the
end. The AIC is an information theoretic criteria and we will
not discuss it further. We restrict ourselves to the following
discussion of the FPE and CAT criteria.

The idea behind the FPE is easy to grasp. Suppose we
plotted the residual estimated prediction error $\hat{p}(L)$ as a function
of L. It would decrease monotonically with L since we are using
more and more filter weights to fit the available T data points.
If we applied the filter to a different sample of the process the
resulting mean square error would on average be greater than $\hat{p}(L)$
since the filter weights would not be tuned to the new sample.
The FPE takes this into account and corrects to obtain an estimate
of the average mean square error. This is the source of the term
in brackets which increases the estimate more and more as L
approaches the available data length, T.

Parzen's CAT criterion has a different basis. Suppose the
process of interest does have an exact convolutional whitening
filter although the filter length required may be very large and
even may go to infinity. One could try to get a best estimate of
that filter. That is the CAT approach. The minimum of CAT(L)
should occur for the L which statistically would give the best
agreement (mean square) between our a(i)'s and the correct whitening
filter. The functional form of CAT(L), unlike that for FPE(L)
discussed above, appears to have no simple heuristic explanation
which follows from the basic idea of the criterion.

These criteria are valuable tools to help with autoregressive spectral analysis methods. Figure 1 shows for example some
results obtained from two time series consisting of 100 samples
from sinusoids with added white noise. They are treated as a
single complex time series and independently as real time series.
Figures (1.b), (1.c) apply to the complex analysis. Plots of the
three criteria, normalized to run from one to zero and labeled as
C, A and F, are shown on (1.b) and the spectrum for the complex
process which resulted from using the filter length indicated by
the common clear minimum of all three criteria is shown in
(1.c). Note that positive and negative frequencies are shown in
all cases. Figures (1.d) through (1.f) summarize the analysis of
the real time series shown as the real part in Figure (1.a). The
differences in the relative minima of A and F at L=6 and L=26 on
Figure (1.e) are two small to be considered significant. The
same is not true of C which has a much stronger minimum for L=6.
In any case Figures (1.e) and (1.f) show the spectra obtained
using L=6 and L=26 respectively. The L=26 choice gives an acceptable result while the L=6 gives only a general smear of low

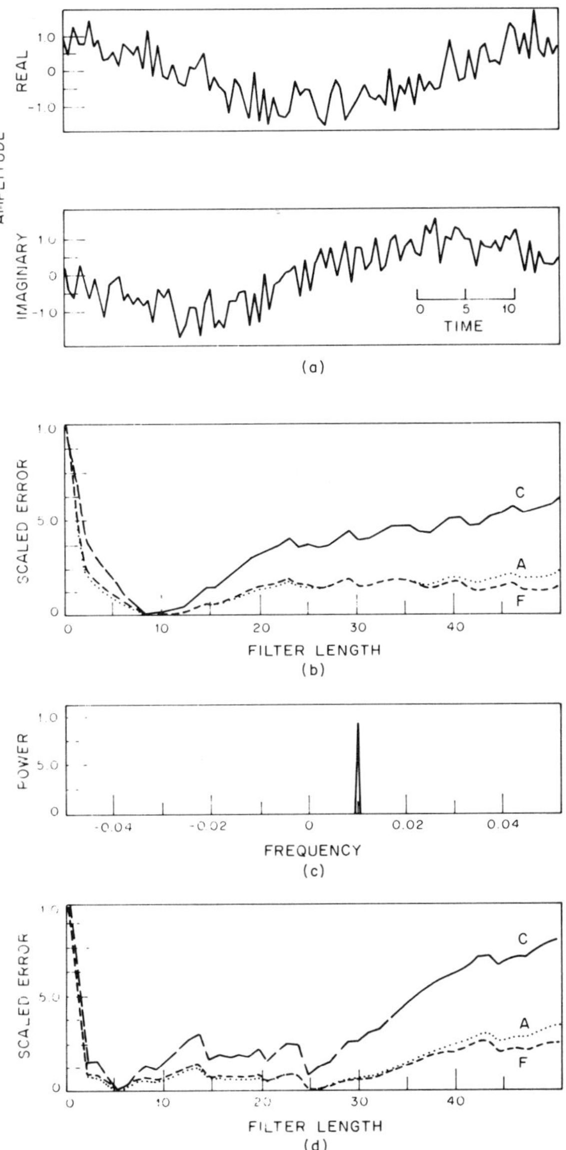

Figure 1.a - 1.j   Signals, order selection criteria, and spectra of sinusoids plus white noise. See text for detailed discussion.

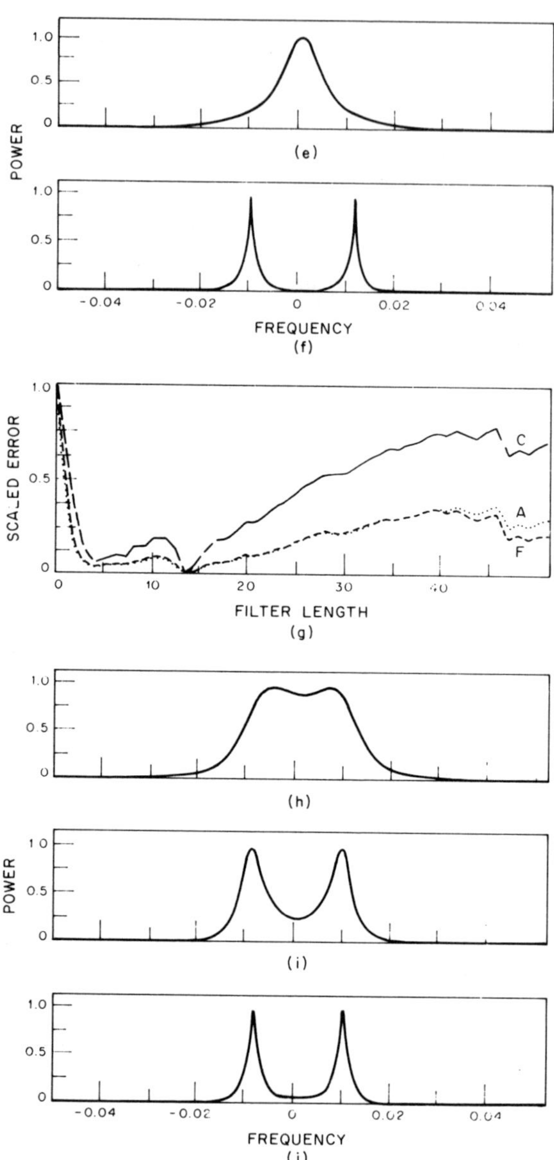

Figure 1.a - 1.j Continued.

frequency energy. Figures (1.g) through (1.j) show the criteria
and some spectra from the real time series identified as imaginary
in Figure (1.a). The three Figures (1.b), (1.i) and (1.j) correspond to the filter lengths L=6, L=13, and L=26.

3. SPECTRAL ANALYSIS FOR A VECTOR TIME SERIES

Suppose we have a time series with vectors rather than
scalars as samples. In this case we are interested in an autocorrelation function with samples which are matrices and we have a
power spectral density matrix for the process. We will not try
to review or survey this general problem area but only present
the basic algorithms for autoregressive estimation of the power
spectral density matrices and an extension of Burg's recursion to
the vector case. At the present time it is only a conjucture
that the Burg extension does force the resulting spectra and
filters to satisfy all theoretical constraints as it seems to do
in the scalar case.

3.1 Multivariate autoregressive spectral estimation

Let $\underline{x}(t)$ be an N vector observed at time t and let it be a
sample of a random process with power spectral density matrix
$P_{xx}(f)$. The power spectral density matrix is NxN. Imagine that
the $\underline{x}$ process can be made white by passing it through a multichannel
filter with the frequency domain response function A(f). A(f) is
here an NxN matrix. The output of A is white in that samples at
different times are not correlated. That is, if $\underline{w}(t)$ is the N-
vector output of A at time t we have $E\underline{w}(t-\tau)\underline{w}^T(t)=0$ for $|\tau|\neq0$.
The symbol E here denotes statistical expectation and T indicates
the complex conjugate transpose operation. For the case $\tau=0$ we
have $E\underline{w}(t)\underline{w}^T(t)=P$ which defines the NxN matrix P. This matrix P
is also the power spectral density matrix of the white noise.
The well known relationships between input and output power
spectral densities of linear filters give us

$$A(f) \, P_{xx}(f) \, A^T(f) = P \tag{3.1}$$

where $\underline{x}$ is the input and A is the filter.

Changing equation (3.1) only slightly **gives**

$$P_{xx}(f) = [A(f)]^{-1} \, P \, [A^T(f)]^{-1} \tag{3.2}$$

which looks like and is a correct matrix version of the important
equation (2.9) for the scalar case. The main difference is that

the order of items on the right side must be correct and a matrix inverse is used. In the scalar case the idea of autoregressive spectral analysis, apart from Burg's algorithm, was to solve equation (2.5) for the whitening filter coefficients and obtain $A(f)$ from equation (2.8) to use in (2.9). The whitening filter is a forward prediction error filter. In the vector case the situation is essentially the same.

Define the forward vector prediction error series as

$$\underline{e}_+(t) = \underline{x}(t) + \sum_{j=1}^{L} a(i)\underline{x}(t-i) \tag{3.3}$$

where the $a(i)$ are the NxN matrix coefficients of the prediction error filter. We select the $a(i)$ as those matrices which minimize $E\underline{e}_+^T(t)\underline{e}_+(t)$. As in the scalar case we try to predict $\underline{x}(t)$ from the previous L samples. The $\underline{e}_+$ series is that part which can't be predicted. For sufficiently large L, the $\underline{e}_+$ series must become white. The equation for the $a(i)$ is

$$[I,a(1),\ldots a(L)] \begin{bmatrix} R(0) & . & . & . & R(L) \\ & & \cdot & & \\ & & \cdot & & \\ & & \cdot & & \\ R)-L) & .. & . & R(0) \end{bmatrix} = [p_a,0\ldots 0] \tag{3.4}$$

where all items are NxN matrices, and p_a is the forward prediction error matrix $E\underline{e}_+(t)\underline{e}_+^T(t)$. This is the needed matrix version of equation (2.5). The subscript "a" on the right side is required since the vector case, unlike the scalar case, does have different forward and backward prediction errors.

The vector autoregressive method is thus substantially the same as the scalar case. The $R(i)$ matrices of equation (3.4) can be estimated from the data and (3.4) solved to get the filter weights. The only remaining complication is that the Levinson recursion must be generalized if it is to be used to solve (3.4). We do that here because it can give substantial computational savings and because we will need the method in the next section which generalizes Burg's algorithm to the vector case.

The generalization of the Levinson recursion involves the coefficients of a backwards prediction error filter. In the scalar case the backward filter coefficients were just a reversed copy of the forward filter so it did not appear explicitly. The backwards filter is defined by the matrices $b(i)$ which minimize $E\underline{e}_-^T(t)\underline{e}_-(t)$ where

$$\underline{e}\,(t) = \underline{x}(t-L) + \sum_{i=1}^{L} b(i)\underline{x}(t-L+1). \qquad (3.5)$$

The b(i) are then the solution of

$$[b(L),\ldots b(1),I] \begin{bmatrix} R(0) & \cdots & R(L) \\ & \vdots & \\ R(-L) & \cdots & R(0) \end{bmatrix} = [0\ldots 0,p_b] \qquad (3.6)$$

where p_b is the backward NxN prediction error matrix.

Now given the solution of

$$\begin{bmatrix} I,a(1)\ldots\ldots,a(L) \\ b(L),\ldots\ldots b(1),I \end{bmatrix} \begin{bmatrix} R(0) & \cdots & R(L) \\ & \cdot & \cdot \\ & & \cdot \\ R(-L) & & R(0) \end{bmatrix} = \begin{bmatrix} p_a,0\ldots,0 \\ 0,\ldots 0,p_b \end{bmatrix} \qquad (3.7)$$

we want to find the solution of the similar set with L increased by one. To see how to do this consider

$$\left\{ \begin{array}{l} \alpha[I,a(1)\ldots a(L),0] \\ \quad + [0,b(L)\ldots b(1),I] \end{array} \right\} \begin{bmatrix} R(0) & \cdots & R(L+1) \\ & \cdot & \cdot \\ & & \cdot \\ R(-L-1) & & R(0) \end{bmatrix} =$$

$$\left\{ \begin{array}{l} \alpha[p_a,0\ldots 0,G_a] \\ \quad +\beta[G_b,0\ldots 0,p_b] \end{array} \right\} \qquad (3.8)$$

where α,β are arbitrary NxN matrices,

$$G_a = R(L+1)+a(1)R(L)\ldots+a(L(R)1) \qquad (3.9)$$

and

$$G_b = b(L)R(-1)+b(L-1)R(-2)\ldots+R(-L-1) . \qquad (3.10)$$

If we select $\alpha=I$ and $\beta=-G_a p_b^{-1}$ we see that we have generated the extended forward prediction error filter. With $\beta=I$ and $\alpha=-G_b p_a^{-1}$ the extended backward filter is obtained. It may be that only the forward filter is of interest but both must be extended to continue the recursion. No simple relation is known between the backward and forward filters or errors but it is known that

$G_a^T = G_b$. Not only can the new values for the extended $a(i)$, $b(i)$ be obtained but we can also get the new values for p_a and p_b. A little algebra with the above equations shows the new values of p_a and p_b to be, in terms of the old values

$$[I - G_a p_b^{-1} G_a^T p_a^{-1}] p_a \qquad (3.11)$$

and

$$[I - G_a^T p_a^{-1} G_a p_b^{-1}] p_b \qquad (3.12)$$

respectively. These are the matrix equivalent of the much simpler $p(L+1) = (1 - |c|^2) p(L)$ for the scalar case.

We note that, as in the scalar case, if the $a(i)$, $b(i)$ are all assumed to be zero for i greater than some value L then (3.9) and (3.10) give a direct mechanism for extending the autocorrelation matrices to larger lags. This can be done using those equations with $G_a = G_b = 0$. As in the scalar case the extended correlation function can be transformed to get a power spectral density matrix but the actual procedure is to use equation (3.2) after the $a(i)$ and p are determined.

We have conducted preliminary tests of multivariate autoregressive spectral estimates. These tests have used direct estimates of the required autocorrelation functions and have also applied the recursion presented in section 3.2 which follows. The results are encouraging and will be presented elsewhere after more experience is gained. Jones (1974) has other examples.

3.2 Vector Burg recursion

The generalization of the Burg recursion given in section 2.2 starts with the definitions

$$\underline{\varepsilon}_+(t) = \underline{e}_+(t) - G_a p_b^{-1} \underline{e}_-(t-1) \qquad (3.13)$$

and

$$\underline{\varepsilon}_-(t) = \underline{e}_-(t-1) - G_b^T p_a^{-1} \underline{e}_+(t) \qquad (3.14)$$

which are motivated by the generalized Levinson recursion. This
is the vector version of equations (2.19) and (2.20). The only
difference is that in order to go from length L to L+1 filters
we must now select the matrix G_a to minimize

$$J = \sum_{t=L+1}^{M} \{\underline{\varepsilon}_+^T(t)\underline{\varepsilon}_+(t) + \underline{\varepsilon}_-^T(t)\underline{\varepsilon}_-(t)\} \tag{3.15}$$

where M is used as the index of the last data point to avoid confusion with the transpose conjugate operator T.

The minimization problem leads to the matrix equation

$$G_a[p_b^{-1}C_{--}p_b^{-1}] + [p_a^{-1}C_{++}p_a^{-1}]G_a =$$

$$- [C_{+-}p_b^{-1} + p_a^{-1}C_{+-}] \tag{3.16}$$

for the unknown matrix G_a where

$$C_{--} = \sum_{t=L+1}^{M} \underline{e}_-(t-1)\underline{e}_-^T(t-1) \tag{3.17}$$

$$C_{++} = \sum_{t=L+1}^{M} \underline{e}_+(t)\underline{e}_+^T(t) \tag{3.18}$$

$$C_{+-} = \sum_{t=L+1}^{M} \underline{e}_+(t)\underline{e}_-^T(t-1) \tag{3.19}$$

and the p_b, p_a are both known for length L. Equation (3.16) can
be solved by treating it as a set of N^2 simultaneous equations.
In fact this particular matrix equation has received much attention
in the past. One approach to it's solution is given by Lacoss
(1976) in a recent note. The entire extension of the Burg algorithm
will be found in Nutall (1976).

The vector equations do reduce to the correct scalar results.
In the scalar case $p_a = p_b$. Identify $G_a p_b^{-1}$ as the c used in the
scalar case and (3.16) becomes

$$c[c_{--}+c_{++}] = -2c_{+-} \tag{3.20}$$

which is the previously obtained corresponding scalar equation.

We must note that a basic proof is missing which is needed
to help justify the extra effort required to apply these extensions
of Burg's algorithms. We note in the scalar case that the constant
c used in each step of the recursion was bounded by one in magnitude.
This is essential since the prediction error for length L+1 given
p, the prediction error for length L, is $(1-|c|^2)p$. Since power
must be positive, $|c|$ must be less than one. In a similar
manner the expressions (3.11) and (3.12) must be non-negative
definite. We are not aware of any proof that this will be satisfied
using the G_a obtained from equation (3.16).

4. ANALYSIS OF WAVE FIELDS

Burg's original work with autoregressive spectral analysis
was not for time series but was for uniform linear arrays of N
seismometers. In that case there is a time dimension and single
space dimension. The sensors take samples of time and space of
the wave field of interest. In the case of a single time series,
Fourier analysis consists of decomposing the data into complex
exponentials, $e^{i2\pi ft}$. In the case of a wave field the generalization of Fourier analysis consists of decomposing the field into
plane waves such as $e^{i2\pi(ft-kr)}$ where k is a spacial frequency
variable referred to as a wavenumber and r is the space variable.
In general k and r are vectors and kr is a scalar product.

Unfortunately, there is not yet any generalization of
Burg's work or the autoregressive approach to the case of non-
uniform arrays or arrays in more than one dimension. However, it
is possible to extend the (misnamed) maximum likelihood method to
the analysis of fields sampled by arbitrary arrays of sensors.
This was done originally by Capon (1969). We cannot here study
the method in detail but wish to complete our presentation with a
brief exposition of the method from the point of view of least
squares unbiased signal processing.

Imagine a set of sensors sampling a wavefield and imagine
that the field contains a plane wave $Ae^{i2\pi(ft-kr)}$. Given f and k
we can design a set of filters such that when they are applied to
the sensors and the outputs are summed we get $Ae^{i2\pi ft}$ from the
plane wave and at the same time minimize the contribution from
the rest of the wave field. The filters will change as a function
of f and k. If we fix f and we map the output power over k we
will obtain what is known as the maximum likelihood estimate of
the wavenumber spectrum for the fixed frequency. This approach

is the same as we took for the analysis of a single scalar time series.

Fortunately there is no need to actually design filters and go through the procedure just outlined. Suppose the NxN power spectral density matrix of the N sensors is known, or has been estimated, at frequency f. Let $[P_{mn}(f)]^{-1}$ be the m,n element of the inverse of that matrix. Capon (1969) showed that the maximum likelihood wavenumber spectral estimate at frequency f is just

$$P_{ML}(f,k) = \left[\sum_{n,m=1}^{N} [P_{mn}(f)]^{-1} e^{i2\pi(k\cdot(r_n-r_m))} \right]^{-1} \qquad (4.1)$$

where r_n is the position vector of the nth sensor. Introducing the complex steering vector

$$E = col(e^{i2\pi k \cdot r_1} \dots e^{i2\pi k \cdot r_N}) \qquad (4.2)$$

we can write (4.1) as

$$P_{ML}^-(f,k) = \frac{1}{E^T P^{-1} E} \qquad (4.3)$$

This looks exactly like the expression for the maximum likelihood spectral estimator for the scalar case except that E is defined in terms of wavenumber and non-uniform sensor locations and the power spectral density matrix takes the place of the autocorrelation matrix. More details will be found in Capon's (1969) paper and other background on noise analysis by arrays will be found in the Lacoss (1969) array analysis paper listed among the references.

5.   ANNOTATED REFERENCES

Anderson, M., On the calculation of filter coefficients for
     maximum entropy analysis, _Geophysics_, _39_, 69-72, 1974.
     - Algorithm for Burg method.

Burg, J. P., Maximum entropy spectral analysis, Ph.D. Thesis,
     Stanford University, 1975.
     - A clear view by an innovative person.

Capon, J., High-resolution frequency-wavenumber spectrum
     analysis, _Proc IEEE_, _57_, 1408-1418, 1969.
     - Basic work on the maximum likelihood method.

Claerbout, J. F., Fundamentals of geophysical data pro-
      cessing with applications to petroleum prospecting,
      McGraw-Hill, 1976.
      - Miscellaneous topics including Burg, filtering,
        scalar and vector Levinson. Includes some programs.

Jones, R. H., Identification and autoregressive spectrum
      estimation, IEEE Trans. on A.C., 19, 894-898, 1974.
      - Multichannel autoregressive methods including
        order determination criteria (FPE and AIC).

Lacoss, R. T., E. J. Kelly and M. N. Toksoz, Estimation of
      seismic noise structure using arrays, Geophysics, 34,
      21-38, 1969.
      - Introduction to frequency-wavenumber analysis.

Lacoss, R. T., Data adaptive spectral analysis methods,
      Geophysics, 36, 661-675, 1971.
      - Study of maximum entropy and maximum likelihood
        methods for scalar time series.

Lacoss, R. T. and A. F. Shakal, More $A_1 E + EA_2 = -D$ and
      $\dot{X} = A_1 X + XA_2 + D, X(0) = C$, IEEE Trans. Automat.
      Cont., AC-21, 405-406, 1976.
      - A minor technical detail concerning the multi-
        variable Burg extension.

Nuttall, A. H., Multivariate linear prediction spectral analysis
      employing weighted forward and backward averaging: a
      generalization of Burg's algorithm. Naval Underwater
      Systems Center Technical Report 5501, 13 October 1976.
      - The same extension was reported by R. H. Jones at the
        Applied Time Series Symposium, May 1976, Tulsa,
        Oklahoma and was developed by Lacoss in unpublished
        notes.

Parzen, E., Multiple time series: determining the order
      of approximating autoregressive schemes, SUNY at
      Buffalo, Computer Science Department, TR-23, July
      1973.
      - CAT reference (obsoletes Parzen in IEEE Trans.
        on A.C., AC-19, 723-729, 1974).

Ulrych, T. J. and T. N. Bishop, Maximum entropy spectral
      analysis and autoregressive decomposition, Rev.
      Geophys. and Space Phys., 13, 183-200, 1975.
      - Excellent overall survey, including many refer-
        ences. Includes some programs.

<u>DISCUSSION</u>

Comment   : L. MEIER

Why in the Berg method is the filter run in the reverse direction
as well as the forward direction?

Reply   : R.T. LACOSS

A stationary random process (scalar) does not know the difference
between positive and negative time directions. On a more mechanical basis going both ways gives a reflection coefficient
$C = -2C_{+-}/(C_{++} + C_{--})$ which is bounded by one and assures stability.
If only forward prediction error is used we get $C = -C_{+-}/C_{--}$,
which can be larger than unity.

CONFIDENCE INTERVALS FOR MAXIMUM ENTROPY SPECTRAL ESTIMATES*

A. B. Baggeroer*

Departments of Ocean and Electrical Engineering
Massachusetts Institute of Technology
Cambridge, Massachusetts, U.S.A. 02139

ABSTRACT. The statistical properties of spectral estimates generated using the Maximum Entropy Method are analyzed using a Wishart Model for the estimated covariance. The transfer function of the estimated regression filter has a form which may be interpreted as a generalization of the student's t distribution and the estimated prediction error has χ^2 distribution. Asymptotic expressions for a large number of observations are derived which are the same as those of Akaike. These expressions allow a direct comparison between the MEM and the Maximum Likelihood Method, where it is shown that while the MEM has a higher peak to background ratio than the MLM, its confidence intervals are correspondingly larger. An example consisting of several closely spaced tones in a background of white noise is used to demonstrate the results.

1. INTRODUCTION

The autoregressive, or MEM, method of spectral estimation has been used extensively in a number of applications including radar, speech and geophysics.(1-4) The principal motivation for its use is the apparently higher resolution when compared with either classical methods, or the alternative "high resolution" Maximum Likelihood Method of spectral estimation.(5) The method is

---

*This work was supported in part by M.I.T. Lincoln Laboratory
(Contract No. F19628-76-C-0002) and by the MIT Research for
Electronics Joint Services Program (Contract No. DAAB07-75-C-1346)

commonly attributed to either Burg or Parzen; but, as pointed out
by Parzen, it is alluded to in several papers concerned with
autoregression analysis. (6,7) Certainly it is related to regression analysis in time series,(8)prediction error filtering in
seismic exploration,(9) predictive encoding in speech,(2) and
innovations and whitening filters in communications and information theory.(10,11) In any event, the method is used extensively.

In evaluating the performance of this method, there are two
principal considerations. First, how well does the implied autoregressive, or all-pole model, fit the class of spectra being
estimated; and second, what is the statistical variability of the
resulting estimates. The first issue has been addressed in a
number of analyses--some related to the problem of covariance
extension, while others concern the goodness of fit in approximating the true spectra.(6,12,13) The second issue has been
studied to a lesser extent, and at the present time this author
is aware of the results of Kromer(14) referenced by Parzen and
the asymptotic ones of Akaike(15) and Berk.(16) In addition, the
Maximum Likelihood Method has been shown to have a χ^2 distribution
when applied to complex processes, such as those which appear in
the estimation of frequency wavenumber functions.(17) In this
note we use some results from regression theory to analyze the
statistical variability of the MEM of spectral estimation. (This
material is an appreciably condensed version of Ref. 29, where
the complete mathematical analysis appears.)

2. THE MAXIMUM ENTROPY SPECTRAL ESTIMATION ALGORITHM

In this section we describe the MEM, or autoregressive spectral
estimation algorithm, and indicate how it is related to regression
analysis. Let $x(n\Delta T)$ be a discrete, stationary, zero mean random
process with covariance $R_x(n\Delta T)$. We assume that an estimate of
this covariance, $\hat{R}_x(n\Delta T)$, is available for $n = 0,1,\ldots,N-1$. This
estimate is used to form an $(N-1) \times (N-1)$, Toeplitz form, covariance matrix, $\hat{\underline{R}}_x$

$$
\hat{\underline{R}}_x \triangleq \begin{bmatrix}
\hat{R}_x(0) & \hat{R}_x(\Delta T) & \hat{R}_x(2\Delta T) & \cdots & \hat{R}_x((N-2)\,\Delta T) \\
\hat{R}_x(\Delta T) & \hat{R}_x(0) & \hat{R}_x(\Delta T) & & \\
& & & \ddots & \hat{R}_x(0)
\end{bmatrix}
$$

and an $N \times 1$ covariance vector $\hat{\underline{r}}_x$

$$
\hat{\underline{r}}_x \triangleq [\hat{R}_x(\Delta T),\ \hat{R}_x(2\Delta T)\ \cdots \quad \hat{R}_x(N-1)\,\Delta T)]^T
$$

Define an N x 1 frequency vector $\underline{E}_N(f)$ as

$$\underline{E}_N(f) \triangleq [e^{-j2\pi f(N-1)\Delta T} \ldots \quad , \ e^{-j2\pi f\Delta T}, \ 1]^+$$

The MEM, or autoregressive spectral estimation algorithm, is specified by

$$\hat{S}_E(f) \triangleq \hat{P}_E(N) \ / \ |\hat{H}(f:N)|^2 \tag{1}$$

where

$$\hat{P}_E(N) = \hat{R}_x(0) - \hat{\underline{r}}_x^{\ T} \hat{R}_x^{-1} \hat{\underline{r}}_x$$

$$\hat{H}(f:N) = \underline{E}_N^+(f) \begin{bmatrix} -\hat{R}_x^{-1} \hat{\underline{r}}_x \\ \hline 1 \end{bmatrix}$$

Alternate implementations, particularly the recursive ones suggested by Burg, (18) also have been suggested. (Asymptotic consistency of the MEM can in fact be demonstrated under mild restrictions whether or not the process is a true autoregressive one.(16))

In our analysis it is necessary to specify the algorithm in a more general form which emphasizes the coupling to regression analysis. In some respects this may be considered as shifting from the "correlation" to "covariance"form, a described in Ref. 2. These generalizations remove the restriction imposed by stationarity on $\hat{R}_x$ and the use of complex sinusoid basis representation, $\underline{E}_N(f)$. The assumptions which we make, however, are the same as those by Capon and Goodman in their analysis of MLM for wavenumber function estimation.(17)

To generalize this analysis, we want to consider the matrix $\hat{R}_x$ and the vector $\underline{r}_x$ to be partitions of a matrix $\hat{\Sigma}_N$, whose probability density we discuss in the next section. As can be seen there, this matrix maybe interpreted as the sample covariance of zero mean, N dimensional, random vector $\underline{x}_N$ with covariance matrix Σ_N. We have

$$\begin{bmatrix} \hat{R}_x & \hat{\underline{r}}_x \\ \hat{\underline{r}}_x^T & \hat{R}_x(0) \end{bmatrix} = \hat{\Sigma}_N = \begin{bmatrix} \hat{\Sigma}_{N-1} & \hat{\underline{\sigma}}_N \\ \hat{\underline{\sigma}}_N^T & \hat{\sigma}_{NN} \end{bmatrix}$$

$T \triangleq$ transpose

$+ \triangleq$ conjugate transpose

where $\hat{\Sigma}_{-N-1}$ is an (N-1)x(N-1) matrix, $\hat{\sigma}_{NN}$ is a scalar. An addition,
let $\underline{\phi}$ be an arbitrary vector which we may identify as the first
N-1 components of $\underline{E}_N(f)$ when applied to the MEM analysis, i.e.

$$E_N(f) = \left[\begin{array}{c} \underline{\phi} \\ \hline 1 \end{array}\right] \begin{array}{l} \} \\ \} \end{array} \begin{array}{c} N-1 \\ \hline 1 \end{array}$$

We then have

$$\hat{S}(\underline{\phi}) = \hat{P}_E(N) \ / \ |\hat{H}(\underline{\phi})|^2 \qquad\qquad (2)$$

where

$$\hat{P}_E(N) = \hat{\sigma}_{NN} - \underline{\sigma}_N^T \hat{\Sigma}_{-N-1} \hat{\underline{\sigma}}_N$$

$$\hat{H}(\underline{\phi}) = 1 - \underline{\phi}^+ \hat{\Sigma}_{-N-1} \hat{\underline{\sigma}}_N$$

The deterministic quantities $S(\underline{\phi})$, $P_E(N)$ and $H(\underline{\phi})$ are similarly
defined when one uses ensemble, Σ_N, instead of estimated ones,
$\hat{\Sigma}_{-N}$. In regression theory, $P_E(N)$ is the mean square error in
estimating the Nth component of x_N given the first N-1 components,
$x_{-N-1} \Sigma_{-N-1}^{-1} \sigma_{-N}$ generates the estimate $(\hat{x}_{-N})_N$, the Nth component of
x_{-N}. Obviously, the error ε_N is given by

$$\varepsilon_N = \underline{x}_N^T \left[\begin{array}{c} -\Sigma_{-N-1}^{-1} \ \sigma_{-N} \\ \hline 1 \end{array}\right]$$

3. PROBABILITY DENSITY MODEL FOR $\hat{\Sigma}_{-N}$

There are a number of methods for estimating a covariance matrix
R_{-x} from sample sequences of a stationary random process; and each
of these methods induces a different form for the probability
density of the resulting estimated covariances. In this section
we introduce a probabilistic model for $\hat{\Sigma}_{-N}$ (and hence by partioning
R_{-x} and r_x), that is frequently used in the statistical literature.
Our basic assumption is that the matrix $M\hat{\Sigma}_{-N}$ has a dimensional
Wishart distribution with M degrees of freedom.(19,20) Since there
are both advantages and disadvantages, it is useful to discuss
the implications of this model. However, first we specify the
form of the Wishart probability distribution.

We define the accumulated covariance as $\hat{S}_{-N} = M\hat{\Sigma}_{-N}$ and we model it
as having an N dimensional Wishart distribution with M degrees
of freedom. It then has a multivariate probability density of
the form

$$P_{S_{-N}}(\underline{X}) = K(N,M) \ \frac{|\underline{X}|^{\frac{M-N-1}{2}}}{|\Sigma_{-N}|^{\frac{M}{2}}} \ e^{-1/2\mathrm{Tr}\,[\Sigma_{-N}^{-1}\underline{X}]} \ , \ \underline{X} \geq 0$$

where $|\ |$ and Tr are, respectively, the determinant and trace operations, and

$$\underline{\Sigma}_N = \frac{1}{M} \, E[\hat{S}_N], \quad K^{-1}(N,M) = 2^{\frac{NM}{2}} \, \pi^{\frac{N(N-1)}{4}} \, \prod_{i=1}^{N} \, \Gamma \left(\frac{M+1-i}{2} \right).$$

In discussing the appropriateness of this probability density for $\underline{\Sigma}_N$, we need to examine some of the issues involved in the measurement methods. Figure 1 illustrates two of the more commonly used methods.

Fig. 1   Comparison of Methods for Covariance Estimation

The most common methods of estimating $R(n\Delta T)$ is to use the sampled "lagged product"

$$R(n \Delta T) \; \alpha \quad x(m \Delta T) \; x \, ((m-n) \, \Delta T)$$

A stationarity structure is imposed directly in that the resulting $R(n \Delta T)$ is function only of the offset variable n regardless of the properties of the sample sequence $x(m \Delta T)$. This method is in some respect computationally inefficient since many of the terms in the sum are highly correlated. It also requires the storage of large arrays which may be prohibitive for small computers. Its most important property, however, is that the stationarity structure imposed so one is dealing with a minimum number of variables for a given total lag extent, N, of the correlation function. The stationarity also makes the results of Toeplitz forms applicable, and these can be used to derive properties of spectral estimates which are always applicable independent of any statistical considerations.

The "pseudo ensemble" method is which we employ in this paper is used in most direct, of FFT, methods of spectral estimation. It also generalizes to wavenumber estimation. It consists of segmenting the time series into M blocks with the length of each block, $N\Delta T$, generally chosen to be at least longer than the correlation extent of the process. Beyond this, the usual considerations of resoltion versus the statistical stability influence the choice of N. These segments are used to calculate an estimate of the covariance matrix

$$R_x(i\Delta T, \, j\Delta T) = \frac{1}{M} \sum_{m=1}^{M} x^m(i\Delta T) \; x^m(j\Delta T) = (\hat{\Sigma}_N)_{ij}$$

with the segments represented as

$$\underline{x}^{m^T} = \left[x^m(\Delta T), \ x^m(2\Delta T), \ \ldots \ x^m(N\Delta T) \right]^T$$

If one can argue that the segment vectors $\underline{x}^m$ are jointly Gaussian and are statistically independent, then $M\hat{\Sigma}_{-n}$ has a Wishart distribution. The Gaussian property must be introduced with the usual arguments in modelling data. The independence can be qualitatively asserted if one makes the segments long compared to the effective correlation extent of the data.

In examining the applicability of this model, the most evident consideration is that the stationarity structure has not been imposed directly. Consequently, the matrix $\hat{\Sigma}_{-N}$ is itself not Toeplitz, although its expected value is Toeplitz. It should be noted that it is seldom imposed directly in most direct, or FFT, methods of spectral estimation, e.g. a simple, unwindowed, average of periodograms can be expressed as

$$\hat{S}_x(f) = \underline{E}_{-N}^+(f) \ \hat{\Sigma}_{-N} \ \underline{E}_{-N}(f)$$

In any event, one could consider that not imposing stationarity upon the estimated covariance detracts from the probabilistic model. The methods which we employ are based upon general regression properties and not specifically autoregressive ones. Our results then may be considered conservative since they use a less restrictive model.

There are also some advantages to this model. First, it can be used in nonstationary situations where one is concerned with a general basic vector $\underline{\phi}$, and not a sinusoidial one, $\underline{E}_{-N}(f)$. Second, it can be extended to arrays and wave vector estimation where nonperiodic element spacing is present, which is commoningly the case. Finally, it is more efficient in that it requires approximately one half the computations of the lagged product procedures.

With these remarks, we simply assert that we are using the Wishart distribution model for the estimated covariance, and we note that this is essentially the same assumption used by Capon and Goodman.(17) The only difference is that we have confined our attention to real processes.

4. STATISTICAL PROPERTIES OF THE MEM SPECTRAL ESTIMATE.

The important aspect in our analysis involves some results from regression theory which are implicit in several texts, e.g. Refs.20 and 21, and explicitly derived by Kshirsagar in Ref. 19. Using these results, we have the following:

i) The matrix $M\hat{\Sigma}_{-N-1}$ has an N-1 dimensional Wishart distribution with M degrees of freedom;

ii)   The vector $\hat{\underline{h}} = \hat{\Sigma}^{-1}_{-N-1} \hat{\underline{\sigma}}_N$ has the conditionally Gaussian distribution $N[\Sigma^{-1}_{-N-1} \underline{\sigma}_N, P_E(N) \hat{\underline{S}}^{-1}_{-N-1}]$ when $\hat{S}_{-N-1}$ is given;

iii)   The scalar $MP_E(N) / P_E(N)$ has a χ^2 distribution with M-N+1 degrees of freedom and is statistically independent of $\hat{\underline{h}}$ and $\hat{\Sigma}_{-N-1}$.

The first result is obvious, and the third result implies that the number, $\hat{P}_E(N)$, in the MEM algorithm which appears as a scale factor has a mean of $(1-\frac{N-1}{M}) P_E(N)$. Note that this scale factor is independent of $\underline{\phi}$ and simply translates the entire estimated spectrum when it is plotted on a logarithmic scale.

As one can observe from Eq. (2), the spectral "features" are determined by the function $|\hat{H}(\underline{\phi})|$. In this section we derive the probability density for the real and imaginary components of $\hat{H}(\underline{\phi})$.

We define $\hat{H}(\underline{\phi}) \overset{\Delta}{=} 1 - \underline{\phi} + \hat{\underline{h}}$

or separating real and imaginary parts

$$\hat{H}_R(\phi) + j\hat{H}_I(\phi) \overset{\Delta}{=} (1 - \underline{\phi}_R^T \hat{\underline{h}}) + j \, \underline{\phi}_I^T \hat{\underline{h}} .$$

Using the properties of Gaussian random vectors and integrating out the conditioning on $\hat{\underline{S}}_{N-1}$, one can demonstrate

$$P_{\hat{H}(\phi)}(H_R, H_I) = \frac{M-N+2}{2\pi} \, |L(\underline{\phi})|^{\frac{1}{2}} \, \frac{1}{(1 + \underline{\Delta}\,\underline{H}^T L(\underline{\phi})\,\underline{\Delta}\underline{H})^{(M-N+4)/2}} \tag{3}$$

where

$$\underline{\Delta}H = \begin{bmatrix} H_R - (1 - \underline{\phi}_R^T \Sigma^{-1}_{-N-1} \underline{\sigma}_N) \\ H_I - (\underline{\phi}_I^T \Sigma^{-1}_{-N-1} \underline{\sigma}_N) \end{bmatrix}$$

and

$$L(\phi) = \left[\begin{array}{c|c} \underline{\phi}_R^T \Sigma^{-1}_{-N-1} \underline{\phi}_R & \underline{\phi}_R^T \Sigma^{-1}_{-N-1} \underline{\phi}_I \\ \hline \underline{\phi}_I^T \Sigma^{-1}_{-N-1} \underline{\phi}_R & \underline{\phi}_I^T \Sigma^{-1}_{-N-1} \underline{\phi}_I \end{array} \right] P_E(N)$$

Equation (3) expresses the fundamental result for describing the probability density structure of the MEM spectral estimator. We

can observe that it has a form which could be interpreted as a
multidimensional generalization of the "noncentral student's t"
distribution.(24)

5. COMPARISON TO AKAIKES' RESULTS AND TO THE MAXIMUM LIKELIHOOD
    METHOD.

Akaikes' paper (Ref. 15) is the principal work on the performance
of regression methods in power spectrum estimation. As such, it
is useful to contrast the analysis introduced here to his. One
of the more interesting results is that both lead to the same
asymptotic distribution for the estimated transfer function, $\hat{H}(\phi)$.
These asymptotic results can also be used to compare the perfor-
manaces the Maximum Entropy and Maximum Likelihood Methods.

The basic assumption in Akaike's paper is that the process is
actually generated by an autoregressive model of known dimensionality, N. No assumptions beyond the existence of some moments
are made about the uncorrelated process exciting the autoregressive model. Using Theorem 2 of Akaike's paper and transcribing
the notation to our we have the following results:

i) $\hat{\underline{h}} - \underline{h} = \hat{\underline{R}}_{-x}^{-1} \hat{\underline{r}}_{-x} - \underline{R}_{-x}^{-1} \underline{r}_{-x}$ is a zero mean, Gaussian random
vector with covariance matrix

$$\sigma^2_{h-h} \simeq \frac{P_E(N)}{M} \underline{R}_{-x}^{-1} = \frac{P_E(N)}{M} \Sigma_{-N-1}^{-1}$$

ii) $\hat{H}(\phi) - H(\phi) = \phi^+(\hat{\underline{h}} - \underline{h})$ is a zero mean, complex Gaussian
random variable with variance

$$\sigma^2_{\hat{H}-H} = \frac{P_E(N)}{M} \underline{\phi}^+ \underline{R}_{-x}^{-1} \underline{\phi} \tag{4}$$

For comparing to our results we use some asymptotic results
involving the Student's t distribution.(25) Two observations are
needed: i) In these examples, $L(\phi)$ is approximately diagonal
so that the density $p_{\hat{H}(\phi)}(H)$ is radially symmetric about its
single mode; ii) under the condition of very large M the density
is very sharply peaked near this mode. In this case we can
approximate the density, integrate out the coordinate orthogonal
to the radius vector of $|H(\phi)|$, and apply the asymptotic results
that the Student's t distribution approaches a Gaussian one as
its order increases. This leads to a Gaussian distribution for
$\hat{H}(\phi)$

where

$$m = H(\underline{\phi})$$

$$\sigma^2_{\hat{H}-H} = \frac{P_E(N)}{M-N+2} \; \underline{\phi}^{+} \, \Sigma^{-1}_{-N-1} \, \underline{\phi} \qquad (5)$$

Comparing Equations (4) and (5) we observe that for large M they are substantially identical.

The above equation can be expressed in a rather interesting way which couples the performance of the Maximum Entropy and Maximum Likelihood Methods. By using matrix identities, it is easy to verify(26)

$$\frac{\sigma^2_{H-H}}{|H(\underline{\phi})|^2} = \frac{1}{M-N+2} \left\{ \frac{S_{MEM}(\underline{\phi})}{S_{MLM}(\underline{\phi})} \right\} - 1$$

This should be contrasted to the results of the Maximum Likelihood estimator where

$$\frac{\sigma^2[S_{MLM}(\phi)]}{|S_{MLM}(\phi)|^2} \simeq \frac{1}{M-N+1}$$

This results allows us to draw some general conclusions in comparing the performance of the MEM and MLM spectral estimators. Typically, the MEM produces spectral features which have a much higher peak to background ratio than the MLM, egs. see Figures 2 and 3 or those of Lacoss in Ref. 5. This is probably its principle attraction for its usage. The drawback, however, is that its confidence intervals are proportionally larger than those of the MLM.

6. PERFORMANCE ANALYSIS EXAMPLE FOR AN MEM ESTIMATOR.

At this point we want to use the results of the previous section to illustrate an example of the performance of an MEM autoregressive spectral estimation. We also want to compare it with the performance indicated by a χ^2 density, for it is used in more conventional spectral estimation procedures as well as the Maximum Likelihood Method. In this analysis we make one assumption regarding the use of the MEM estimator. In most spectral estimation problems, one is principally concerned with the salient features and not the absolute scale of the estimated spectrum. In particular, one usually is concerned with determining if an observed spectral peak is truly representative of the process or if it is only a noise spike introduced in the estimation procedure. In our analysis, therefore, we do not consider the scaling of the overall estimate introduced by the random variable $\hat{P}_E(N)$. Since it is independent of $\hat{H}(\phi)$, we focus solely upon the variations $|\hat{H}(\underline{\phi})|^{-2}$.

For characterizing the performance we want to find the confidence intervals in dB such that

$$\Pr\left[10\ \log_{10}\ \frac{|H(\underline{\phi})|^2}{|\hat{H}(\phi)|^2} > L_u(\alpha)\right] = \frac{\alpha}{2} \tag{6a}$$

and

$$\Pr\left[10\ \log_{10}\ \frac{|H(\underline{\phi})|^2}{|\hat{H}(\phi)|^2} < L_\ell(\alpha)\right] = \frac{\alpha}{2} \tag{6b}$$

so that

$$\Pr\left[L_\ell(\alpha) < 10\ \log_{10}\ \frac{|H(\underline{\phi})|^2}{|\hat{H}(\underline{\phi})|^2} < L_u(\alpha)\right] = 1 - \alpha$$

In order to illustrate the results of the previous sections the results were applied to a simple example. The ensemble covariance was taken to be

$$\underline{\Sigma}_{-N} = \sum_{k=1}^{3} \sigma_k^2\ \underline{E}_{-N}(f_k)\ \underline{E}_{-N}^{+}(f_k) + N_o \underline{I}_{-N}$$

with

$$\sigma_1^2 = 100. \qquad\qquad f_1 = 2.\ \text{Hz}$$

$$\sigma_1^2 = 10. \qquad\qquad f_2 = 1.5\ \text{Hz}$$

$$\sigma_3^2 = 1. \qquad\qquad f_3 = 1.\ \text{Hz}$$

$$N_o = 1 \qquad\qquad \Delta T = 0.1\ \text{sec}$$

Figures 2 and 3 illustrate respectively the Maximum Likelihood and Maximum Entropy Spectral Estimates for N = 24, corresponding to covariance intervals and 2.4 sec. The most evident characteristics are that the MLM has less resolution but better sidelobe structure when compared with the MEM. In fact, the MEM does not always indicate the presence of a tone in that it is masked by a sidelobe strucuture as in Figure 3. Some of the sidelobe structure in the MEM may be suppressed by using the covariance extensions concept introduced by Burg (12) and Pusey (18).

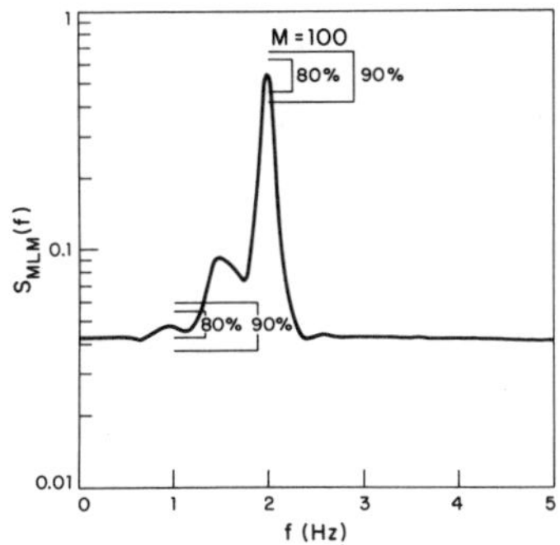

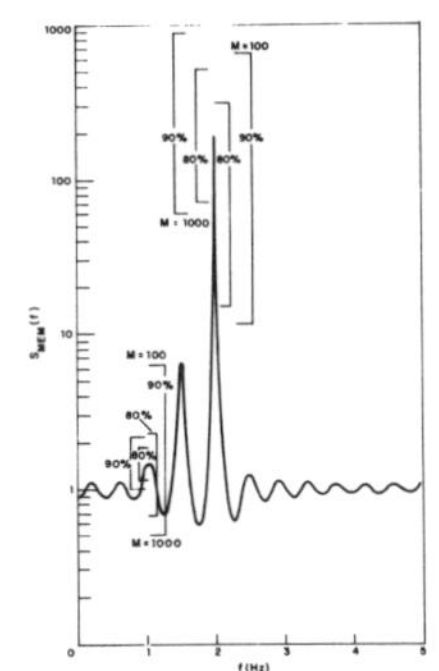

Fig. 2 MLM Spectral Estimate   Fig. 3 MEM Spectral Estimate

Figures 4 (a),(b) illustrate the $\alpha = 0.1$ (90%) and 0.2(80%) confidence intervals specified by Eqs. (6a) and (6b) for the 24-point estimator at 1 and 2 Hz, respectively. Note that there are two scales, the dB scale for $L_u(\alpha)$ and $L_\ell(\alpha)$ and an albegraic one for $10^{-L_u(\alpha)/20}$ and $10^{-L_\ell(\alpha)/20}$. The most distinguishing aspect of these confidence intervals is the maximum of the upper limit $L_u(\alpha)$. This is related to the probability of observing $\hat{H}(\underline{\phi})$ near the origin, so it is particular relevant to the occurrence of spurious peaks in the estimated spectra using the MEM procedure. At small M the density is so broad that it is very significant near the origin.

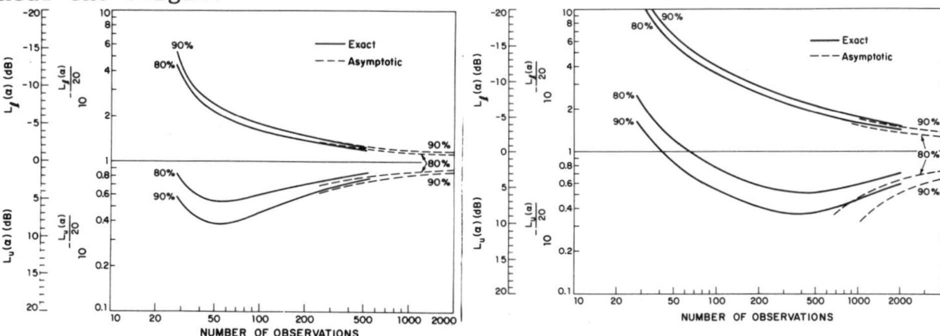

Fig. 4a Confidence Intervals for MEM Spectral Estimate
              (a) @1 Hz   (b) @2 Hz

In comparing the confidence intervals at the two frequences, it is evident that the estimate at the stronger tone at 2 Hz requires many more observations for the same percentage error. This is in contrast to both the MLM and conventional methods where the percentage error is a constant. The difference is introduced by the matrix $\underline{L}(\underline{\phi})$ which determines the sharpness of the density.

SUMMARY.

The probability density and confidence intervals for regression
spectral estimators have been derived where a Wishart model for
the estimated covariance has been employed. These results were
obtained by using two theorems in multivariate analysis for
Wishart probability densities. The principal result is that the
transfer function of the predictor error, or regression, filter
has a density which may be interpreted as a generalization of the
Student's t distribution. When asymptotic limits are taken, the
results are the same as Akaikes.

These results and some examples allow one to compare the performance of the MEM and MLM. They indicate that while the peaks
of the MEM estimator are larger than MLM. the variance on them
is also larger. For finite M there is also an algebraic
dependence on the tail of the density for the MEM analysis vs.
the exponential dependence in the MLM analysis. In addition,
the experience of this author and other suggests that the MEM
has better resolution than the MLM; however,its sidelobes
structure is larger.(4,5)

In summary, the MEM (or regression) has better resolution but
is a noiser estimate than the MLM. This tradeoff, is however,
well established in spectral analysis.(22)

PRINCIPAL SYMBOLS USED

Ensemble quantities and estimates of them are distinguished by the
use of the superscript $^\wedge$ on the estimated quantities.

$\underline{h}$	: predictor weighting vector
$H(f)$	: transfer function of prediction error, or regression, filter
$H(\underline{\phi})$	: generalized transfer function of prediction error, or regression filter
$\underline{R}_x$	: $(N-1) \times (N-1)$ covariance matrix of the process $x(n\Delta T)$
$\underline{r}_x$	: $(N-1)$ vector containing the covariance elements $R_x(\Delta T), \ldots R_x((N-1)\Delta T)$
$\underline{\Sigma}_N$	: $N \times N$ generalized covariance of the random vector, $\underline{x}_N$
$\underline{\Sigma}_{N-1}, \sigma_{-N}, \sigma_{NN}$	: partitioned elements of the matrix $\underline{\Sigma}_N$
$\underline{S}_N$	: accumulated matrix of $\underline{\Sigma}_N$, $\underline{S}_N = M\underline{\Sigma}_N$
Tr	: trace operation
$\mid \; \mid$	: determinant operation

$\underline{\phi}$: generalized basis vector

$\underline{E}(f)$: frequency basic vector

REFERENCES

1. L. C. Pusey, "High Resolution Spectral Estimates," Technical Note 1975-7, Lincoln Laboratory, M.I.T. (21 January 1975), DDC AD-A006406.

2. J. Makhoul, "Linear Prediction: A Tutorial Review," Proc. IEEE 63, 561-580 (1975).

3. D. R. Brillinger, "An Empirical Investigation of the Chandler Wobble and Two Proposed Excitation Processes," Bull. Inst. Stat. Inst. 39, 413-434 (1973).

4. D. Porter, A. Baggeroer, and M. Briscoe, "Estimates of Internal Wave Spectra Using Autoregressive Analysis," Contribution No. 3541, Woods Hole Oceanographic Inst. (May 1975).

5. R. T. Lacoss, "Data Adaptive Spectral Analysis Methods," Geophysics 36, 601-675(1971), DDC AD-734104.

6. J. P. Burg, "The Relationship Between Maximum Entropy and Maximum Likelihood Spectra," Geophysics 37, 375-376 (1972).

7. E. Parzen, "Multiple Time Series Modeling," in Multivariate Analysis, P. Krishnaian, Ed. (Academic Press, New York, 1970), pp. 389-407.

8. T. W. Anderson, The Statistical Analysis of Time Series, (Wiley, New York, 1971).

9. E. A. Robinson, "Predictive Decomposition of Time Series with Application to Seismic Exploration," Geophysics 32, 418-484 (1967).

10. N. Wiener, The Extrapolation, Interpolation and Smoothing of Stationary Time Series with Engineering Application (Wiley, New York, 1949).

11. T. Kailath, "The Innovations Approach to Detection and Estimation Theory," Proc. IEEE 58, 680-695 (1970).

12. L. Pusey and A. Baggeroer, "The Role of the Stationarity Equation in Spectral Analysis and Wave Propagation," Proc. 1974 Int'l Information Theory Symposium, University of Notre Dame, October 1974.

13. E. Parzen, "Some Recent Advanced in Time Series Modeling," IEEE Trans. Automatic Control AC-19, 723-730 (1974).

14. R.E. Kromer, "Asymptotic Properties of the Autoregressive Spectral Estimator," PhD disseration, Department of Statistics, Stanford University, 1969.

15. H. Akaike, "Power Spectrum Estimation Through Autoregressive Model Fitting," Annuals of the Institute of Statistical Mathematics, Tokyo (1969) pp. 407-419.

16. K. Berk, "Consistent Autoregressive Spectral Estimates," Annals of Statistics, Vol 2. No. 3 (1974) pp. 489-502.

17. J. Capon and N. R. Goodman, "Probability Distributions for Estimators of the Frequency-Wavenumber Spectrum," Proc. IEEE 58, 1785-1786 (1970), DDC AD-723788.

18. J.D. Burg, "New Concepts in Power Spectrum Estimation," paper presented at 40th Annual Int'l SEG Meeting, New Orleans, Nov. 1970.
19. A. Kshirsagar, Multivariate Analysis (Marcel Dekher, Inc. New York, 1972).
20. S. S. Wilks, Mathematical Statistics (Wiley, New York, 1964).
21. T. W. Anderson, Introduction to Multivariate Statistical Analysis (Wiley, New York, 1952)
22. R. B. Blackman and J. W. Tukey, The Measurement of Power Spectra (Dover, New York, 1952)
23. C. Bingham, M. D. Godfrey, and J. W. Tukey, "Modern Techniques of Power Spectrum Estimation," IEEE Trans. Audio Electroacoustics AU-15, 56-66 (1967).
24. J. S. Bendat and A. C. Piersol, Random Data: Analysis and Measurement Procedures (Wiley, New York, 1971).
25. Handbook of Mathematical Functions, M. Abramowitz and I. Stegun, Eds. (National Bureau of Standards, AMS-55, U.S. Government Printing Office, 1964).
26. H. L. Van Trees, Detection, Estimation, and Modulation Theory, Pt. 1 (J. Wiley, New York, 1968).
27. T. Kailath, Fredholm Resolvent, Wiener-Hopf Equations and Riccati Differential Equations, IEEE Transaction on Information Theory, Vol. IT-16, pp. 276-288, May 1970.
28. J. P. Burg, "The Relationship Between Maximum Entropy Spectra and Maximum Likelihood Spectra," Geophysics 37, pp. 375-376 (1976).
29. A. B. Baggeroer, "Confidence Interval for Regression (MEM) Spectral Estimates," accepted for publication IEEE Information Theory Transactions.

TEAM DECISION THEORY IN THE OPTIMAL CONTROL OF A COMMUNICATION
NETWORK*

R. Zoppoli

Istituto di Elettrotecnica, University of Genoa, Genoa,
Italy

ABSTRACT. Measurement devices (observers) and controllers are
considered as the cooperating agents of two distinct subteams
interconnected by a communication network. Observers are given
the task of controlling the data flow entering the network. The
optimal decision rules of the agents are derived via team–theory.

1. OBSERVING AND CONTROLLING SUBTEAMS IN STATIC DECISION PROCESSES

An organization can be considered as a set of decision
makers acting on a common process but not sharing the same information, and pursuing goals that, in general, may be different. A
team is defined as an organization, in which decision makers
share different information but, as quoted in [1] , have "common
interests", and then cooperate to the accomplishment of a common
objective.

One of the main reasons for this "dispersion of information
and authority" [2] stems from the presence of real communication
networks (characterized by costs for messages, noises and inter-
ruptions, etc.) which interconnect posts where information is
handy and posts where decision actions are exerted on the controlled system. Measurement posts and controlling posts constitute

---

* This work has been supported by the National Council of Research
  of Italy (C.N.R.).

G. Tacconi (ed.), *Aspects of Signal Processing, Part 2*, 631-638. *All Rights Reserved.*
Copyright © 1977 by D. Reidel Publishing Company, Dordrecht-Holland.

the nodes of a graph which defines the topology of the communication network.

Now a question naturally arises: is it advisable "to give intelligence" to measurement posts in order to allow them to decide whether it is convenient to send the acquired information to controlling posts, to select what data are worth transmitting and, in general, to define the communication interchange? Since the answer may turn out to be positive, as will be clear later on, we shall consider two distinct subteams: a subteam of k "non-classical" decision makers OA 1, ..., OA k (observing agents), who are given the task of controlling the flow of observed data, and a subteam of n "classical" decision makers CA 1, ..., CA n (controlling agents), who take decision actions on the system. In a static process, the observing subteam plays first, then the controlling subteam follows. To be more specific, some definitions need to be introduced.

<u>Definition 1</u>: Let the vector $s_j \in R^{q_j}$ be the information variable observed by OA_j. s_j is related to the random vector $x \in X$ (the so-called state of the world) by a vector-valued function

$$s_j = \zeta_j(x), \quad j = 1, \ldots, k \tag{1}$$

x is comprehensive of all possible observation and communication noises. Then $\zeta \triangleq (\zeta_1, \ldots, \zeta_k)$ is the observation structure of the team.

On the basis of the vector s_j observed, OA_j selects the transmission procedure by means of a decision vector h_j. For example, $h_j = 1$ ($h_j = 0$) might mean that a message is sent (is not sent) to a unique central CA. Then we have

<u>Definition 2</u>: Let $h_j \in \Omega_j \subset R^{t_j}$ be the decision action by means of which OA_j controls the data flow to the CAs. Consider the decision functions

$$h_j = \varphi_j(s_j) = \varphi_j \left[\zeta_j(x) \right], \quad j = 1, \ldots, k \tag{2}$$

$\varphi \triangleq (\varphi_1, \ldots, \varphi_k)$ is the decision rule (or strategy) of the observing subteam.

<u>Definition 3</u>: Let the vector $y_i \in R^{p_i}$ be the message received by DM i from the observing subteam. Consider the functions

$$y_i = \eta_i(x, h), \quad i = 1, \ldots, n \tag{3}$$

where $h \triangleq (h_1^T, \ldots, h_k^T)^T$. Then $\eta \triangleq (\eta_1, \ldots, \eta_n)$ is the communication structure of the team.

For computational reasons, it is convenient to assume vectors h_j to be received by CA i without errors. This assumption is realistic enough since in general Ω_j is composed of a small amount of elements and h_j can be easily protected by means of suitable codes. Thus we have

Definition 4: Let $u_i \in R^{m_i}$ be the control action exerted on the system by CA i. Consider the decision functions

$$u_i = \gamma_i(y_i, h) = \gamma_i\left[\eta_i(x,h),\ h\right],\ i = 1, \ldots, n \tag{4}$$

$\gamma \triangleq (\gamma_1, \ldots, \gamma_n)$ is the decision rule of the controlling subteam.

Suppose now the common goal of the team to consist in minimizing an expected cost

$$\bar{J}(\varphi, \gamma) \triangleq E\left[J(x, \varphi_1, \ldots, \varphi_k, \gamma_1, \ldots, \gamma_n)\right] \tag{5}$$

where expectation is taken with respect to x. Cost J, functions ς_j, η_i and probability density $p(x)$ are assumed to be known to all decision makers. Then we can state the following

Definition 5: The subteam decision rules φ^o, γ^o are optimal if

$$\bar{J}(\varphi^o, \gamma^o) \leq \bar{J}(\varphi, \gamma),\ \forall\ \varphi_j \in \phi_j,\ \forall \gamma_i \in \Gamma_i \tag{6}$$

where ϕ_j, Γ_i are the classes of admissible decision rules.

Remark 1: observe that all parameters and functions characterizing the statistics, the information structures and the cost of the decision process are known to all agents. Then, each agent is provided with enough " a priori" information to compute his own optimal decision rule. In this sense, we will say that the team is informationally decentralized in the on-line observations, is not informationally decentralized in the process structure, and is computationally self-sufficient. It follows that no coordination mechanism is required among the agents, nor among the agents and a computation center, to determine local decision rules. See also [2] and [3] for related definitions.

Suppose now a fixed observation structure to be acting that corresponds to a constant strategy $\hat{\varphi}_j \in \phi_j$ (i.e., $\hat{\varphi}_j$ does not depend on s_j). Then we have

Definition 6: The nonnegative quantity

$$\bar{J}\left[\hat{\varphi}, \gamma^o(\hat{\varphi})\right] - \bar{J}(\varphi_0, \gamma_0), \tag{7}$$

where $\gamma^o(\hat{\varphi})$ is the optimal strategy relating to $\hat{\varphi}$, is defined as the expected value of task decentralization (EVTD) for the given constant observation structure [4].

EVTD is the maximum cost one is willing to pay in order to substitute $\hat{\varphi}$ with an "intelligent observing subteam".

2. CONSTANT OBSERVATION STRUCTURES

Well known results are available for constant observation structures. To simplify notations, let us drop strategy φ in cost (5), and consider only optimization of γ. η is now comprehensive of both observation and communication structures. Unless for very simple cases, inequality (6) is not very useful for computing γ^o. A constructive definition is the following.

Definition 7: a decision rule γ^* is person-by-person satisfactory (p.b.p.s.) if, for any i,

$$\bar{J}(\gamma_1^*,\ldots,\gamma_i^*,\ldots,\gamma_n^*) \le \bar{J}(\gamma_1^*,\ldots,\gamma_i,\ldots,\gamma_n^*) \tag{8}$$

Clearly, conditions (8) are necessary, but not sufficient, for optimality. By specializing the cost function J, we have the following

Theorem 1 [5]: if $J(x, u_1,\ldots, u_n)$ is convex and differentiable with respect to $u \triangleq (u_1^T, \ldots, u_n^T)^T$ for each $x \in X$, then a p.b.p.s. decision rule is optimal. $\qquad\square$

An explicit solution can be given for the optimal team strategy γ^o under LQG assumptions. Namely, assume that

$$y_i = \eta_i(x) = H_i\, r + \varepsilon_i, \quad i = 1, \ldots, n \tag{9}$$

where $r, \varepsilon_1, \ldots, \varepsilon_n$ are mutually independent Gaussian random vectors with p.d.f. $p(r) = N(0,R)$, $R>0$, $p(\varepsilon_i)=N(0,E_i)$, $E_i > 0$ ($i = 1, \ldots, n$). Assume that

$$J = u^T Q\, u + 2\, u^T D\, r, \quad Q = Q^T > 0 \tag{10}$$

Then we have the following

Theorem 2 [5]: the unique optimal decision rule for the LQG team problem defined above is given by the linear control laws

$$u_i = \gamma_i^o(y_i) = F_i\, y_i, \quad i = 1, \ldots, n \tag{11}$$

where matrices F_i are the unique solution of the linear system

$$Q_{ii}F_i(H_i R H_i^T + E_i) + \sum_{j \neq i} Q_{ij}F_j RH_i^T = -D_i RH_i, \quad i = 1, \ldots, n \qquad (12)$$

Matrices Q_{ij}, D_i are partitioned blocks of suitable dimensions derived from matrices Q and D, respectively. □

Remark 2: it is easy to show that the control law (11) does not exhibit a separation property (see, for instance, a counter-example in [6]). This means that the optimal solution for the agents is not obtained by multiplying the best conditional estimates of r by the gain matrices of the deterministic solution.

Extensions from static to dynamic teams have been discussed by Ho and Chu [7]. However, analytical results are presently available only for the so-called LQG one-step-delay sharing information pattern (see [8], [9] also for references), in which the agents exchange without noises the observations taken at the preceding stage. Actually, this case can be shown to be reducible to a static team problem.

3. DERIVATION OF OPTIMAL DECISION RULES FOR THE OBSERVING SUBTEAM

Let us consider the general case presented in Section 1 (intelligence is given to the observing subteam), and let us try to extend this problem to an N-stage dynamic decision process. The structure of the communication network plays a central role in solving the problem. A possible classification of structures, for which growing difficulties must be overcome to derive the agents' optimal decision rules, may be the following: 1) point-to-point communication link ($k=1$, $n=1$); 2) star-shaped network with a central controller ($k>1$, $n=1$); 3) star-shaped network with a central observer ($k=1$, $n>1$); 4) bipartite communication network ($k>1$, $n>1$). Some results will now be presented for the first two cases.

Consider a point-to-point communication link in which OA observes the state x_i of a linear dynamic system

$$x_{i+1} = Ax_i + Bu_i + \xi_i, \quad i = 0, 1, \ldots, N-1 \qquad (13)$$

through a linear noisy observation channel $s_i = Hx_i + \varepsilon_i$. The initial state x_0, and noises ξ_i, ε_i are mutually independent, Gaussian with $E(x_0) = \alpha$, $E(\xi_i) = 0$, $E(\varepsilon_i) = 0$ and known covariance matrices. OA is given the task of deciding whether to send observation s_i to CA ($h_i = 1$) or not ($h_i = 0$). Messages are noise-free (possibly penalized by costs c_i) but stochastically interrupted with probability 1-p. The process cost is given by

$$J = \sum_{i=0}^{N-1} \left[\| u_i \|^2_P + ch_i \right] + \| x_N \|^2_V \tag{14}$$

where $P = P^T > 0$, $V = V^T \geq 0$. CA is assumed to send OA back a message
by means of which he acknowledges (or does not) the correct
reception of s_i. Let I^1_i and I^2_i be the information sets of OA and
CA at stage i, respectively. Then we have the following
<u>Theorem 3</u> [10] : CA's optimal decision rule is given by

$$u^o_i(I^2_i) = - L_i \mu_i, \quad i = 0, 1, \ldots, N-1 \tag{15}$$

where $\mu_i \triangleq E(x_i | I^2_i)$ and L_i are the controller's gain matrices of
LQ standard optimization problems [11]. OA's optimal decision
rule takes on the form

$$h^o_i(I^1_i) = f_i(\lambda_i, z_i, \Sigma^p_{2i}), \quad i = 0, 1, \ldots, N-1 \tag{16}$$

where $\lambda_i \triangleq \hat{x}_i - A\mu_{i-1} - Bu_{i-1}$, $z_i \triangleq K_{2i}\left[s_i - H(A\mu_{i-1} + Bu_{i-1})\right]$,

$\Sigma^p_{2i} \triangleq cov(x_i | I^2_{i-1})$, $\hat{x}_i \triangleq E(x_i | I^1_i)$ and K_{2i} is CA's Kalman filter

gain when no interruption occurs at stage i. The sequence of
functions f_0, f_1, ..., f_{N-1} can be derived by solving an optimization problem with a dynamic programming structure which is
given in [10].   □

 Consider now a static team composed of a subteam of OAs
and a central CA interconnected by a star-shaped communication
network. Without loss of generality, only two agents, OA 1 and
OA 2, are given the task of observing and transmitting random
vectors r_1 and r_2 , respectively, to CA. r_1 , r_2 are n-dimensional,
mutually independent, zero-mean with known p.d.f.. Observations
and transmissions are noise-free (see [10] for the noisy case).
Transmissions are penalized by costs c_1 , c_2 . Then modify cost
(10) as follows

$$J = c_1 h_1 + c_2 h_2 + u^T Q u + 2u^T \tilde{D}(r_1 + r_2), h_1, h_2 \in \{0,1\} \tag{17}$$

 It is immediately seen that CA's optimal control law is
given by $u^o = - Q^{-1} \tilde{D}(h_1 r_1 + h_2 r_2)$. On the contrary, only necessary
conditions for optimality of PAs' strategies can be given. More
specifically, p.b.p.s. strategy pairs φ^*_1 , φ^*_2, which are candida-
tes for optimality, may be derived from the following

<u>Theorem 4</u> [4]: the p.b.p.s. strategy pairs φ_1^*, φ_2^* (if they exist) are given by

$$h_j^*(r_j) = 1\left(\left\|r_j - k_j\right\|_S^2 - c_j - \left\|k_j\right\|_S^2\right), \quad j = 1,2 \tag{18}$$

where $1(.)$ is the unit step function and $S \triangleq \tilde{D}^T Q^{-1} \tilde{D}$. k_1, k_2 are n-dimensional vectors which must satisfy the following system of 2n (nonlinear) equations

$$k_i = \mathop{E}_{r_j}\left[h_j^*(r_j) r_j \right] \triangleq f_i(k_j), \quad i = 1, 2, \ j = 1,2, i \neq j \tag{19}$$

$\square$

Theorems 3 and 4, although solving rather particular problems, give useful elements toward the building of a more comprehensive theory of the optimal control of data flow in large scale systems. From Theorem 3, it can be observed (see [10] for a more detailed discussion) that "backward messages" from CA to OA are essential to insure the controller a separation property. Theorem 4 suggests a coordination mechanism among PAs based on the p.b.p. satisfactoriness criterion. Unfortunately, optimal solutions can be obtained only via a numerical approach. Extension of this latter result to the dynamic case, as well as generalization to the other communication networks previously defined, are presently under investigation.

REFERENCES

1. J. Marschak, R. Radner, <u>Economic Theory of Teams</u>, Yale University Press, New Haven and London, 1972.
2. F.N. Bailey, Decision Processes in Organization, to be published in <u>Recent Contributions to Large Scale Systems</u>, ed. by R. Saeles.
3. K. Miyasawa, Decision Processes in Organizations, <u>Conference on Design and Management of Large Organizations</u>, IIASA, Baden, Austria, July 1973.
4. P.P. Puliafito, R. Zoppoli, Optimal Control of Data Flow from Peripheral to Central Units in an Information Structure, <u>Proc. IFAC Symposium on Large Scale Systems</u>, Udine, Italy, June 1976, 479-487.
5. R. Radner, Team Decision Problems, <u>Ann. Math. Statist.</u>, vol. 3, Sept. 1962, 857-881.
6. N.R. Sandell, M. Athans, Solution of Some Nonclassical LQG Stochastic Decision Problems, <u>IEEE Trans. Automatic Control</u>,

vol. AC–19, April 1974, 108–116.

7. Y.C. Ho, K.C. Chu, Team Decision Theory and Information Structures in Optimal Control Problems – Part I, _IEEE Trans. Automatic Control_, vol. AC–17, Feb. 1972, 15–22.

8. T. Yoshikawa, Dynamic Programming Approach to Decentralized Stochastic Control Problems, _IEEE Trans. Automatic Control_, December 1975, 796–797.

9. B.Z. Kurtaran, A Concise Derivation of the LQG One–Step – Delay Sharing Problem Solution, _IEEE Trans. Automatic Control_, December 1975, 808–810.

10. G. Casalino, F. Davoli, P.P. Puliafito, R. Zoppoli, Models of Communication Networks in Large Scale Systems, 3rd Italian–Polish Conference on _Modern Applications of Mathematical Systems and Control Theory to Economics and Technology_, Białowieza, Poland, May 1976. To be published also in _Ricerche di Automatica_.

11. M. Aoki, _Optimization of Stochastic Systems_, Academic Press, New York and London, 1967.

TOPOLOGICAL STRUCTURE IN GROUP TRANSFORMS

M.L. Retter

Department of Nuclear Physics, Oxford University
Keble Road, Oxford, England.

ABSTRACT.   Fast transform techniques are based upon discrete group
theory, and non-abstract group representations are associated
with symmetry and structures.   In this paper we emphasize the
topological structures implied by the underlying group symmetries,
and attempt to give them some physical reality.   A class of tensor
network configurations is considered, which perform the complete
array real-time transformation.   The 'lattice processor' is a
generic term we apply to these lattice structures and we illustrate
them with specific Transformation and Reverberation models
(T_2^3, R_3^1).   Some philosophical remarks on these R- and T- lattices,
suggest they may be of value in parallel processing, or as a model
for space-time, in the brain, or elsewhere.

1.   INTRODUCTION

The fast transform techniques such as the FFT, are based upon
discrete group theory, and group theories are associated with
structure and symmetries.   The fast algorithms can thus be associated with a structure such as the familiar signal-flow networks.
Here we wish to give some physical insight into these totally
parallel array processing configurations, which serial hardware
algorithms seem to avoid.   The most appropriate way of describing
these tree graph structures is matrix and tensor algebras.   The
essential point about the fast transform is that there is a
reduction in the number of operations, and this is reflected by
the sparseness of the matrices in the matrix representation.   We
interpret these matrices as connectivity matrices and where there
is an entry in a matrix ^rA at element (ij) then there is a link
between cell i plane r, and cell j plane (r+1).   We use the term

G. Tacconi (ed.), Aspects of Signal Processing, Part 2, 639-647. All Rights Reserved.
Copyright © 1977 by D. Reidel Publishing Company, Dordrecht-Holland.

cell or node to refer to a node in the graph or lattice representation, and r shows the index of the matrix decomposition,
or plane in the lattice.

The magnitude of the matrix entries show the weight given to
the value of cell i at the cell j. These weights are powers
of $e^{2\pi j/N}$ for the FFT, but here we consider only real simple
weights of ±1, and $\sqrt{2}$, and so stress the structural representation.
Primarily we consider the Walsh and Haar transforms over 8 and
8x8 elements. Consider the following transformation:-

$$F_i = G_{ij}\, f_j = A^3 f,\quad A = \begin{bmatrix} 1\ 1 & & & 0 \\ & 1\ 1 & & \\ & & 1\ 1 & \\ & & & 1\ 1 \\ & & 0 & \\ 1\text{-}1 & & & \\ & 1\text{-}1 & & \\ & & 1\text{-}1 & \\ & & & 1\text{-}1 \\ 0 & & & \end{bmatrix} \tag{1}$$

f_j is an input vector of N points and N is composite (i.e. 2^n).
The sparse matrix form of A makes G the Walsh transform matrix
with Kronecker ordering. If we now permute the nodes and there-
for the matrix elements, we arrive at the fast Walsh transform

$$F_i = {}^2A^1A^0Af = G_{ij}f_j \tag{2}$$

where ^2A, ^1A and ^0A are merely the permuted symmetric forms of A
(given in ref. 1), and so lead to the well known (in place)
tree-graph diagram of figure (1). We use the left superscript to

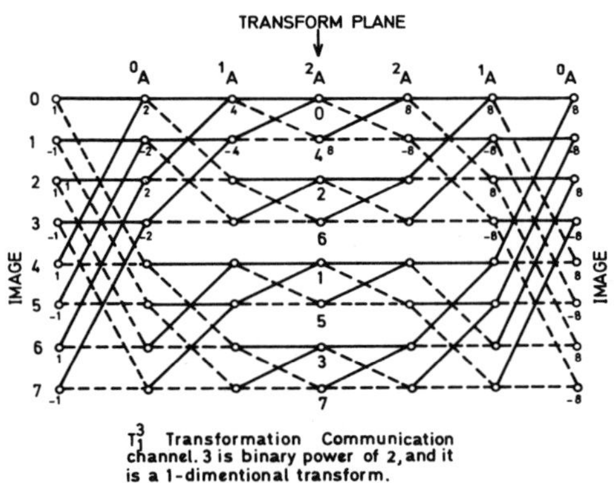

T_j^3 Transformation Communication
channel. 3 is binary power of 2, and it
is a 1-dimentional transform.

Fig. 1

avoid confusion with the power and because all weights are the
same except in sign, we use dotted lines to indicate at which
nodes the difference is taken. The unit lattice transform element out of which this figure is constructed is the Hadamard

matrix $H_1 = \begin{bmatrix} 1 & 1 \\ 1 & -1 \end{bmatrix}$. The forward and reverse transform of Fig. 1 is described by the following equation

$$f = (^0A \ ^1A \ ^2A)(^2A \ ^1A \ ^0A) f \tag{3}$$

the ^iA when normalized are sparse symmetric unitary matrices, and we may regard the diagram as a section of a communication channel.

1.1 Classification

We classify the networks using the following scheme. Where we have the value at a node propagated to a new node outside the set, we refer to this as a Transmission or T-lattice. Where the results replace the contents of a node in the same set, we denote this as a Reverberation or R-lattice. To these we give two indices, the power is the power of 2 which gives N, and the subscript gives the dimensionality of the transform. So Figures (1) and (2a) are T_1^3-lattices, while (2b) is an R_2^2-lattice which is equivalent to a

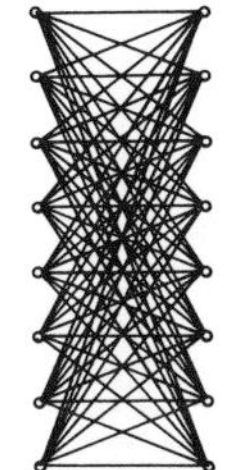

T_1^3 – lattice, 64 links
totally connected
symmetric graph

Fig. 2(a)

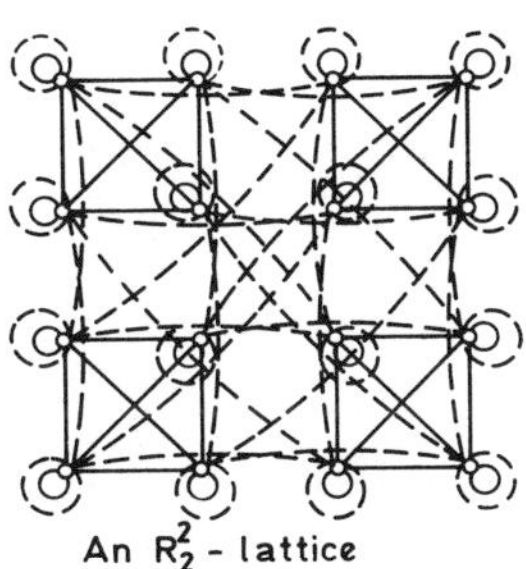

An R_2^2 – lattice

Fig. 2(b)

base 4 transform, where all links are bidirectional and the solid lines are active first, then the dotted mesh second, or vice-versa. Figure (2a) illustrates the fact that G_{ij} is not sparse, having N^2 elements, requiring a lattice with N^2 links.

Figure (3) illustrates an R_3^1-lattice and performs the same transform as Figure (1) except the nodes now have the topological form of a 3-D cube. The 64 links show, as the cube vertices communicate, how each node transmits its value to its neighbours. There are two states, the function value (a, b, c...) at each vertex, and its Walsh transform "over the cube", at each vertex. It is a unit radix-8 transform and we could think of it (dynamically) as oscillating between function and transform spaces. If the output and input nodes of Figure (2a) came together then it would be an R_1^3-lattice topologically equivalent to Fig. 3. However, to show the fundamental identity between the n-dimensional and composit 1-D transforms one should identify

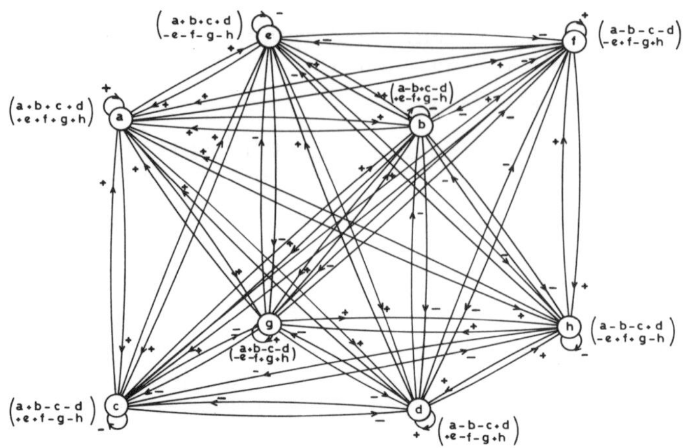

Figure 3   A Walsh - Fourier R^1_3 - lattice processor

Figure (1) with Figure (3) but with fewer links. Three 1-D transforms performed in sequence along the edges of the cube. This
happens because the transform is separable symmetric and implies
the G-matrix has another (tensor) decomposition, $G = H_1^{\otimes 3}$

2. THE 'LATTICE-PROCESSOR'

These lattices can be regarded as physical devices if each
node is a digital element that sums incoming, and transmits
outgoing pulse trains. We could use a binary rate multiplier
or stochastic computer which represents numbers by pulses, the
density of which in time (or mean time) represents the magnitude
of the number, and accuracy thus depends on time of observation.
Here we consider signal networks as composed of arrays of such
elements (called 'metrons' in Reference (1)) which combine incoming pulse streams and propogate the result as a new pulse
stream to the next layer of elements. We give the generic
term 'lattice-processor' to communication networks composed of
these elements. Input and output elements are the same for the
R-lattice, so that one transitiónal state has to be localised in
time, before the next ^rA links become active. If there is no
normalization and pulses are not lost then the pulse rate can
increase by a factor of α, which is the base or radix, or
valency, between stages or states of the metrons. Networks
like 1 and 2(a) have been discussed in recent years by Harmuth
in his book, and in relationship to acoustic imaging[3], where
thenode elements are operational amplifiers.

2.1 Some Examples

The structural interpretation of the sparse matrix representation
of the transform theory, we now discuss in relation to a specific

physical model, built to illustrate the 2-dimensional real-time FFT structure. Figure 4 illustrates a T_2^3-lattice processor which piecewise performs the transform in the plane, where the transform steps for each dimension are interleaved. We write the 2-D transform as follows

$$\hat{\chi} = {}^2A\ {}^1A\ {}^OA\ (\chi)\ {}^OA^t\ {}^1A^t\ {}^2A^t \tag{4}$$

where χ is a two-dimensional 8x8 array image and t stands for the transpose which we omit (along with matrix indices) since the ${}^iA$ are all symmetric. The left super index again denotes the appropriate permutation of the Good-Yates matrix A, of equation (1) . Let us note in passing that (4) can equally well be written as

$$\hat{\chi}_{il} = G_{ij}\ \chi_{jk}\ G^t_{kl} \quad \text{or} \quad \hat{\chi}^l_i = G^{kl}_{ij}\ \chi^k_j \tag{5}$$

where the G represent matrix or tensor point functions which we will identify with a specific structure. Now to complete the reverse transform, we repeat the unitary transform to give

$$\chi = {}^OA\ {}^1A\ {}^2A\ {}^2A\ {}^1A\ {}^OA\ (\chi)\ {}^OA\ {}^1A\ {}^2A\ {}^2A\ {}^1A\ {}^OA \tag{6}$$

This corresponds to the double transform of Figure (1) but now in 2 dimensions, where again all matrices are sparse and independent of the others. The tree graph structure to be associated with it, is not so simple, but the fast algorithm $(N^2 \rightarrow N\log N)$ is necessarily reflected by the sparseness of the lattice. The matrices specify the couplings made between planes of metron cells, represented here by the 8x8 array of holes in the perspex planes. Rewriting (6) we have

$$\chi = {<}^OA{<}^1A{<}^2A{<}{<}^2A{<}{<}^1A{<}{<}^OA(\chi){>}^OA{>}{>}^1A{>}{>}^2A{>}^2A{>}^1A{>}^OA{>} \tag{7}$$

The brackets show in what order the matrices are wired starting at the bottom of the model. ${}^OA$ is wired vertically using the 128 strings (2 per hole) and then ${}^OA$ is applied in the second dimension, horizontally, and so on. The model shows the double transform of equation (7), where input and output patterns χ (the coloured beads) have the same form. The brackets delineate how they are ordered in space, and the inner 6 pairs <> yield the scattered transform at plane 7 from the bottom.

The reverse transform, planes 8, 9, 10 is condensed because the matrices have been combined. If we write (7) using the tensor notation of (5) we have

$$\chi^{p'}_{q'} = {}^OG^{p'j'}_{q'i'}\ {}^1G^{i'n'}_{j'm'}\ {}^2G^{m'r}_{n's}\ {}_2G^{sm}\ {}_2G_{rn}\ {}^1G^{ni}\ {}^1G_{mj}\ {}^OG^{jq}\ {}^OG_{ip}\ \chi^p_q \tag{8}$$

where the ${}^iG$ are the factors of G and represent the ${}^iA$ or combined ${}^iA$ matrix factors ${<}^jA|\ldots\ldots|^kA^t{>}$. The contravariant and

covariant indices are taken to apply to the vertical or horizontal dimension, and to ease physical construction i=j=k.
The last three tensor operators at left of (8) show the reverse
transform where wiring density should double (α=4) because we
now have a base 4 unit transform, with all 4 cells in plane r
mapped to all 4 on the next plane. To reduce complexity the
available 8 strings were crossed on the cube faces. There is
no significance in the colour of the strings used, and no means
of indicating weighting or normalization. However, the T -lattice
processor represents a communication network that recovers the
input after 9 operators have been applied.

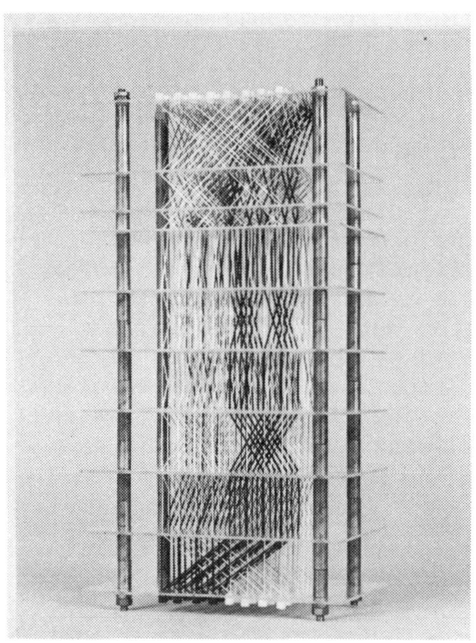

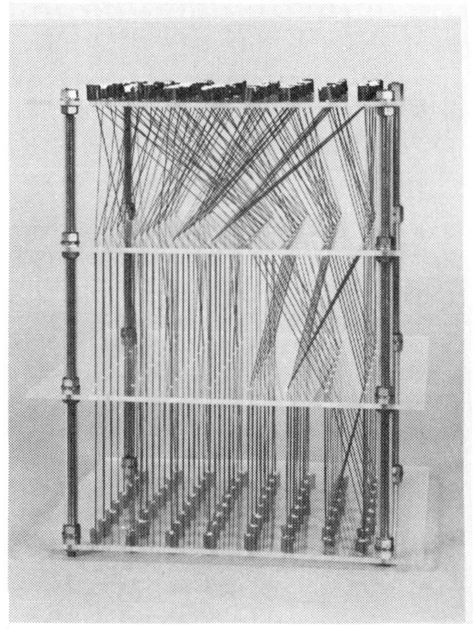

Fig.4.
Model of a Walsh-Fourier
T_2^3-lattice Processor

Fig. 5.

A Haar T_2^3- lattice

Another example of a T_2^3-lattice is shown in Figure (5). This
represents the structure for the Haar transform, though the
wiring density is half what it should be. As before we write,

$$F_i = H_{ij}\, f_j = {}^2H\, {}^1H\, {}^0H\, f \tag{9}$$

but the ${}^iH$ are no longer permutations of the same matrix although
they are still sparse.

$$^0H = A; \quad {}^1H = \begin{bmatrix} 1 & 1 & & \\ & 1 & 1 & 1 \\ 1 & -1 & & \\ & 1 & -1 & \\ & & & \sqrt{2}\ \sqrt{2}\ \sqrt{2}\ \sqrt{2} \end{bmatrix} ; \quad {}^2H = \begin{bmatrix} 1 & 1 & & \\ 1 & -1 & & \\ & & \sqrt{2} & \\ & & & \ddots \\ & & & \sqrt{2} \end{bmatrix} \tag{10}$$

The two-dimensional forward transform that is illustrated by this figure we can write as follows

$$\widehat{\chi}_j^i = {}^2H^{ik}_{jl}\ {}^1H^{km}_{ln}\ {}^0H^{mp}_{nq}\ \chi_q^p \tag{11}$$

The nature of the Haar transform means that the valency at the nodes is not constant within a given plane.

The geometrical symmetry introduced in these models is of secondary importance since the fundamental unit cross-over lattice network is to be thought of as a topological link only. It is more difficult to imagine an R^3_3 lattice where the 3-D Walsh transform on an 8x8x8 cubic lattice is written

$$\widehat{M}_{stu} = {}^2G^{pqr}_{stu}\ {}^1G^{lmn}_{pqr}\ {}^0G^{ijk}_{lmn}\ M_{ijk} \tag{12}$$

The ^iG tensor operators are structures of the form of Figure (3) which couple nodes on a cubic lattice that are (2^i) nodes apart. Again, a more general transform does not need to have i=j=k=1. i.e.

$$^iG^{pqr}_{stu} = {}^jG^p_s\ {}^kG^q_t\ {}^lG^r_u$$

In contrast to the T-lattice, the G operators in an R-lattice must be considered as operating throughout the lattice and localized and ordered in time, whereas the reverse applies to the T-lattices.

2.2 Properties of the Lattice-processors

The sparseness of the matrix representations leads naturally to a sparse lattice representation, and the T^3_2-lattice of Figure (4) could have been wired like 2(a), but in 2-dimensions with 64^2 links. We increase the number of planes and therefore nodes required, but reduced the number of links.

An integral property of these tensor lattice transforms is that information is scattered by what we call 'lexicographic-inversion'. If we consider the image or frequency space as an ordered set, then the dual (or transform) space is permuted, or can be regarded as a pseudo-random permutation. It can be easily shown that if the scattered space is partially destroyed in space, or in time by malfunction of the metron, then dispersed noise is introduced in the dual space. This property is common with holography where the image can be recovered from the incomplete hologram and incomplete key wave. These lattices are

specified by remarkably little information, and their topological symmetry is manifest by emphasizing the geometrical
symmetry in the models. The way the threads are crossed is
arbitrary and topologically one could have a T_2^n-lattice
with circular symmetry rather than square, which maps $r, \theta \rightarrow x, y$.

Such structures can support a 'spectrum' of different
transformations, depending on the link weights at the lattice
metrons. For example, the Haar transform may apply in one
dimension over n_1 elements, Walsh in a second over n_2
elements and so on[4]. Time variable transformations are another
possibility. The R- and T-configurations have been considered
in a biological connection, by Von Foerster where they are
called action and interaction neural nets. An interesting
property of the R_m^n-lattice is that the transform of an 1-
dimensional image (where 1 < m) will replicate throughout the
remaining (m-1) dimensions of the lattice. We have an information replicating automaton. If we consider a cubic lattice of
side 2^3, and apply an image to one face, then apply the three
operators of equation (12), M will be present along the third
dimension, so that the scattered transform irradiates very
quickly into the entire lattice volume.

3. <u>CONCLUDING PHILOSOPHICAL REMARKS</u>

The main purpose of this paper is to give some physical insight
into a class of topological structures, which we term the
'lattice-processor'. In a brief way we deal with the lattice
representation of the group theory behind fast transform
theory. We illustrate these ideas with two 'string and perspex'
T_2^3-lattice processor models, which can be regarded as an
extension of signal flow networks into the space-time domain,
where information is distributed both spacially and temporally
in these tensor networks. It comes about as a direct result
of the duality (or separability) of the n-D and composite 1-D
discrete transform, of which the FFT is the prime example[6].
The lattice processor was conceived as an abstract and deter-
minate model for cortical structure and function[7], but it may
have wider applications.

It has been suggested that the brain functions in a holo-
graphic manner or does Fourier transforms. Here, then, are
structures which will support asynchronous generalized space-
time holography[8,9], and the real-time FFT, so that there is
indeed justification for looking for such approximate topological structures in the brain[10]. The optic Chiasma is one
example of the cross-over unit lattice network. However, since
genetics can grow an eye-lens system, it is not impossible
nature long ago has found cleverer and more economical structures.

The concepts of tensors and topology in the brain are not new
[11,12]. The development of sonar systems must clearly be
related both to the inhomogenous propagation theories in
the environmental media, besides various hydraphone sensor
geometries, but and also to the inboard transformation and
decision networks which process that environment. These
lattices or 'fibre-bundles' may be of value in other space-
time areas such as particle physics. The extension of
matrix-mechanics to real-time tensor-mechanics may yield some
unifying concepts of time in the particle, brain, and automaton.
The Fourier integral theory is so fundamental to all physics,
that any links of the discrete theory to a class of topological
lattices is intriguing, and worth deep consideration, and
perhaps might lead to some ideas on the quantisation of space-
time.

ACKNOWLEDGMENTS

The author thanks T. Harris for his construction of the
Haar T_2^3-lattice.

REFERENCES

1.   M.L. Retter, A structural approach to Modern Integral
     Transform Theory, 1975. Hatfield Poly. W.F. Symp.
2.   B.R. Gaines, Stochastic Computing Systems. Vol. 2. Advances
     in Information Systems Science, Ed. J. Tou, Plen. 69.
3.   H.F. Harmuth, J. Kamal, S.S.R. Murthy. Two dimensional
     Spatial Hardware Filters for Acoustic Imaging. 1974
     Walsh Symp. Edts. H. Schreiber, G. Ferrell.
4.   H.C. Andrews, C.L. Patterson, Outer Product Expansions
     and their Uses in Digital Image Processing. IEEE Trans.
     on Computers, Vol. C-25 No. 2, pp 140-148.
5.   H. Von Foerster, Circitry of Clues to Platonic Ideation.
     Symp on Art. Intell. Plenum 1962, Ed. C.A. Muses.
6.   I.J. Good, The Interaction Algorithm,J. Royal Stat. Soc.
     1958, B20, pp 361-372; also 1960 B22 pp 372-375.
7.   M.L. Retter, A Look at the Eye-Brain Super-Complex.
     Unpublished note 1974. Oxford Nuclear Physics.
8.   Y. Morita, Y. Sakurai, Holography by Walsh Waves, U.S.
     Walsh Funct. Symp. 1973, pp 122-126.
9.   J.W. Carl, M. Kabrisky, Playing the Identification Game
     with Walsh Functions, 1971, U.S. W.F. Symp, pp 203-209.
10. M. Eden, Visual Image Processing in Animals and Man.
     NATO ASI, 1976, Digital Image Processing and Analysis.
11. J.W. Lynn, Tensors on the Brain,Electronics and Power
     July, 1970, pp 268-270.
12. E.C. Zeeman, Topology of the Brain, MRC Conference H.M.S.O.
     1965 (Maths. and Compt. Science in Biology).

SOME SIGNAL PROCESSING ASPECTS IN MEDICAL ULTRASOUND

C.T. Lancée and N. Bom

Department of Experimental Echocardiography,
Erasmus University, Thoraxcenter, Rotterdam,
the Netherlands

INTRODUCTION. Echo techniques are being used with a wide spectrum
of applications to-day. Most applications are in the low frequency sound range. However, for diagnostic purposes pulses in the
1 to 10 MHz frequency range are being used. Obviously, important
parameter differences exist when low frequency marine sonar is
compared to medical ultrasound. Differences include target structure; attenuation; noise sources; beam pattern requirements;
and particularly in cardiac imaging, the need for real time.
All these aspects influence the signal handling heavily.

In two-dimensional cardiac imaging optimal resolution and therefore narrow sound beams are important, this leads to focusing
methods. A further requirement is that the line density on the
display should be preferably high.

Whereas the above quoted signal handling is carried out on-line,
off-line processing may be used for image enhancement. As a first
step towards automation the schematic reconstruction of a cardiac
image resulting from a line to line brightness connectivity method will be shown. The paper will finish with a few remarks on
future signal processing aspects for general medical ultrasound
applications.

PARAMETER COMPARISON

a. Near field/far field

Whereas in most naval sonar applications the acoustic aperture
to target distance ratio is small, the opposite occurs with dia-

gnostic ultrasound. In general the beam pattern is made up of
two distinct regions. The near field or fresnel zone, and the
far field or Fraunhofer zone. In cardiac imaging echoes will
be created at reflecting structures in both sound beam regions.
Distance D (the distance of gradual transition from near field
to far field) may be defined by $D \simeq r^2/\lambda$ where λ is wave length.
As a typical example, a transducer element with a diameter of
1 cm, and a frequency of 3 MHz, has a corresponding D of $\simeq$ 5 cm.
This is about one third of the maximal range.

b. Absorption

Sound absorption may be explained as the effect of a timelag
between a variation in sound pressure and the associated variation
in density in the medium. This absorption increases with frequency and, in the Megahertz range, is an important limiting factor.
As a rough estimate the attenuation in soft tissue for a plain
wave is in the order of 2dB/MHz/cm. Thus, at 5 MHz and for an
echo returning from the posterior cardiac wall at 10 cm depth,
the attenuation is approximately 100 dB. The absorption in bone
is even higher. This is the reason for selection of "low" ultrasound frequencies if the sound has to propagate through bone
as happens in neurological applications.

c. Main differences

Underwater sonar	Medical ultrasound
Purpose: to localize a limited number of targets as precise as possible. Target is in the plane of scanning a very small part of this plane. Range is kilometers. Scanning time is seconds. Image presentation is slow. Propagation medium is water. Low attenuation. Target reflectivity moderately variable. Maximum power limited by cavitation. Noise defined by thermal effects and reverberation.	Purpose: to record all echo producing structures along a small area around the beam axis of each beam. Target is the complete plane of scanning. Range is centimeters. Scanning time is μsec. Flicker free dynamic image 50 frames/sec. Propagation medium is biological tissue. High attenuation. Target reflectivity highly variable. Max. power limited by safety requirements. "Noise"defined by strong reflectors outside main beam.

e. Need for hard wired processing

In order to create a complete image of the area of interest with

a repetition frequency of 50 frames/sec, one has an effective
time of 20 msec. At a maximum range of 16 cm this yields a possible number of scans of 100. By using this time scale the full
capacity of video techniques can be used. For instance the real
time registration of moving images is of great importance to
medical purposes. Apart from real time video processing the
resulting image quality is defined by the direct processing at
the "front" part of the system. Since there is practically no
time left between scans all the processing has to be carried
out in real time. Therefore the transducer array design together
with the beam forming circuitry has to be optimal. As a result
all possible processing is accomplished by hard wired electronics
such as lumped constant delay lines, summing amplifiers, gain
controlled amplifiers, etc.

TWO DIAGNOSTIC ULTRASOUND SYSTEMS

a. A catheter system

In the literature various catheter tip ultrasound systems are
mentioned. In these systems each element is used for transmission
and reception. Similar to radar the echoes may be displayed on
corresponding lines on an oscilloscope in PPI fashion (Plan
Position Indicator). In order to obtain directionality each
element must be large compared to wave length. Since a catheter
is small, this has limited the number of elements. With phase
compensation and simultaneous use of subgroups of elements,
directional beams can be made sequentially in many directions
with a high speed, resulting in a high frame rate.

Optimum results will be obtained with a narrow main beam and
small side lobes. Apparently an array with a large number of
elements should be aimed for. This has the additional advantage
of an equally high number of "viewing" directions. These considerations and the construction limitation, led to the selection

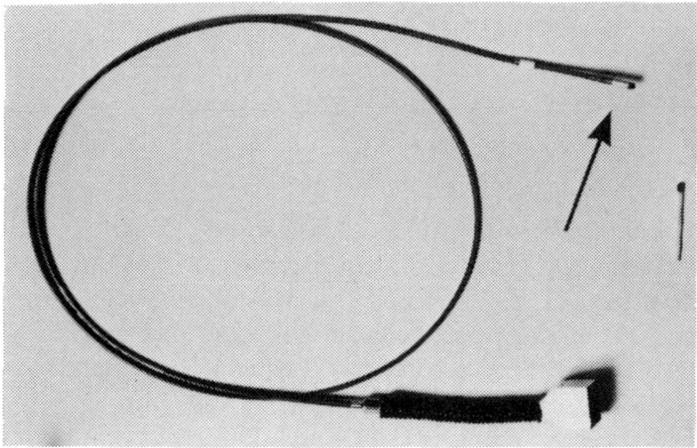

fig 1: Cardiac cath-
eter with acoustic
32-element array on
tip.

of a 32-element catheter tip with a steering in sub-groups of
8 single elements and phase corrected beam forming in 32 directions. A photograph of the catheter is shown in figure 1.

b. Multiscan system

The principle of the non-invasive linear array or multiscan system is based on the use of 20 parallel single acoustic elements
mounted in a row. All acoustic elements have a separate excitation stage and reception circuit. Each element in sequence transmits a short acoustic pulse into the tissue and receives the
returning echoes. With the multiscan a cardiac cross-section
is visualized in real time. The cross-sectional size is 8 x 16
cm. Each frame consists of 40 echo lines. To date the multi
element system has been introduced into clinical practice in
over 15 countries. The present application is mainly directed
towards the fast and easy accumulation of data or cardiac size,
shape and motion patterns.

ON-LINE PROCESSING

a. Line density on display

The presentation of two-dimensional acoustic images on any display
will be optimal with the highest possible line density on the
display. Since a depth of 16 cm (corresponding with one line)
requires a travel time of approximately 200 μsec, the number
of lines will remain limited if moving images are to be displayed.
In most situations increase of line density would require more
acoustic elements and/or electronics. Since the parallel lines
of a linear array system match well with video display techniques
(where line duration is in the order of 64 μsec) increase of
line density becomes possible by digital interpolation techniques.
An example of various line densities in a cardiac image is shown
in figure 2. It should be emphasized that the line density (or
resolution) on the display in no way reflects the acoustical
resolution.

b. Lateral resolution study

Image distortion caused by beam shape may be separated in a)
distortion due to beam width and b) distortion due to side lobes.
Whereas beam width will cause local ambiguity, side lobes may
cause reflecting structures to be malpositioned on the display.
It is thus always important to aim for a narrow beam width in
the main axis direction as well as a low sensitivity in the other
directions. A study on optimization of the lateral resolution

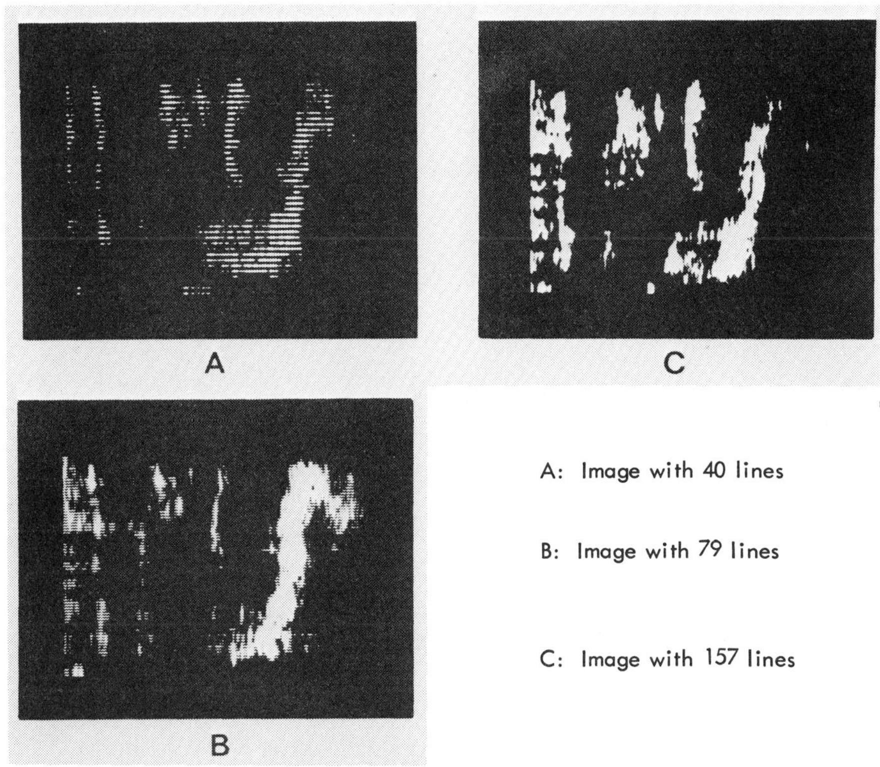

fig 2: The effect of increase in line density on the same cardiac image.

allowed the comparison of various configurations. The sound field was calculated for various transducer geometries where miniature elements were considered as combined line sources. Calculated results were thereafter compared with measured data from corresponding test configurations in a water tank. In figure 3 a comparison of measured data for summation of focus zones versus single focus zones is shown with the beam width as a function of depth. Note: the term beam width is defined as the measure for the total lateral trajectory of a wire reflector satisfying the condition that the received echo signal amplitude is greater than -10 dB relative to the signal strength when the reflector is positioned on the main axis. This study resulted in an optimal configuration which will be presented during the lecture.

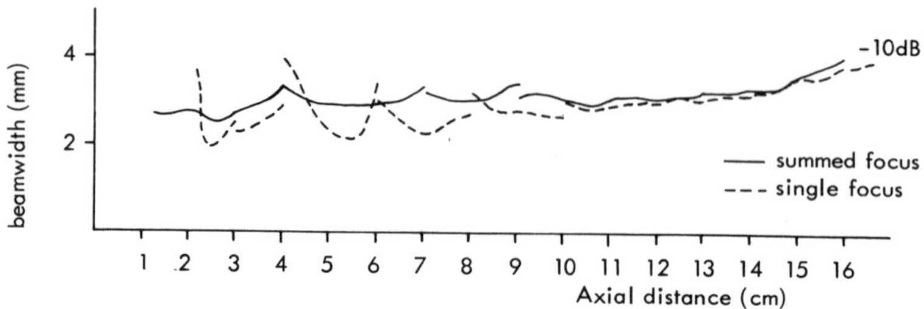

fig 3: Measured beam width as function of depth

c. Side lobe suppression

It is a well known characteristic of arrays that when the distance between the array elements is greater than $\frac{1}{2}$ λ, there will
be a grating lobe. Due to the technical limitations in the construction of the ultra miniature circular phased array the inter-
element minimum spacing is in the order of 1 λ. For the actual
array of 32 elements the antenna pattern shows a pronounced
grating lobe at approximately 55 degrees. As a result of the
requirement of real time processing the only possible methods
to reduce this grating lobe are: 1. Amplitude distribution.
2. Phasing. 3. Non-linear processing. Mostly because of the low
number of elements (8) per array the methods 1 and 2 fail. A
method based on differences in arrival time at the individual
elements has been used to suppress the incoming echoes from grating directions.

d. Dynamic range compression

Pulse echo systems for the heart require a wide dynamic range.
Apart from the attenuation and beam broadening which can be corrected for, there remains a wide variation in reflectivity. In
order to visualize as much echo information on a display as possible with only a limited dynamic range, dynamic range compression may be considered. In practice the situation is difficult,
since due to this wide variation in echo strength throughout
the scanned area the side lobe echoes may even be stronger than
the main lobe echoes. Therefore one should aim for the highest
possible main lobe -side lobe ratio. Preliminary investigations

showed that dynamic range compression, while keeping the main
lobe/side lobe ratio constant, is possible. This is accomplished
by using an array of small elements, while each element has its
own dynamic range compression prior to the beam forming circuitry. This approach enables the use of array techniques for optimization of the directivity pattern combined with an optimal matching of the echo information onto the display characteristics.

OFF-LINE PROCESSING

The purpose of off-line computer processing of two-dimensional
images is the study of filtering methods, aperture correction,
the effect of change in line density on the display, differentiation of the signal, etc. Also much effort has been directed
towards schematic contour reconstruction of the two-dimensional
images following a line to line brightness connectivity search.
Most applications are intended for further image enhancement
and structure recognition purposes.

a. The data registration

For the off-line processing a system has been developed which
allows the recording of up to 20 seconds sequential images at
the necessary high bit rate for subsequent data processing. The
method employs a PDP-11 E-10 computer (16K memory). Real time
recording of the moving images on a digital disc utilizes direct
memory access. It offers the possibility to store 50 digitized
pictures per second with 16 intensity levels. This requires a
bit rate of 1 Mbit/sec for the system. With the 1,2 Mword (16
Bit) disc a maximum storage of 20 seconds results.

b. Image filtering

Since filtering of two-dimensional sequential images is a time
consuming action the filtering has been limited to the space
domain. The fourier domain has been used only to establish the
characteristics of various filters. The filters were to decrease
the image distortion due to beam width in the vertical direction
and similarly to increase the axial resolution by shortening
the received echoes. A first approach with a linear digital high
emphasis filter proved not optimal since background intensity
increased and contrast decreased. The latter was caused by the
decreasing intensity of the saturated parts of the image.

A second approach used a non-linear digital filter, based on
first and second derivative of intensity as function of position.

With this filter method some improvement of the two-dimensional
images has been obtained.

c. Connectivity procedure

After image enhancement with filtering as described above the
cardiac structures may be reconstructed in a semi-automatic way.
With such a technique it eventually may become possible to
calculate cardiac size; stroke volume, etc. Image quality however
is not high enough for a completely automatic pattern recognition
at this time.

In principle the cardiac structures are reconstructed following
the intensity from line to line in a connective way. Four parameters are still to be selected manually. These are a) the intensity of a structure and b) the maximum intensity deviation
allowed, c) the maximum allowed "width" of a structure and d)
the maximum vertical distance between lines were two seperate
echoes are still assumed to be created by the same structure.
In the above described off-line filtering and cardiac image reconstruction techniques, only a first approach has been made
mainly based on an empirical and intuitive approach.

CONCLUDING REMARKS

a. Future processing possibilities

As mentioned earlier there is a great need for extremely fast
data-processing in the field of real time ultrasonic imaging.
Apart from dedicated analogue processing in a rather classical
way, technology of today indicates a tremendous increase of
possibilities within the near future. Extremely fast correlation
and convolution techniques may become possible by new products
such as: a) Surface acoustic wave devices, b) Charge coupled
devices, c) Opto-electronic and opto-acoustic devices, d) New
transducer materials and e) Various integrated circuits. As a
result of these developments it should be possible in future
to construct at moderate costs optimally scanned arrays yielding
high resolution plus high fidelity cross-sectional images in
real time.

b. Synthetic aperture system

Practically all developments in the medical ultrasound area are
induced by preceding investigations in either sonar, radar or
N.D.T. techniques. One of such techniques is the synthetic
aperture system used in airborne radar applications.

In combination with side looking radar antennas S.A.S. enables
high resolution mapping of areas. Instead of a lateral trajectory
of one antenna accomplished by motion a similar effect can be
obtained by using a linear array. By subsequent scanning from
each position within the array and storage of the echo signals
in a fast memory, high resolution imaging can be achieved. In
this way parallel processing at many elements in a large aperture is replaced by repeated processing of one element followed
by fast combinational scanning of the memory content. The development of such techniques may provide a new powerful diagnostic
instrument to the medical profession.

RECHERCHE ET CARACTERISATION DE DEPENDANCES LINEAIRES ENTRE
SIGNAUX PAR ANALYSE INTERSPECTRALE. APPLICATIONS

J.L. LACOUME

Centre d'Etude des Phénomènes Aléatoires et Géophysi-
ques (CEPHAG), Equipe de Recherche associée au CNRS.
B.P. 15, 38040 GRENOBLE-CEDEX (France)

RESUME. Nous présentons une analyse théorique du problème de la
recherche de dépendances linéaires stationnaires entre deux signaux. Cette étude nous permet de montrer, dans l'hypothèse d'une
certaine régularité des densités spectrales de puissance (dsp) des
signaux, que la densité spectrale de puissance croisée conduit à
l'estimateur optimal. Dans le cas de signaux décalés en temps,
de manière différente selon la fréquence, nous étudions l'estimateur du temps de retard et nous présentons un exemple d'application.

1. <u>INTRODUCTION</u>.

Dans de nombreux problèmes de mesure ou d'estimation, l'observation est constituée par N (N≥2) signaux temporels, connus sur la
durée -T, +T. Le but poursuivi par l'expérimentateur est de re-
chercher si ces N signaux dépendent d'une cause unique ou s'ils
sont issus de plusieurs causes indépendantes. De nombreuses études
ont été consacrées à ce type de situation décrit de manière géné-
rale. Nous allons ici particulariser les hypothèses, ce qui nous
conduira à une solution théorique du problème de l'estimation op-
timale de la phase. Nous présenterons ensuite des exemples d'application pratique.

2. <u>POSITION DU PROBLEME ETUDIE</u>.

Pour simplifier le formalisme, nous nous limiterons à l'étude de
signaux à 2 composantes. Par ailleurs, nous supposons que la liaison éventuelle entre les signaux est linéaire et stationnaire (filtrage). Le problème général est celui du choix entre le cas où il

existe une liaison (hypothèse H_1 et celui où il n'existe pas de liaison (hypothèse H_o) et, dans l'hypothèse H_1 celui de l'estimation des paramètres du filtre. Nous supposons donc :

- <u>Hypothèse H_o</u> : $r_1(t) = B_1(t)$ $-T < t < +T$

                       $r_2(t) = B_2(t)$ $-T < t < +T$

$B_1(t)$ et $B_2(t)$ sont des f.a. centrées, gaussiennes, de dsp $\gamma_{B_1}(\nu)$ et $\gamma_{B_2}(\nu)$ <u>indépendantes</u>

- <u>Hypothèse H_1</u> : $r_1(t) = S(t) + W_1(t)$ $-T < t < +T$

                       $r_2(t) = \mathcal{F}[S(t)] + W_2(t)$ $-T < t < +T$

où

     $S(t)$: f.a. gaussienne, centrée, de dsp $\gamma_S(\nu)$

     $W_1(t)$, $W_2(t)$: gaussiens, centrés, indépendants

     $\mathcal{F}$ est une transformation linéaire stationnaire que nous représentons par son gain complexe :

$$G(\nu) = |G(\nu)| e^{i\phi(\nu)}$$

Un élément fondamental du modèle est le "degré de connaissance" dont on peut disposer sur $S(t)$ et $G(\nu)$. Nous écartons le cas où $S(t)$ est parfaitement connu ($S(t)$ certain), mais nous serons amenés à faire des hypothèses sur la forme de $S(t)$. Disons que l'échelle des fréquences étant fixée par $\nu_o = 1/2T$ (2T durée du signal) nous serons conduits à supposer $S(t)$ localement blanc (ou à dsp localement connue, ce qui revient au même par blanchiment). Pour le moment, précisons que $S(t)$ est localement blanc si sa dsp $\gamma_S(\nu)$ varie peu sur $K\nu_o$ où K est un entier de l'ordre de 5 à 10. Simultanément, nous supposons que le filtre $\mathcal{F}$ est localement constant, c'est-à-dire que son gain complexe varie peu sur la même plage de fréquence.

3. <u>RESOLUTION THEORIQUE DU PROBLEME.</u>

3.1 <u>Description du modèle.</u>

<u>Sous H_o</u> : $r_1(t) = B_1(t)$ $-T < t < T$

                $r_2(t) = B_2(t)$ $-T < t < T$

B_1, B_2 gaussiens, centrés, indépendants.

$$\gamma_{B_1}(\nu) = \gamma_{B_2}(\nu) \qquad \text{dsp}$$

<u>Sous H_1</u> : $r_1(t) = S(t) + B_1(t)$ $-T < t < +T$

                $r_2(t) = \mathcal{F}[S(t)] + B_2(t)$ $-T < t < +T$

$S(t)$, B_1, B_2 gaussiens, centrés, indépendants, de dsp $\gamma_S(\nu)$.

$$\gamma_{B_1}(\nu) = \gamma_{B_2}(\nu)$$

$\mathcal{F}$ filtre linéaire stationnaire (fls). Gain complexe : $G(\nu)e^{j\phi(\nu)}$

- S, B_1 et B_2 sont supposés <u>localement blancs</u> soit, la dsp de ces f.a. varie peu sur $K\nu_o$ où K entier ~ 5 à 10 . $\nu_o = 1/2T$.

- $\mathcal{F}$ <u>localement constant</u> : le gain complexe de $\mathcal{F}$ varie peu sur $K\nu_o$.

3.2 <u>Base de KARHUNEN-LOEVE</u>.

Supposons S , B_1 , B_2 blancs et $G(\nu)$ constant.

$$\underline{R}(t) = \begin{pmatrix} r_1(t) \\ r_2(t) \end{pmatrix}$$

$$\gamma_{\underline{R}|H_o} = N_o \underline{\underline{I}} \quad ; \quad \gamma_{R|H_1} = \begin{pmatrix} N_o+N_S & G\,N_S \\ G^*N_S & |G|^2\,N_S+N_o \end{pmatrix}.$$

Les v.p. de $\gamma_{\underline{R}|H_1}$ sont $\lambda_1 = N_o+N_S(1+|G|^2)$; $\lambda_2 = N_o$

Les vecteur propres

$$V_1(\nu) = \frac{1}{\sqrt{1+|G|^2}}\begin{pmatrix} 1 \\ G \end{pmatrix} \quad ; \quad V_2(\nu) = \frac{1}{\sqrt{1+|G|^2}}\begin{pmatrix} -G^* \\ 1 \end{pmatrix}$$

Une base de KARHUNEN-LOEVE est dans ce cas formée de deux suites :

$$\underline{\phi}_{1_K}(t) = V_1(K\nu_o)e^{2\pi jK\nu_o t} \quad ; \quad \underline{\phi}_{2_K}(t) = V_2(K\nu_o)e^{2\pi jK\nu_o t}$$

où $\nu_o = 1/2T$.

Nous montrons en annexe que cette base reste approximativement de KARHUNEN-LOEVE si S, B_1 et B_2 sont localement blancs et $G(\nu)$ localement constant. La valeur de l'approximation a en particulier été chiffrée dans [1].

3.3 <u>Fonctionnelle du rapport de vraisemblance</u>.

Calculons la conditionnellement à $G(\nu)$:

$$\Lambda[R(t)|G(\nu)] = \frac{P_{R|H_1,G}(r)}{P_{R|H_o}(r)}$$

Posant $L_G = \text{Log}[\Lambda(r(t)|G(\nu))]$ on obtient facilement avec $B_1(t)$, $B_2(t)$ bruits blancs gaussiens de dsp N_o

$S(t)$ fonction aléatoire gaussienne de dsp $\gamma_S(\nu)$

$$r_{1_K} = 1/\sqrt{2T}\ \langle e^{2\pi j\nu_o Kt}|r_1(t)\rangle \quad ; \quad r_{2_K} = 1/\sqrt{2T}\ \langle e^{2\pi j\nu_o Kt}|r_2(t)\rangle$$

$$L_G = L_1 + \sum_K A(K\nu_o)\big[|r_{1_K}|^2+|G(K\nu_o)|^2|r_{2_K}|^2+G(K\nu_o)r_{1_K}r_{2_K}^*+G^*(K\nu_o)r_{1_K}^*r_{2_K}\big]$$

où $\quad L_1 = \frac{1}{2}\sum_K \text{Log}\,\dfrac{N_o}{N_o+\gamma_S(K\nu_o)[1+|G(K\nu_o)|^2]}$

$$A(K\nu_o) = \frac{-\gamma_S(K\nu_o)}{2\,N_o[N_o+\gamma_S(K\nu_o)(1+|G(K\nu_o)|^2)](1+|G(K\nu_o)|^2)}$$

A partir de cette expression de L_G nous allons traiter les problèmes d'estimation de la phase de $\,^G\mathcal{F}$.

3.4 Estimateur optimal de la phase du filtre $\mathcal{F}$.

Ecrivons : $G(\nu) = |G(\nu)|\,e^{i\phi(\nu)}$. L'estimateur du maximum de vraisemblance de ϕ est solution de

$$\left.\frac{\partial L_G}{\partial\phi}\right|_{\hat{\phi}_{MV}} = 0$$

Soit : $\displaystyle\sum_K i\,A(K\nu_o)\,|G(K\nu_o)|\,[r_{1_K}r_{2_K}^*\,e^{\,i\hat{\phi}_{MV}(K\nu_o)} - r_{1_K}r_{2_K}^*\,e^{\,-i\hat{\phi}_{MV}(K\nu_o)}]=0$

Supposons $\phi(\nu)$ constant sur $K_1\nu_o$, $K_2\nu_o$; alors

$$e^{\,2i\hat{\phi}_{MV}(K\nu_o)} = \sum_{K_1}^{K_2} r_{1_K}^*\,r_{2_K}\Bigg/\sum_{K_2}^{K_2} r_{1_K}\,r_{2_K}^* = J \,.\, J \ \text{peut être obtenu :}$$

- directement par la formule ci-dessus : moyenne des sorties de (K_2-K_1) filtres de largeur ν_0 ;

- comme la moyenne temporelle de la sortie d'un filtre de largeur de bande $(K_2-K_1)\nu_o$;

- comme une filtrée de l'estimation de la fonction de corrélation par un filtre centré à $((K_1+K_2)/2)\cdot\nu_0$ et de réponse impulsionnelle

$$h(t) = \text{sinc } \pi(K_2-K_1)\nu_o t$$

J est donc une estimée de la phase de la dspc de $R_1(t)$ et $R_2(t)$. Dans les conditions indiquées ci-dessus, l'estimateur MV de la phase du filtre $\mathcal{F}$ est donc la phase d'une estimée de la dspc.

4. EXEMPLE D'APPLICATION : ESTIMATION DE RETARDS.

Nous avons étudié expérimentalement le cas suivant :
$\mathcal{F}$ = filtre retard pur sur des bandes de fréquence disjointes.

Soit : $\displaystyle G(\nu) = \sum_i G_i\,\pi_{\nu_{1_i}}(\nu-\nu_{2_i})\,e^{\,2i\pi\nu\tau_i}$

Ceci correspond à des signaux reçus en 1 et 2, décalés de

τ_1 dans $\nu_{11}\pm\nu_{21}$; τ_2 dans $\nu_{12}\pm\nu_{22}$, etc...

Nous admettons que les bandes de fréquence dans lesquelles le retard est constant sont assez grandes vis-à-vis de la bande élémentaire ν_o = 1/2T   (filtre localement constant).

4.1   Estimation de la phase ou du retard. Définition. Propriétés.

L'estimateur optimal de la phase a été calculé comme transformée de FOURIER d'un estimateur de la fonction d'intercorrélation (fig. 1) [2].

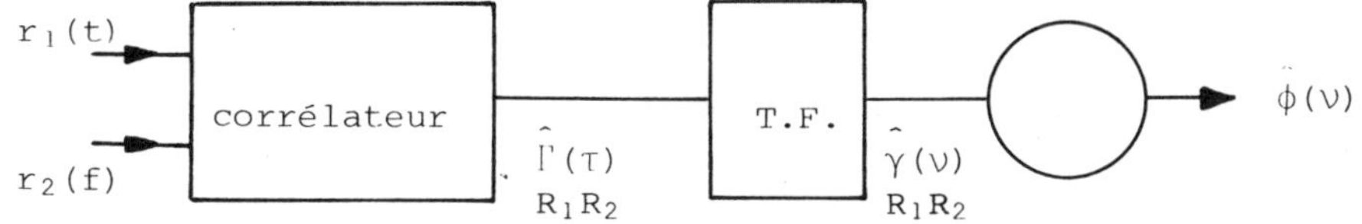

Fig. 1

Les conditions expérimentales sont définies par :

2T = temps d'intégration pour le calcul de l'intercorrélation
   (ou ν_o = 1/2T   bande passante élémentaire d'analyse)

$K\nu_o$:bande passante du filtre équivalent à la T.F.

K, de l'ordre de 10, définit la zone sur laquelle les signaux sont
   supposés "constants".

Avec les hypothèses faites on peut montrer que l'estimateur de la phase est non biaisé. La variance de cet estimateur est [4] :

$$\sigma^2{}_{\delta\phi} = \frac{g(K)}{K|c|^2}$$

où $|c| = \dfrac{|\gamma_{12}(\nu)|}{\sqrt{\gamma_1(\nu)\gamma_2(\nu)}}$ est la dspc normée de 1 et 2 (coefficient de cohérence)

et   g(K) est défini en annexe 2.

Pour la mesure de τ on a : $\phi = 2\pi\nu\tau$

En faisant une régression linéaire sur   2N+1   filtres de largeur $\Delta\nu$ on obtient approximativement

$$\sigma_{\delta\tau} = \frac{\sigma_{\delta\phi}}{2\pi L \cdot \sqrt{2N-1}\ \Delta\nu}   \quad\text{où}\quad L^2 = 2[1+2^2+ \ldots + N^2]$$

4.2   Résultats expérimentaux [3].

Sur la figure 2 nous présentons la mesure d'écarts de temps entre des pulsations magnétiques (Pc1) reçues en 2 stations SV – SO distantes de 1 000 km. Nous avons représenté la dsp, le module de la dspc et le coefficient de cohérence (c(ν)), la phase de la dspc.

Les courbes de phase sont tracées pour plusieurs retards $\tau_1, \tau_2, \ldots$ introduits lors du traitement et l'on cherche sur ces courbes des paliers horizontaux (méthode de 0). L'examen des courbes de phase montre des zones très variables associées à des valeurs de $|c(\nu)|$ assez faibles. Dans cette situation on doit conclure que la partie correlée est très faible. Il apparaît par contre des paliers : trois sont visibles sur la courbe (b), mettant en évidence trois bandes de fréquences dans lesquelles notre modèles s'applique et pour lesquelles le retard est mesurable. Enfin, la courbe (a) fait apparaître un palier très net sur lequel nous pouvons vérifier le calcul de précision de la mesure de temps faite au 4.1. Sur ce palier :

$$|c(\nu)| \sim 0,5$$

2N+1 = 11 (nombre de filtres occupant le palier)

$\Delta\nu = 0,02$ Hz (largeur de bande d'un filtre).

Avec ces valeurs : $\quad \sigma_{\delta\phi} = 0,19$ rd = 10° $\quad ; \quad \sigma_{\delta\tau} \simeq 0,05$ s.

On vérifie tout d'abord que l'amplitude des fluctuations de la phase est bien de 20° environ. Enfin, les temps $\tau_1, \tau_2 \ldots$ diffèrent de 0,06 s et l'on voit bien que cette valeur est la limite de précision de mesure du temps.

5. CONCLUSION.

Nous avons montré que dans le cas particulier de l'estimation d'un écart de temps, ou de la mesure du déphasage dans le cas de signaux et d'un filtrage localement constants, la phase de la dspc est l'estimateur MV de la phase ou du retard. Ce résultat prend uniquement en compte comme connaissance a priori le caractère localement constant, son domaine d'application est donc très vaste. Cet estimateur est non biaisé et sa variance se calcule à partir du coefficient de cohérence des 2 signaux. Les résultats quantitatifs présentés ont enfin été vérifiés dans un cas réel d'application.

REFERENCES.

1. Covariance between FOURIER coefficients representing the time waveform observed form an array of sensors. - W.S. HODGKISS, L.W. NOTTE, J. Acoust. Soc. Am. _59_, 3, March 76.

2. Etude de la précision de l'estimateur de la densité spectrale croisée, obtenue par transformation de FOURIER de la fonction de corrélation - A. KNOB, Rapport CEPHAG 55/75.

3. Movement and distribution of the energetic particles which generate Pc1. F. GLANGEAUD et al, (3.coll. Geophysics, Amsterdam 76).

ANNEXE 1   Base de KARHUNEN-LOEVE approximative.

Le problème concerne les corrélations éventuelles entre les composantes du processus sur des vecteurs de base différents.

Soit : $r_{1_K} = <\phi_{1_K}|R>$; $r_{2_K} = <\phi_{2_K}|R>$

$$E[r_{iK_1},r^*_{jK_2}] = \int_{-\infty}^{+\infty}\phi^*_{iK_1}(\nu)\,\phi_{jK_2}(\nu)\,\gamma_B(\nu)\,d\nu$$

où $\phi_{iK_2}(\nu)=TF[\pi_T \cdot \phi_{iK}]$; $\phi_{jK_2}(\nu)=TF[\pi_T \cdot \phi_{jK_2}]$

Cette intégrale est toujours petite car

1) si K_1 très différent de K_2 : $\phi_{iK_2}(\nu)\phi_{iK_1}(\nu) \sim 0$

2) si K_1 voisin de K_2 dans la zone où le produit $\phi_{iK_2}(\nu)\phi_{iK_1}(\nu)$ est important, on a $\gamma_z(\nu) \sim$ constante et $\int \phi_{iK_1}(\nu)\,\phi_{jK_2}(\nu)\,d\nu \sim 0$. Des évaluations chiffrées précises sont données dans [1].

ANNEXE 2   Calcul du biais et de la variance de l'estimateur de $\hat{\phi}(\nu)$. On a posé :

$$R_1(t) = S(t) + B_1(t) \qquad\qquad -T < t < T$$
$$R_2(t) = \mathcal{F}[S(t)] + B_2(t) \qquad -T < t < T$$

$$\hat{\gamma}(\nu)=TF[\hat{\Gamma}_{S,\mathcal{F}S}(\tau)]+TF[\hat{\Gamma}_{SB_2}(\tau)]+TF[\hat{\Gamma}_{B_1\mathcal{F}S}(\tau)]+TF[\hat{\Gamma}_{B_1B_2}(\tau)] \qquad (A1)$$

1/ <u>Biais</u> : avec les hypothèses faites l'estimateur est approximativement <u>sans biais</u>.

2/ <u>Variance de</u> $\hat{\gamma}_{12}(\nu)$: $\hat{\gamma}_{12}(\nu)$ est un nombre complexe. Nous calculerons sa matrice de covariance en posant :

$\hat{\gamma}_{12}(\nu)= (R+iJ)z$; z : nombre complexe de module 1 et proportionnel à $\gamma_{12}(\nu)$

$$\underline{\underline{\Gamma}} = \begin{pmatrix} \Gamma_{RR} & \Gamma_{RJ} \\ \Gamma_{JR} & \Gamma_{JJ} \end{pmatrix}$$

Les calculs faits dans [2] montrent que $\Gamma_{RJ} = \Gamma_{JR} = 0$; $\Gamma_{JJ} = g(K)\,\Gamma_{RR}$ où $g(K) = 0{,}1$ pour $K = 11$. On montre [2] que :

$$\Gamma_{RR} = \frac{f(K)}{K}\,[\,|G|^2\gamma_S^2 + N_o\,\gamma_S + N_o|G|^2\gamma_S + N_o^2\,],\ \text{ce qui nous conduit,}$$

en posant $|c|^2=|\gamma_{12}|^2 / \gamma_1\gamma_2$ à

$$\sigma^2_{\delta\phi} = \frac{g(K)}{K|c|^2}$$

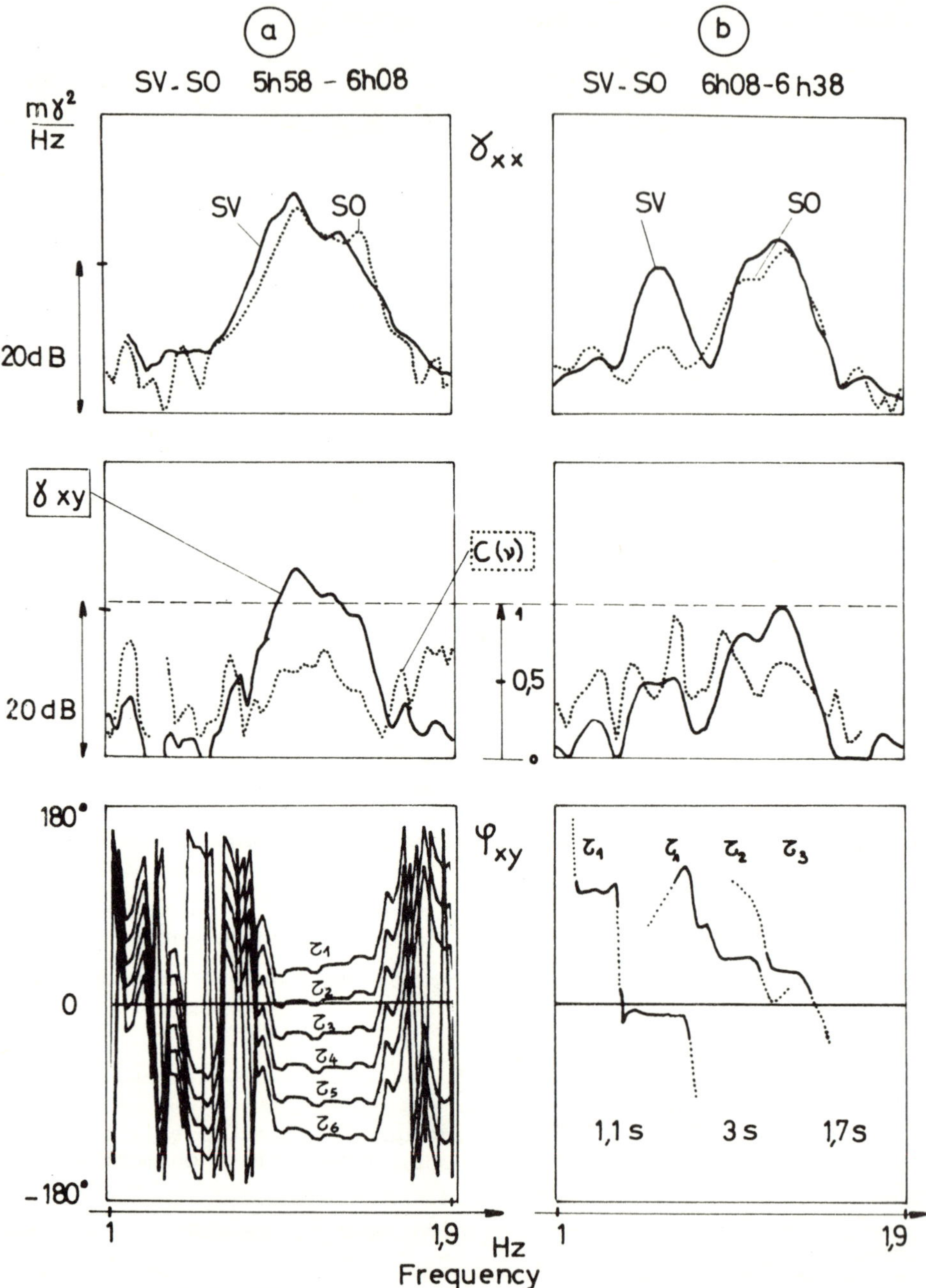

Fig. 2 : Dsp, module de la dspc et de la dspc normée, phase de
la dspc de pulsations magnétiques reçues en 2 sta-
tion distantes de 1 000 km.

S U B J E C T 6

MODERN PROCESSOR ARCHITECTURE AND TECHNIQUES

LINEAR SIGNAL PROCESSING ARCHITECTURES

H. J. Whitehouse and J. M. Speiser

Naval Undersea Center
San Diego, CA 92132

ABSTRACT. A large portion of the computational load for many signal processing problems consists of the computation of linear transforms. For time-invariant linear transforms such as cross convolution or matched filtering, the transversal filter provides a highly parallel computational module with high throughput and minimal control overhead. This paper will show how similar computational modules can be configured to provide similar computational advantages for a large class of time-variant linear transforms including one-dimensional and multi-dimensional discrete Fourier transforms and one-dimensional and two-dimensional discrete cosine transforms. Furthermore, time-variant transform modules may be combined to implement high capacity time-invariant linear transforms. The implementation of these techniques using surface acoustic wave (SAW) and charge coupled device (CCD) technology permits the real-time solution of several important signal processing problems including image data compression, spectrum analysis, convolutional array scanning and beamforming. Advanced digital and integrated analog/digital architectures will permit these fast processing techniques to be extended to high accuracy and adaptive processing tasks.

INTRODUCTION

Linear transforms play an important role in science and engineering. They occur in several very natural ways: First, the system being studied may be modeled either exactly or approximately by a linear input-output relationship. Secondly, a linear transform may be used to describe the system in a preferred coordinate system in order to diagonalize the system's input-output relationship. Thirdly, a linear transform may be used to diagonalize a quadratic form associated with a system, such as the stored energy in a linear dynamical or electrical system, or the covariance matrix of a random vector. While these concepts have a long history in the literature of linear algebra and its applications to physics and mathematical statistics, their application to modern signal processing problems frequently imposes a very heavy computational load if the processing must be performed in real time. Operations which are often required include spectral analysis, matched filtering, crosscorrelation, beamforming, adaptive filtering, and data compression. These operations usually require both significant amounts of memory and many multiplications and additions. This paper will describe how linear memory technologies such as surface acoustic waves (SAW) and charge coupled devices (CCD) can be efficiently exploited to provide highly parallel computational modules to perform the above operations with the high speed, low cost, and minimal control overhead required for real time signal processing applications.

THEORETICAL BACKGROUND

Linear transforms

A general linear transform is given in equation (1), where the output y is represented in terms of the input x and the transform kernel k.

$$y(t) = \int k(t, u)\, x(u)\, du \tag{1}$$

Such a transform can arise in several ways: It may represent the input-output relationship of a system which we wish to study such as a communication channel, an antenna, a transmission line, a scatterer, or a filter. Alternatively, it may represent a transformation which we wish to apply to a signal which is given to us.

Subject to minor regularity conditions, the transform can be diagonalized if and only if it commutes with its adjoint, [1], i.e. $K K^* = \int k(t, z)\, k^*(u, z)\, dz = \int k^*(z, t)\, k(z, u)\, dz = K^* K$. Such an operator is called normal.

Two important classes of linear operators are normal: the shift-invariant operators and the self-adjoint operators. A linear operator is shift-invariant if its kernel is of the form $k(t, u) = h(t - u)$, and hence the output is the convolution of the input with the impulse response.

The covariance function of a stochastic process, $R_z(t, u) = E\, z(t)\, z^*(u)$ may also be viewed as the kernel of a linear operator. Since $R_z(u, t) = R^*_z(t, u)$, the operator is self adjoint, and hence normal. The transform defined by the eigenvectors of the covariance function is called the Karhunen-Loeve transform [2], and provides a representation of the process in terms of uncorrelated random variables. That is if $z_n = \int z(t)\, e_n^*(t)\, dt$, then the z_n's are uncorrelated, and $z(t) = \Sigma\, z_n\, e_n(t)$.

Transforms using transversal filters

The transversal filter provides a high speed implementation of convolution with a fixed reference function, allowing an efficient realization of transform with shift-invariant kernels [3]. In order to address a wider class of transforms, this paper will show how several techniques of factoring a transform kernel into a product of simpler ones may be combined with advanced signal processing technologies.

For a shift-invariant linear transform, the kernel is of the form $k(t, u) = h(t - u)$, and the input-output relation is the convolution shown equation (2).

$$y(t) = \int h(t - u)\, x(u)\, du \tag{2A}$$

If x and h are bandlimited to $(-W, W)$, then so is y, and convolution integral of (2A) may be replaced by its sampled equivalent (2B), where $x_n = x(n/2W)$, $y_n = y(n/2W)$, and $h_n = h(h/2W)$.

$$y_s = \Sigma\, h_{s-n}\, x_n \tag{2B}$$

The sampled data convolution of equation (2B) may be implemented by a transversal filter as shown in Fig. 1. A new sampled convolution output is provided each time the input shifts one tap position to the right. so that the throughput rate is equal to the reciprocal of the intertap delay. In effect, the transversal filter combines a large number of multipliers with minimal control overhead because of the simple serial data flow. Two limitations are evident however: Only a fixed transform is provided, and the transform kernel must be shift-invariant.

Many signal processing problems require transforms which are not shift-invariant. The most important of these is the Fourier transform given in equation (3).

$$X(f) = \int x(t)\, e^{-i2\pi ft}\, dt \tag{3A}$$

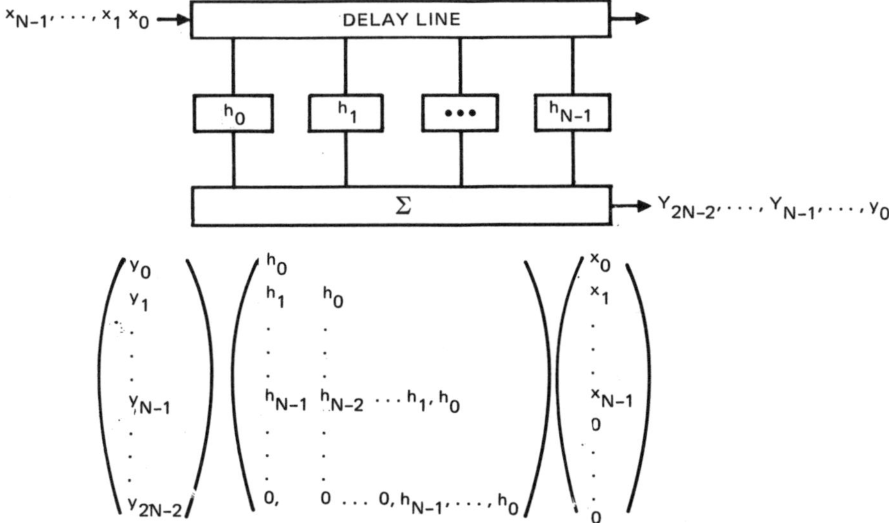

Figure 1. Transversal Filter

$$X_k = \sum_{n=0}^{N-1} x_n\, e^{-i2\pi nk/N} \tag{3B}$$

The Fourier transform is useful because it provides a decomposition into sinusoids — and sinusoids are the eigenvectors of every shift-invariant linear transform. The convolution theorem may therefore be regarded as a diagonalization by an eigenvector decomposition. However, the Fourier transform itself is not a shift-invariant transform, and must be put in somewhat modified form to provide a transversal filter implementation. A suitable substitution is $-2ft = (f-t)^2 - f^2 - t^2$, or the corresponding identity using the variables n and k [4]. The result of this substitution is shown in equation (4).

$$X(f) = e^{-i\pi f^2} \int x(t)\, e^{-i\pi t^2}\, e^{i\pi(f-t)^2}\, dt \tag{4A}$$

$$X_k = e^{-i\pi k^2/N} \sum_{n=0}^{N-1} x_n\, e^{-i\pi n^2/N}\, e^{i\pi(k-n)^2/N} \tag{4B}$$

Equation (4) corresponds to computing the Fourier transform by a premultiplication with a chirp, a convolution with a chirp, and a postmultiplication by a chirp as shown in Fig. 2. If the convolution is performed with a transversal filter, then one new transform coefficient is computed each time the data shifts one tap position down the delay line.

An alternate substitution to reduce the continuous Fourier transform to a convolutional architecture is $f' = \ln f$, $t' = \ln t$, or $f = e^{f'}$, $t = e^{t'}$, as shown in equation (5), assuming $x(t) = 0$ for $t < 0$.

$$X(e^{f'}) = \int_{-\infty}^{\infty} x(e^{t'})\, e^{-i2\pi\, e^{f'+t'}}\, e^{t'}\, dt' \tag{5A}$$

$$X(0) = \int x(t)\, dt \tag{5B}$$

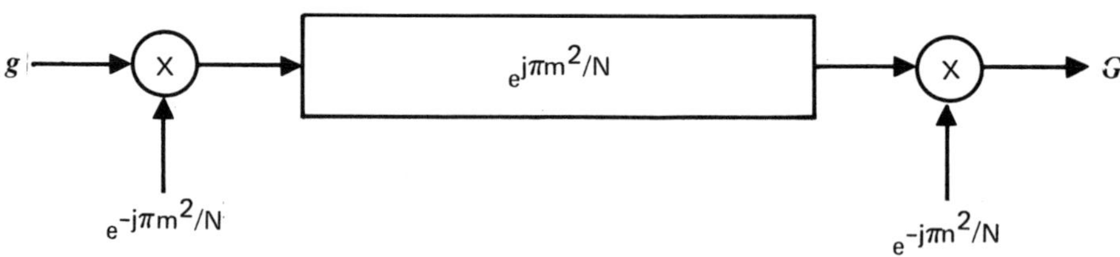

Figure 2. Chirp-Z-Transform Implementation of the DFT

An analogous substitution for the discrete Fourier transform of equation (4B) is available if the transform block size, N is a prime number [5]. In that case, there is as integer R, called a primitive root for N, which plays the role of a base for a table of logarithms in the integers modulo N. That is, if n and k are nonzero modulo N, one can write $n = R^{n'}$ (modulo N), $k = R^{k'}$ (modulo N), and $nk = R^{n'+k'}$ (modulo N). The integer n' is called the index of n with respect to the primitive root R, and is analogous to a logarithm to the base R [6].

The discrete Fourier transform representation analogous to (5A & D) is given in (5C & D).

The zero frequency term must be computed separately, as given in (5B & D). Since every nonzero integer modulo N appears exactly once in the sequence 1, R, $R^2 \ldots R^{N-1}$, this sequence may be regarded as a permutation of $1, \ldots N - 1$.

$$G_{R^{k'}} = g_0 + \sum_{n'=1}^{N-1} g_{R^{n'}} \, e^{-i2\pi R^{n'+k'}/N} \tag{5C}$$

$$G_0 = \sum_{n=0}^{N-1} g_n \tag{5D}$$

The interpretation of (5D) is that the discrete Fourier transform in permuted order may be calculated by permuting the input data, convolving with a permuted complex sinusoid, and adding a connection, g_0. The corresponding architecture is shown in Fig. 3.

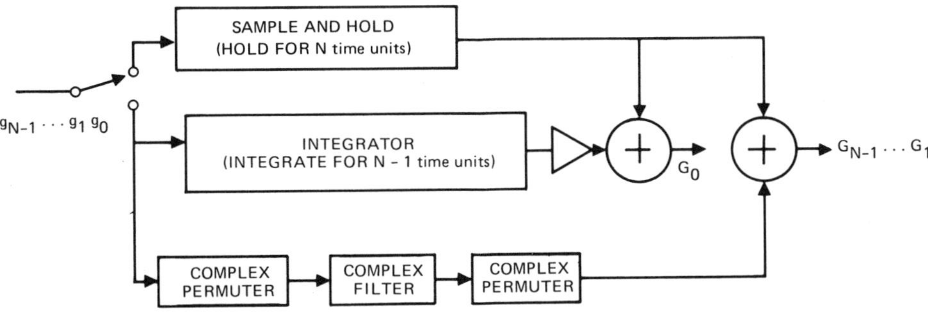

Figure 3. Prime Fourier Transform Architecture

A concept for an analog permutation memory using capacitors as storage elements and field effect transistors as switches is shown in Fig 4.

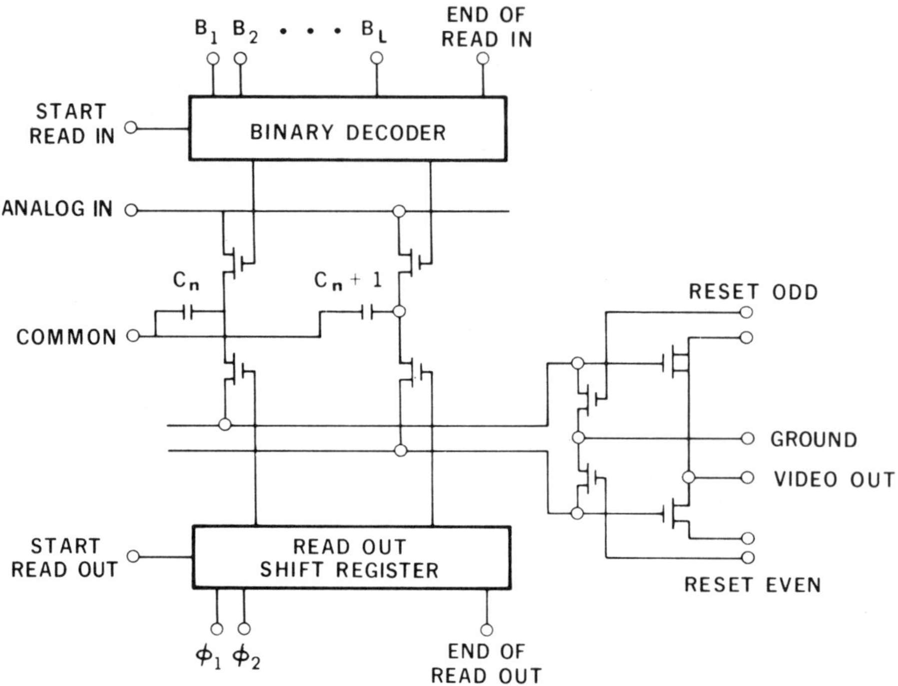

Figure 4. Analog Permutation Memory

Data Compression

Multidimensional data fields may have a large number of degrees of freedom: the product of the number of samples in each of the dimensions. If the samples are individually quantized for transmission, this results in a data rate proportional to the product of the number of degrees of freedom and the number of bits of quantization per sample. Frequently this forces a compromise between the data channel bandwidth and the quality of transmission, either requiring an excessively high transmission bandwidth or a coarse quantization which limits the quality of the data.

Encoding

Memoryless encoding of the data is optimal only when the data samples are statistically independent. Any set of random variables with a given covariance matrix may be converted by a unitary linear transformation into uncorrelated random variables. If the original random variables were jointly Gaussian, then the uncorrelated random variables will be statistically independent Gaussian random variables, and a subsequent memoryless encoding will be optimal. If the original random variables are not jointly Gaussian, the transformation eliminates only linear dependences. In any case, the Karhunen-Loeve transform results in uncorrelated variables with variances equal to the corresponding eigenvalues of the data covariance matrix. If a small number of bits are used to code the variables having small variance, the overall bit rate can be reduced for a given mean-square error in the data reconstructed from the encoded data [2].

For a second order stationary random process, the sinusoids are an asymptotic approximation to eigenvectors when the coding block length is long compared to the correlation length of the process. Even when the exact eigenvectors are sinusoids, as in the case of a process with exponential covariance function, the sinusoids may be nonuniformly spaced in frequency, thus limiting the accuracy with which the discrete Fourier transform approximates the exact Karhunen-Loeve transform [7]. The problem is caused by the boundaries produced by a finite block length. In effect, the DFT treats the data as periodically extended and will require high frequency coefficients if there is a difference between the left and right boundary values, since such a difference corresponds to a jump discontinuity in the periodic extension of the function. A modification of the DFT to eliminate the jump discontinuity is provided by the Discrete Cosine Transform. Alternatively, recent studies have shown that the effect of the boundary values may be compensated for by filtering the boundary values, subtracting the filter output from the original signal and then exactly diagonalizing the difference process by means of sinusoids, uniformly spaced in frequency [8].

Discrete cosine transform

Two different types of discrete cosine transform (DCT) are useful for reduced redundancy television image transmission [9]. Both are obtained by extending the length N data block to have even symmetry, taking the discrete Fourier transform (DFT) of the extended data block and saving N terms of the resulting DFT.

The "Odd DCT" (ODCT) extends the length N data block to length $2N - 1$, with the middle point of the extended block as a center of even symmetry. Since the DFT of a real, even sequence is a real, even sequence, the ODCT is its own inverse if a normalized DFT is used. The "Even DCT" (EDCT) extends the length N data block to length 2N, with a center of even symmetry located between the two points nearest the middle. For example, the odd length extension of the sequence A B C is C B A B C, and the even length is C B A A B C. In both cases, the symmetrization eliminates the jumps in the periodic extension of the data block which would occur if one edge of the data block had a high value and the other edge had a low value; in effect it performs a sort of smoothing operation with no loss of information. It will be noted that the terms "odd" and "even" in ODCT and EDCT refer only to the length of the extended data block in both cases the extended data block has even symmetry.

Both types of DCT may be implemented using a serial access architecture similar to that previously described for the Chirp-Z transform (CZT) implementation of the DFT.

Let the data sequence be $g_0, g_1, \ldots g_{N-1}$. The ODCT of g is defined as

$$G_k = \sum_{n=-(N-1)}^{N-1} g_n \, e^{\frac{-i2\pi nk}{2N-1}} \quad \text{for } k = 0, 1, \ldots, N-1 \tag{6}$$

where

$$g_{-n} = g_n \text{ for } n = 0, 1, \ldots, N-1. \tag{7}$$

By straightforward substitution it may be shown that

$$G_k = 2 \, \text{Re} \sum_{n=0}^{N-1} \tilde{g}_n \, e^{\frac{-i2\pi nk}{2N-1}} \tag{8}$$

where $\tilde{g}$ is defined by equation (9).

$$\tilde{g}_n = \left\{ \begin{array}{l} 0.5 \, g_0, \, n = 0 \\ \quad g_n, \, n = 1, \ldots, N-1 \end{array} \right\} \tag{9}$$

The identity (10) may be used to obtain the CZT form of the ODCT shown in equation (11) and Figures 5 and 6.

$$2nk = n^2 + k^2 - (n-k)^2 \tag{10}$$

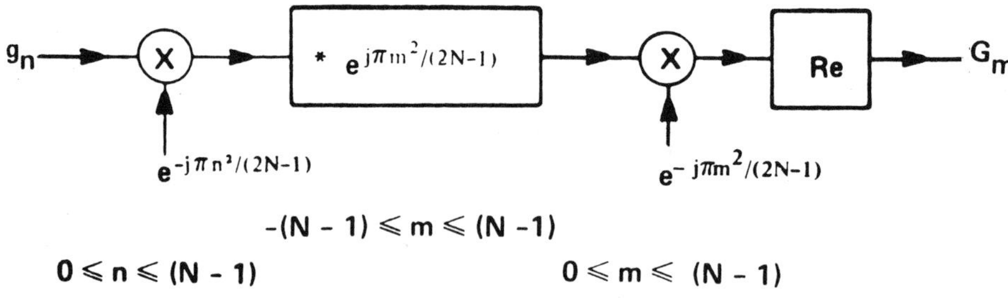

Figure 5. Serial Access Implemenataion of the ODCT

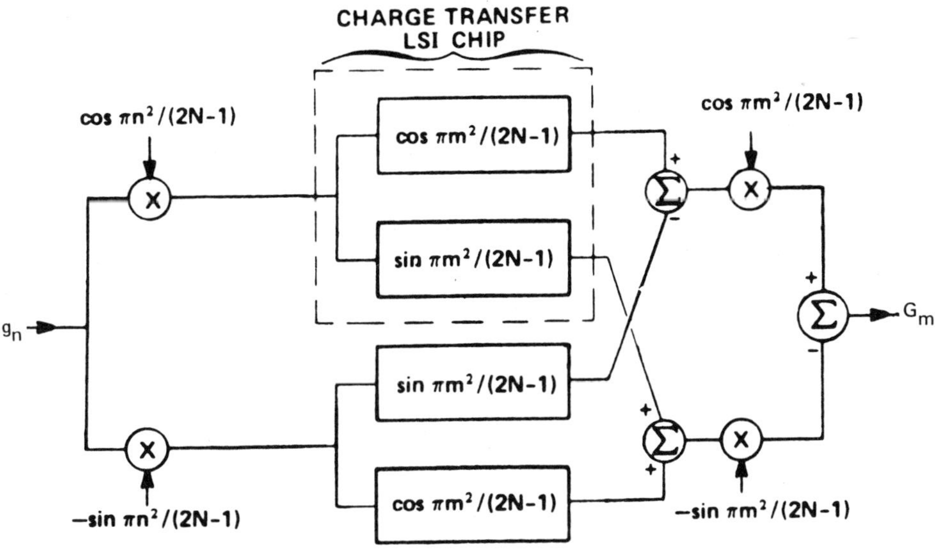

Figure 6. Charge Transfer Implementation of DCT.

$$G_k = 2\,\text{Re}\left\{e^{\frac{-i\pi k^2}{2N-1}}\sum_{n=0}^{N-1}e^{\frac{-i\pi n^2}{2N-1}}\,g_n\,e^{\frac{i\pi(n-k)^2}{2N-1}}\right\} \qquad (11)$$

The EDCT of g is defined by equation (12), where the extended sequence is defined by equation (13).

$$G_k = e^{\frac{-i\pi k}{2N}}\sum_{n=-N}^{N-1}g_n\,e^{\frac{-i2\pi nk}{2N}}\quad \text{for } k = 0, 1, \ldots, N-1 \qquad (12)$$

$$g_{-1-n} = g_n \quad \text{for } n = 0, 1, \ldots, N-1 \qquad (13)$$

If the mutually complex conjugate terms in equation (12) are combined, then equation (14) results. Equation (14) may be viewed as an alternate way of defining the EDCT.

$$G_k = 2\,\mathrm{Re}\left\{ e^{\frac{-i\pi k}{2N}} \sum_{n=0}^{N-1} g_n\, e^{\frac{-i2\pi nk}{2N}} \right\}$$

$$= \sum_{n=0}^{N-1} g_n \cos\left[\frac{2\pi(n + 0.5)k}{2N}\right] \tag{14}$$

Equation (14) may be put in the CZT format given in equation (15)

$$G_k = 2\,\mathrm{Re}\left\{ e^{\frac{-i\pi k}{2N}}\, e^{\frac{-i\pi k^2}{2N}} \sum_{n=0}^{N-1} g_n\, e^{\frac{-i\pi n^2}{2N}}\, e^{\frac{i\pi(n-k)^2}{2N}} \right\} \tag{15}$$

Prime cosine transform

This section discusses a high speed implementation of the odd discrete cosine transform (ODCT) which eliminates the multipliers required in implementations based on the chirp-Z transform. The discrete cosine transform is useful for television data compression since its basis vectors closely approximate those of the optimum Karhunen-Loeve transform for exponentially correlated data [7, 8].

The ODCT is defined as the first N Fourier coefficients of the length 2N – 1 even extension of the data, assuming that the data consists of N real values. This is shown in equation (16).

$$G_k = \sum_{n=-(N-1)}^{N-1} g_n\, e^{-i2\pi kn/(2N-1)} \qquad \text{for } k = 0, 1, \ldots, N-1 \tag{16}$$

where

$$g_{-n} = g_n$$

In order to be able to use a variant of Rader's Prime Transform algorithm [5] we assume that P = 2N –1 is a prime. Note that the data block length, N, need not be a prime, as shown in the Table.

It was shown above that if P is a prime, the discrete Fourier transform (DFT) of length P can be implemented using a circular convolution of length P – 1 together with two analog permuter memories of length P. It will be shown here that the symmetry of the extended data in equation (16) permits the size of the circular convolution and the permuter memories to be reduced by a factor of two.

Table 1. Selected Primes and the corresponding OCDT lengths and filter lengths. (The filter lengths shown assume that the data is not recirculated or reread into the filter.)

P (prime)	$N = (P + 1)/2$ (data block length)	$2N - 3 = P - 2 =$ filter length
31	16	29
61	31	59
127	64	125
251	126	249
257	129	255

For each prime P, there is an integer R, called a primitive root of P, such that the residues of $R, R^2, \ldots R^{P-1}$ are all distinct modulo P, and include every non-zero residue modulo P [6]. Therefore, for each integer n not congruent to zero mod P, n can be represented uniquely as a power of R modulo P, say $n = R^{n'}$(mod P). This representation is useful because it allows us to replace multiplication by addition in the exponent of the DCT, equation (16), and DCT thus reduces the DCT to a circular correlation, as shown in equation (17) analogous to equation (5C) for the DFT.

$$G_{R^{k'}} = g_0 + \sum_{n'=-(N-2)}^{N-1} G_{R^{n'}}\, e^{\frac{-i2\pi R^{n'+k'}}{2N-1}} \tag{17}$$

Since zero does not have an index (logarithm) with respect to the primitive root, the zero frequency point in the transform must be computed separately, as shown in equation (18).

$$G_0 = \sum_{n=-(N-1)}^{N-1} g_n = g_0 + 2 \sum_{n=1}^{N-1} g_n \tag{18}$$

The interpretation of equation (2) is that the DFT coefficients in permuted order are obtained by adding g_0 to the circular crosscorrelation of a permuted sinusoid with a permuted version of the data points excluding g_0. For the special case of the DCT, the symmetry of the extended data allows us to replace the complex exponential by a cosine, as shown in equation (19).

$$G_{R^{k'}} = g_0 + \sum_{n'=-(N-2)}^{N-1} g_{R^{n'}} \cos\left(2\pi R^{k'+n'}/(2N - 1)\right) \tag{19}$$

It will now be shown that the permuted data and the permuted cosine have
periodicity $N - 1$, so that the circular correlation of length $P - 1 = 2N - 2$ can be
reduced to a circular correlation of length $N - 1$. First, note that both the extended data and the cosine function are even. Let h_n be any even sequence. It will be
shown that h_{R^S} has period $N - 1$, where the subscript R^S is reduced modulo $P = 2N - 1$.

It is well known in number theory that $R^{(P-1)/2} = -1 \pmod P$ [6]. In our case,
$(P - 1)/2 = (2N - 2)/2 = N - 1$. Therefore $h_{R^S + (N-1)} = h_{R^S R^{N-1}} = h_{-R^S} = h_{R^S}.$

Using this periodicity property applied to equation (19) lets us write the
ODCT in shorter form, as shown in equation (20).

$$G_{R^{k'}} = g_0 + 2 \sum_{n'=1}^{N-1} g_{R^{n'}} \cos(2\pi R^{n'+k'}/(2N - 1)) \tag{20A}$$

$$G_0 + g_0 + 2 \sum_{n=1}^{N-1} g_n \tag{20B}$$

The circular correlation required for equation (20A) may be implemented
by a transversal filter of length $2(N - 1) - 1 = 2N - 3$, with tap weights of
$\cos(2\pi R^S/(2N - 1))$, for $s = N - 2, \ldots 1, 0, 1, 2, \ldots N - 2$.

The architecture of the transform is shown in Figure 7, and is virtually identical to that of the prime Fourier transform, except that the prime cosine transform
need only permute real data and filter the permuted real data with a filter having

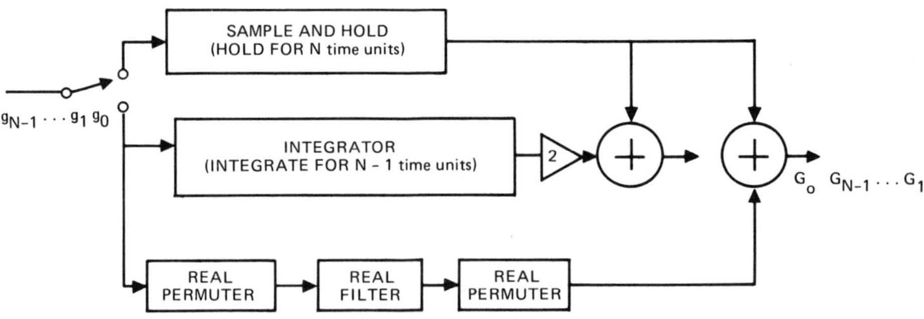

Figure 7. Prime ODCT architecture.

real weights. The prime cosine transform is thus considerably simpler to implement than a prime Fourier transform of the same block length.

TWO-DIMENSIONAL DFT

The DFT of a two-dimensional array, $g(n_1 n_2)$ may be computed by successive applications of the one-dimensional DFT. This concatenation is readily observed by writing the expression for the two-dimensional DFT as shown in Equation (21).

$$G(k_1, k_2) = \sum_{n_2=0}^{N_2-1} \sum_{n_1=0}^{N_1-1} g(n_1, n_2) \, e^{\frac{-j2\pi n_1 k_1}{N_1}} \, e^{\frac{-j2\pi n_2 k_2}{N_2}} \tag{21}$$

In two dimensions the concatenation property of the DFT can be exploited along with auxiliary memory to compute the two-dimensional transform by successively computing the CZT of rows of the input signal matrix and then using the auxiliary memory as a row to column transformation, i.e., transposing the partial Fourier transform matrix, and computing the two-dimensional Fourier transform with a second CZT.

An alternative method of computing the two-dimensional DFT is to use linear congruential scanning of the data. The two-dimensional discrete Fourier transform system will use an input scanning device and a one-dimensional discrete Fourier transform device as shown in Figure 8. The two-dimensional transform block size N_1 by N_2 is chosen such that N_1 and N_2 are relatively prime integers (i.e., they

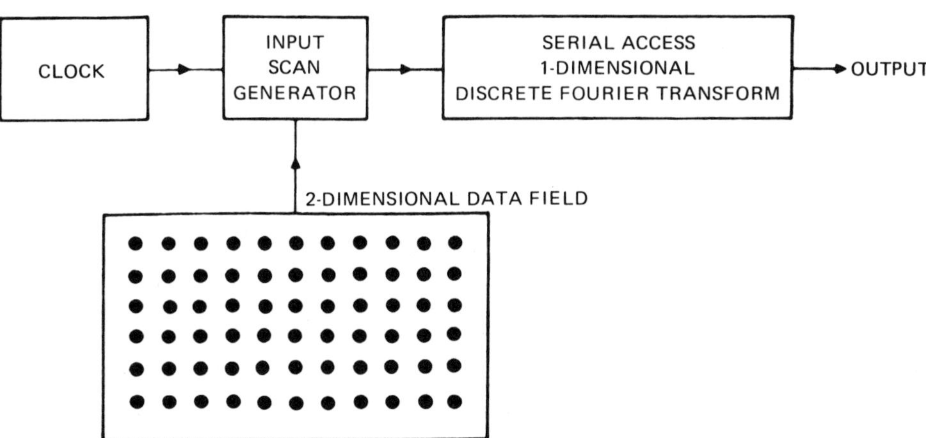

Figure 8. Two-dimensional Discrete Fourier Transform Device

must have no common divisor). The one-dimensional Fourier transform device has a block length of $N = N_1 N_2$. The purpose of the input scanning device is to so order the input data that the one-dimensional Fourier transform of the length $N_1 N_2$ serial data is identical to an N_1 by N_2 two-dimensional Fourier transform of the N_1 by N_2 input data samples. If desired, an output scanning device may also be used to provide the transform output points in normal order. The required scan may be derived from the representation [10] of a one-dimensional discrete Fourier transform matrix as a direct product matrix. The one-dimensional DFT in equation (22) is equivalent to the two-dimensional DFT in equation (23) when $N = N_1 N_2$ and N_1, N_2 are relatively prime.

$$F_k = \sum_{n=0}^{N-1} f_n \, e^{\frac{-j2\pi nk}{N}} \quad \text{for } n, k = 0, 1, \ldots, N-1 \tag{22}$$

$$G(k_1, k_2) = \sum_{n_1=0}^{N_1-1} \sum_{n_2=0}^{N_2-1} g(n_1, n_2) \, e^{-j2\pi \left(\frac{n_1 k_1}{N_1} + \frac{n_2 k_2}{N_2} \right)} \tag{23}$$

for $n_1, k_1 = 0, 1, \ldots N_1 - 1$ and $n_2, k_2 = 0, 1, \ldots N_2 - 1$.

In order to make the two transforms equivalent, it is necessary to find a pair of one-to-one functions $n(n_1, n_2)$ and $k(k_1, k_2)$ such that

$$\frac{nk}{N} = \frac{n_1 k_1}{N_1} + \frac{n_2 k_2}{N_2} \quad \text{(Modulo 1)} \tag{24}$$

or

$$nk = n_1 k_1 N_2 + n_2 k_2 N_1 \quad \text{(Modulo } N_1 N_2) \tag{25}$$

This may be accomplished by letting

$$n(n_1, n_2) = n_1 N_2 + n_2 N_1 \quad \text{(Modulo } N) \tag{26}$$

$$k(k_1, k_2) = k_1 U_1 N_2 + k_2 U_2 N_1 \quad \text{(Modulo } N) \tag{27}$$

where the constants U_1 and U_2 are the solutions of

$$N_2 U_1 = 1 \quad \text{(Modulo } N_1) \tag{28}$$

$$N_1 U_2 = 1 \quad \text{(Modulo } N_2) \tag{29}$$

Equations (27) and (28) will have solutions if and only if N_1 and N_2 are mutually prime. Under this condition, the mappings described by equations (25) and (26) will satisfy the requirement of equation (23). The linear congruential scan prescribed by Equation (25) may also be used to perform two-dimensional convolution or crosscorrelation using an ordinary one-dimensional transversal filter or crosscorrelator.

TWO-DIMENSIONAL CZT

Once the data is in a two-dimensional format, with simultaneous serial access to all the rows, it may be transformed in the "horizontal" direction by the structure shown in Figure 9. A transversal filter with multiple input taps may be used to access a column of the partially transformed output in a single shift time of the partial transform device. With appropriate coding of the parallel input column access device, it may also perform the discrete chirp premultiplication and the discrete chirp convolution of a DFT in the "vertical" direction. A complete two-dimensional CZT architecture is shown in Figure 10, and the required coding for the column access filter device is shown in Figure 11. The complex arithmetic may be implemented as described previously.

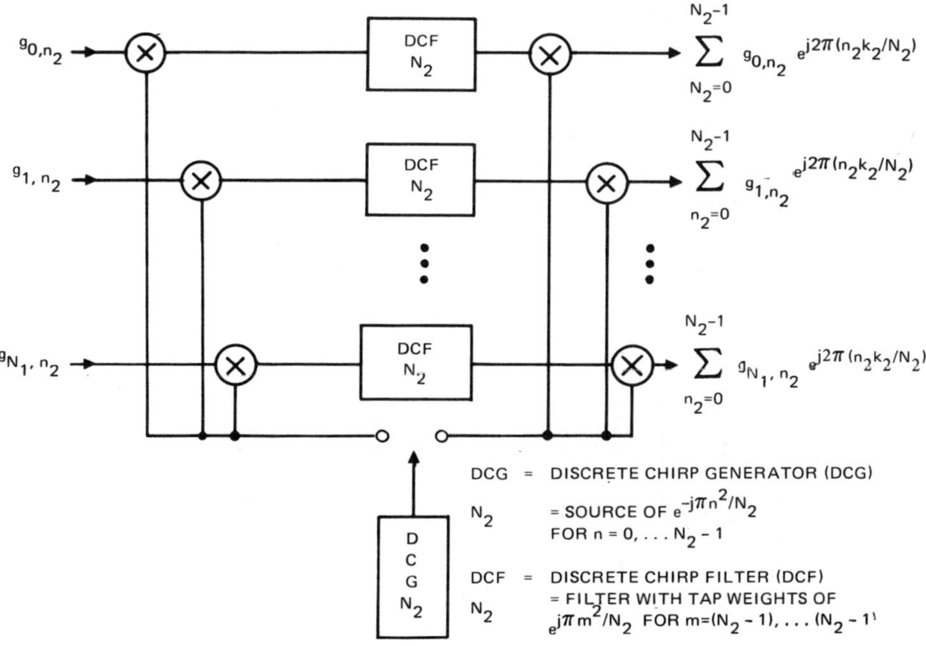

Figure 9. Two-dimensional Partial Chirp-Z Transform

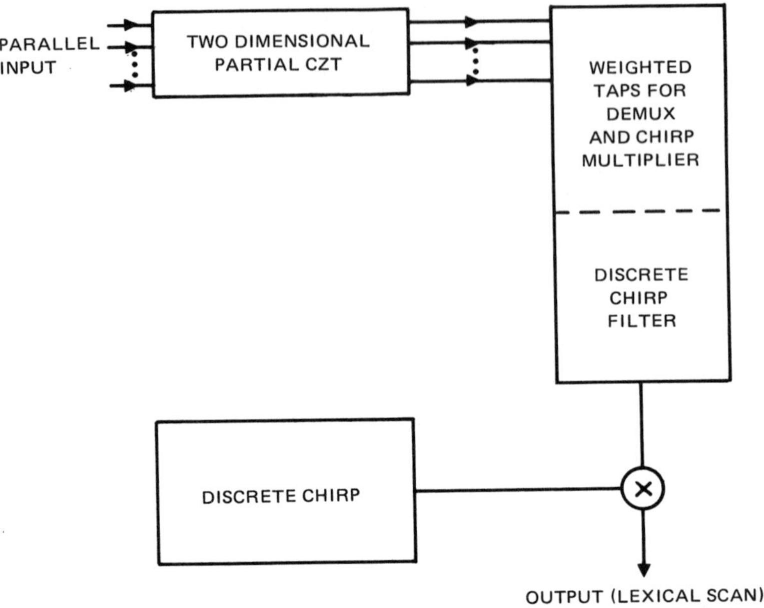

Figure 10. Hybrid Implementation of Two-dimensional CZT.

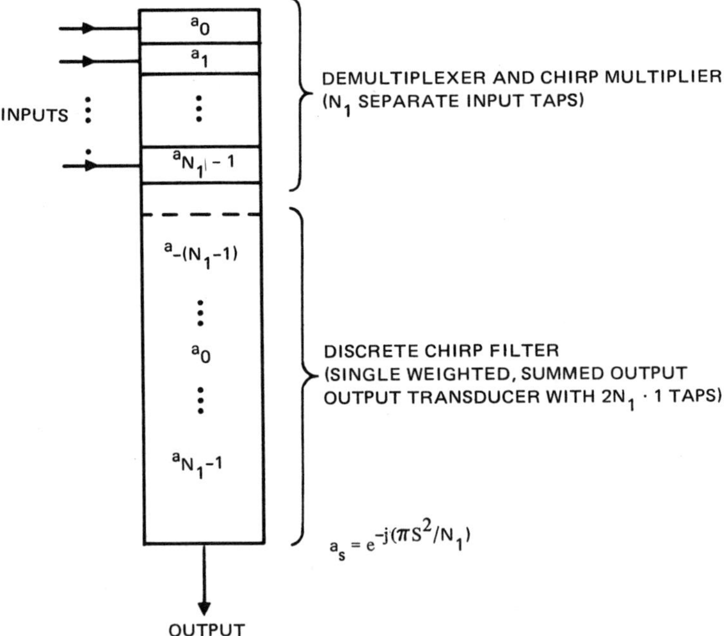

Figure 11. Tap Weights and Structure for Combined Demultiplexer, Chirp Multiplier, and Discrete Chirp Filter.

MODULAR CZT

Two methods for combining N_2 chirp-Z transform (CZT) modules of length N_1 to perform a discrete Fourier transform (DFT) of length N_1N_2 are described. The first method uses an auxiliary parallel-input, parallel-output DFT device of size N_2 and allows the transform of size N_1N_2 to be performed in the same time required for a single CZT module to perform a size N_1 transform. The second method uses an auxiliary parallel-input, serial-output DFT device of size N2. If the second method is implemented entirely in a single technology, such as with CCDs, it performs the size N_1N_2 transform in N_2 times the amount of time required for a single CZT module to perform a size N_1 transform.

A one-dimensional discrete Fourier transform may be written as a partial transform of a doubly subscripted representation of the data, followed by a pointwise multiplication, followed by a second partial transform [11] as shown in equations (30) – (35).

$$G_k = \sum_{n=0}^{N-1} g_n \, e^{\frac{-j2\pi kn}{N}} \quad \text{for } k = 0, \ldots, N-1 \text{ and } N = N_1N_2 \tag{30}$$

Let

$$n = n_1N_2 + n_2 \quad \text{for } k_1, n_1 = 0, \ldots, N_1 - 1 \tag{31}$$

$$k = k_1 + k_2N_1 \quad \text{for } k_2, n_2 = 0, \ldots, N_2 - 1 \tag{32}$$

$$G_{k_1+k_2N_1} = \sum_{n_1=0}^{N_1-1} \sum_{n_2=0}^{N_2-1} g_{n_1N_2+n_2} \, e^{\frac{-j2\pi(k_1+k_2N_1)(n_1N_2+n_2)}{N_1N_2}} \tag{33}$$

$$G_{k_1+k_2N_1} =$$

$$\sum_{n_1=0}^{N_1-1} \sum_{n_2=0}^{N_2-1} g_{n_1N_2+n_2} \, e^{\frac{-j2\pi k_1n_1}{N_1}} \, e^{\frac{-j2\pi k_2n_2}{N_2}} \, e^{\frac{-j2\pi k_1N_2}{N_1N_2}} \tag{34}$$

$$G_{k_1+k_2N_1}$$

$$= \sum_{n_2=0}^{N_2-1} e^{\frac{-j2\pi k_2 n_2}{N_2}} e^{\frac{-j2\pi k_1 n_2}{N_1 N_2}} \sum_{n_1=0}^{N_1-1} g_{n_1 N_2 + n_2} e^{\frac{-j2\pi k_1 n_1}{N_1}} \tag{35}$$

Figures 12 and 13 show modular CZT implementations which follow from equation (35). The individual CZT subsystems shown in Figures 12 and 13 would be similar to the CZT implementations previously described. The parallel DFT required for the second partial transform in Figure 13 may be implemented as combination of summers and attenuators. Unfortunately, a parallel DFT implementation of this type becomes unwieldly if the dimension N_2 is very large.

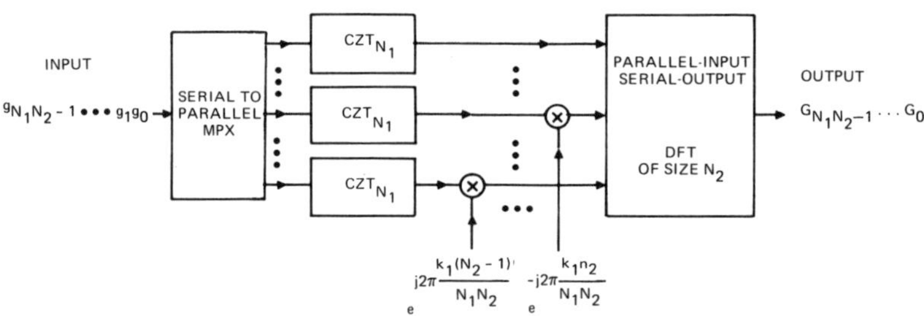

Figure 12. Organization of Modular CZT

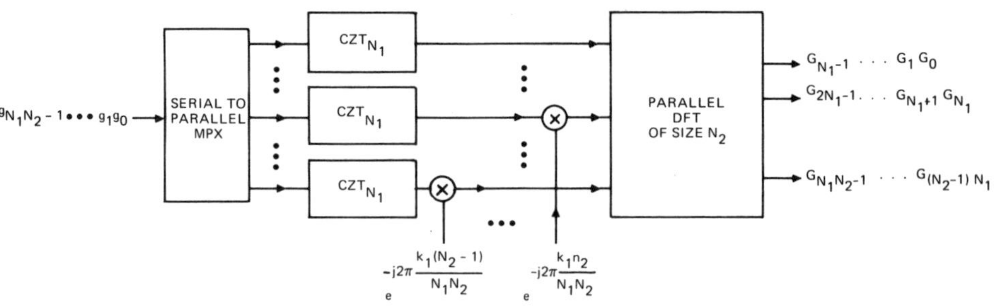

Figure 13. Alternate Organization of Modular CZT

Parallel input discrete Fourier transform implementations may be derived from the identities of equations (36) – (38).

$$H_k = \sum_{n=0}^{M-1} e^{\frac{-j2\pi kn}{M}} h_n \tag{36}$$

$$kn = \frac{1}{4}\left\{ (k+n)^2 - (k-n)^2 \right\} \tag{37}$$

$$H_k = \sum_{n=0}^{M-1} e^{\frac{-j\pi(k+n)^2}{2M}} e^{\frac{j\pi(k-n)^2}{2M}} h_n \quad \text{for } k = 0, \ldots, M-1 \tag{38}$$

If the factors in equation (38) are interpreted as two waves propagating in opposite directions relative to the function to be transformed, it may be seen that the structure of Figure 14 also performs a discrete Fourier transform with speed comparable to that of a CZT.

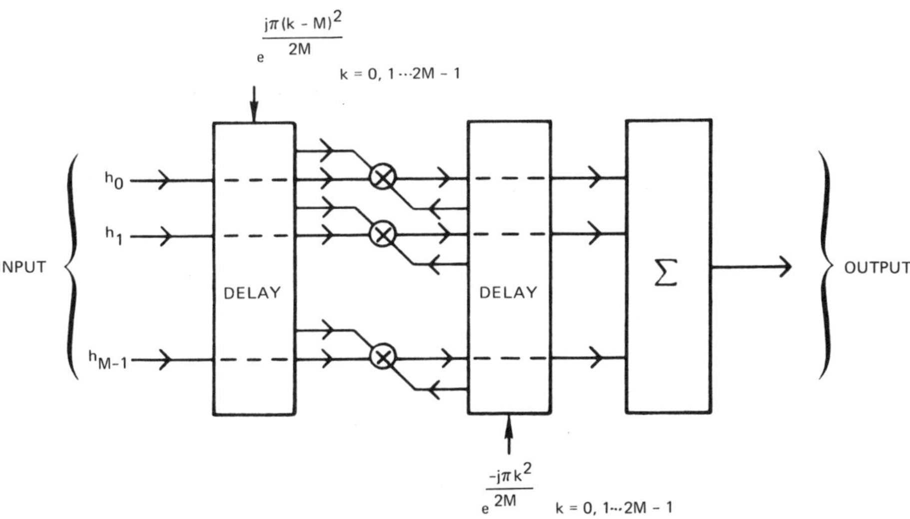

Figure 14. Parallel-Input, Serial-Output CZT Using Multi-Port Convolver

BEAMFORMING

The geometry of an arbitrarily spaced receiving array is shown in Fig. 15, where R_n is the vector from the origin to the position of the nth array element, and R is the vector from the origin to the assumed source location. The delays required for a delay-and-sum beamformer are proportional to the distances from the source to the various elements: $d_n = \|R - R_n\|$. Let g_n denote the complex amplitude of the signal at the nth array element. The signal is assumed to have been decomposed via a temporal Fourier transform so that it is only necessary to form beams separately for each frequency. For a signal component with frequency corresponding to wavelength λ, the desired beamformer output in the far field approximation is given in equation (39), where U is a unit vector in the same direction as R.

$$\sum_{n=0}^{N-1} g_n \, e^{-i2\pi \, [U, R_n]/\lambda} \tag{39}$$

For a line array, this is given by equation (40)

$$G(\theta) = \sum_{n=0}^{N-1} g_n \, e^{-i2\pi(x_n \cos\theta)/\lambda} \tag{40}$$

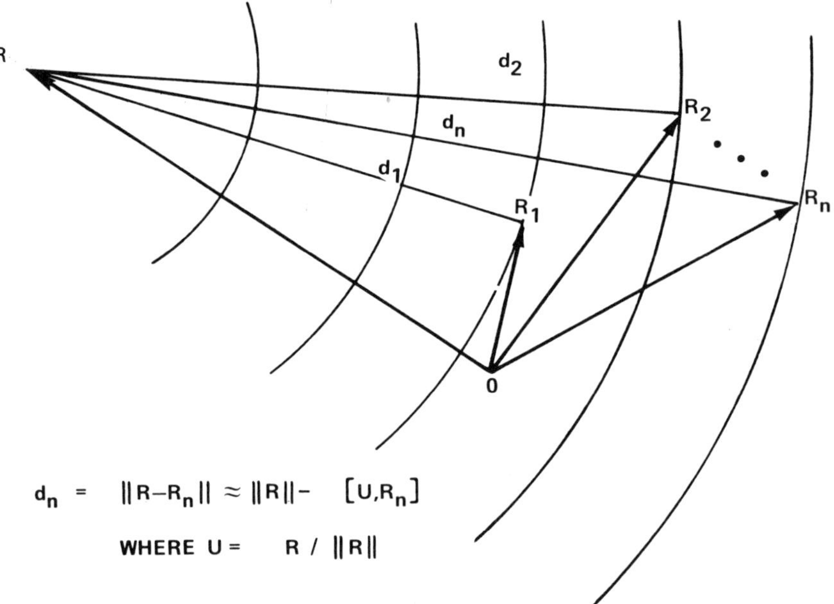

Figure 15. Propagation Geometry for Nonuniformly Spaced Array

In the special case of a uniformly spaced line array shown in Fig. 16 the above equation reduces to the expression shown in equation (40A).

$$G(\theta) = \sum_{n=0}^{N-1} g_n \, e^{-i2\pi (n\,d\,\cos\theta)/\lambda} \tag{40A}$$

The DFT defined by equation (41) therefore calculates the beamformer output at angles θ_k satisfying $\cos\theta_k = (k/N)\,(\lambda/d)$.

$$G_k = \sum_{n=0}^{N-1} g_n \, e^{-i2\pi nk/N} \tag{41}$$

In other words, a two-dimensional temporal-spatial Fourier transform can form beams [12] in the nonuniformly spaced look directions given by equation (42)

$$\theta_k = \cos^{-1}\,(k\lambda/Nd) \tag{42}$$

Note that if the multiport convolver of Fig. 14 is used to perform the spatial transform, and continuous chirps are used to drive its serial input ports, then it evaluates the spatial transform at all frequencies and hence forms beams in all look directions.

Frequency domain beaming for geometrically spaced line arrays and uniformly spaced circular arrays

These two seemingly dissimilar classes of arrays have been grouped together, because a single, simple beamformer architecture suffices for both of them: A temporal Fourier transform, followed by a spatial crosscorrelation or crossconvolution at each temporal frequency shown in Fig. 17.

For the geometrically spaced line array, let the nth element be located at $x_n = d\,r^n$. Then the desired beamformer output is given by

$$G(\theta) = \sum_n g_n \, e^{-[i2\pi x_n \cos\theta]/\lambda} = \sum_n g_n \, e^{-i[2\pi d\,r^n \cos\theta]/\lambda} \tag{43}$$

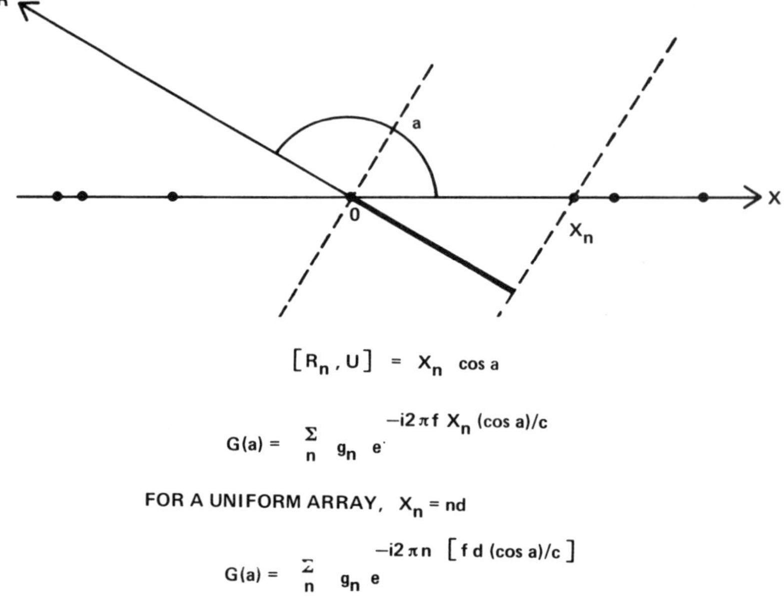

Figure 16. Line Array Beamforming

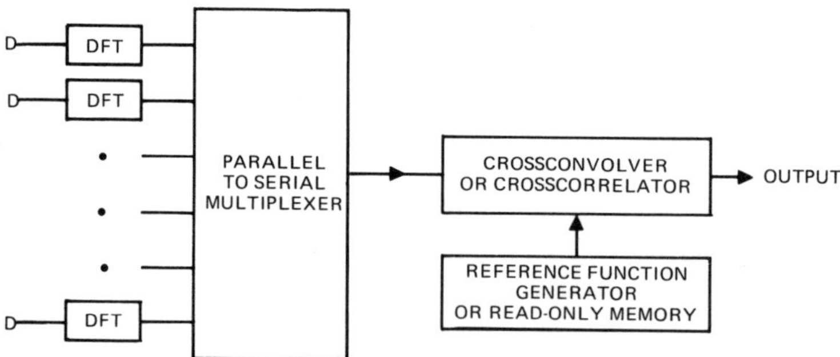

Figure 17. Frequency Domain Beamformer for Geometrically Spaced Line Array or Uniformly Spaced Circular Array.

If the kth beam is formed in the direction θ_k satisfying $\cos \theta_k = a\, r^k$, then the beamformer output is a crosscorrelation:

$$G(\theta_k) = \sum_n g_n\, e^{-[i2\pi da\, r^{n+k}]/\lambda}$$

$$= \text{aperiodic crosscorrelation of } g_n \text{ with } e^{-[i2\pi da\, r^n]/\lambda} \tag{44}$$

The analysis of the circular array beamformer is equally simple. Let the nth element in the circular array be located at $(x_n, y_n) = (R_0 \cos(2\pi n/N), R_0 \sin(2\pi n/N))$. The inner product of a unit vector in the θ direction with the nth element position vector is $R_0 (\cos \theta \cos(2\pi n/N) + \sin \theta \sin(2\pi n/N)) = R_0 \cos(\theta - (2n/N))$. If the kth beam is steered in the direction $\theta_k = 2\pi k/N$, then the beamformer output becomes a discrete circular convolution:

$$G(\theta_k) = \sum_{n=0}^{N-1} g_n\, e^{-i2\pi R_0 \cos((2\pi/N)(k-n))/\lambda} \tag{45}$$

The crossconvolution or crosscorrelation may of course be either performed directly with a CCD or surface acoustic wave crosscorrelator, or may be performed as a DFT, followed by a pointwise multiplication, followed by an inverse DFT.

If the tap spacings in the multiport convolver are allowed to be nonuniform, it can also perform the spatial processing required for a nonuniformly spaced line array: Simply make the tap spacing proportional to the element spacing in the array. This concept may be extended to provide beamforming for nonuniformly spaced two-dimensional arrays by mixing nonlinear chirps in triple product convolvers with nonuniformly spaced taps to provide the required beamforming phase shift. Equations (46-51) give the derivation of the required phase terms from the appropriate nonlinear chirps and tap positions and the resulting beamformer architecture is shown in Fig. 18.

$$G(u) = \sum_n g_n\, e^{\frac{-i2\pi}{\lambda}(x_n \cos u + y_n \sin u)} \tag{46}$$

Let

$$a(t) = e^{-iK_1 \cos K_2 t} \tag{47}$$

$$b(t) = e^{-iK_1 \sin K_2 t} \tag{48}$$

Then

$$G(u) = \sum_n = g_n \, a(t + z_n/c) \, a(t - z_n/c) \, b(t + w_n/c) \, b(t - w_n/c) \tag{49}$$

where the instantaneous look direction is $u = K_2 t$ and the tap positions are

$$z_n = (c/K_2) \cos^{-1} (x_n/M) \tag{50}$$

$$w_n = (c/K_2) \sin^{-1} (y_n/M) \tag{51}$$

M may be any number greater than $\max_n (\max (|x_n|, |y_n|))$

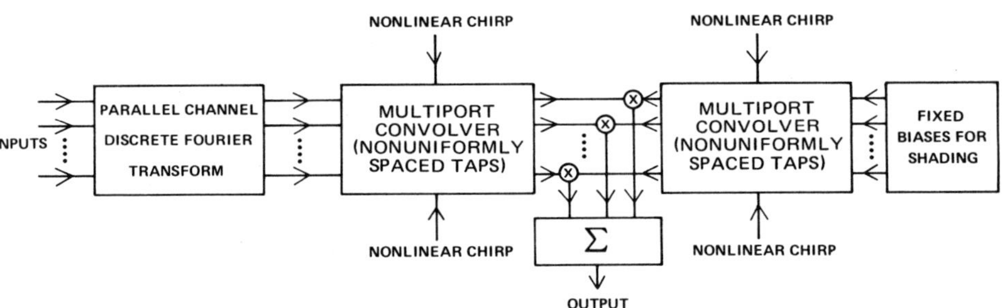

Figure 18. Architecture for nonuniform two-dimensional array beamformers

EXPERIMENTAL BACKGROUND AND CURRENT METHODS

Many types of transversal filter implementations may be easily accomplished using sampled data techniques [3]. Each offers some advantages and some disadvantages in any particular application, although they are all architecturally similar.

Digital correlators

The first device considered will be the Large Scale Integrated Circuit (LSI) version of a conventional discrete component digital correlator. In its simplest form a binary correlator consists of two shift registers, many exclusive ORs and a resistive summing network. Such a correlator can be assembled from conventional medium scale integrated circuit logic modules or can be obtained in LSI form.* The LSI unit shown in Fig. 19 is composed of two 64-bit shift registers, 64 exclusive ORs, and an analog network which sums the outputs of the exclusive ORs. Each shift register can be operated independently at rates up to 20 MHz; the outputs from corresponding stages in the two shift registers are the two inputs of an exclusive OR, so that the correlator output represents the number of agreements minus the number of disagreements for the contents of the two shift registers, and changes whenever new information is shifted into either register [13].

The binary versus binary digital correlator provides a direct solution when a matched filter is required for a binary code. More general signal processing tasks requiring multilevel versus multilevel convolution are more simply addressed via sampled analog transversal filters such as surface acoustic wave devices (SAWs) and charge coupled devices (CCDs).

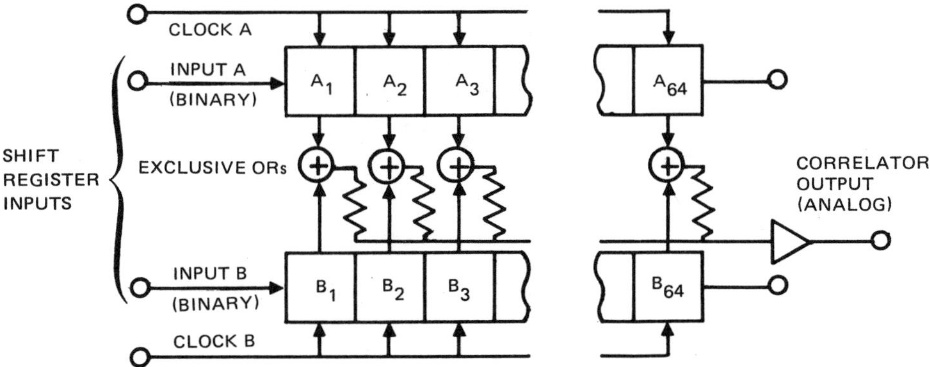

Figure 19. Binary Correlator

*TRW Systems Group, One Space Park, Redondo Beach, CA 90278.

SURFACE ACOUSTIC WAVE DEVICES

Surface Acoustic Wave (SAW) devices [14] can accept either analog or sampled analog input and the output is analog. A substrate of piezoelectric material is polished on one face and a pattern of aluminum or other conductor is deposited by photolithographic techniques. In its simplest configuration, sets of interdigitated finger electrodes are spaced at the sampling rate distance through the use of the relationship $d_s = C_R t_s$, where d_s is the tap spacing, C_R is the Rayleigh wave velocity, and t_s is the sampling increment. For a typical substrate of ST-cut quartz, $C_R \sim$ 3mm/μsec and for a typical sampling increment of 150 nsec, $d_s = 0.45$ mm and a 150 point DFT can be implemented by recirculating convolution in an active length less than 75 mm (3 in.) on a 100 mm (4 in.) substrate. A simple real transversal filter is shown in Figure 20 along with the method of establishing the tap weights. A CZT complex filter [15] is shown packaged in Figure 21. A similar device has been used as the premultiplication and postmultiplication reference function generator and double-balanced mixers have been employed as the multipliers.

Physical limitations on the size of available substrates limit the available DFT size. For an active length of substrate $L_s = 150$ mm and a surface wave velocity $C_R = 3$ mm/μsec the maximum transform size N_{max} for a data rate F_s is approximately $N_{max} = L_s F_s / C_R \approx 50 F_s$ with F_s in megaHertz. Current SAW transversal filters operate at sample rates from 1 to 100 MHz with carrier frequencies from 5 to 500 MHz with typical fractional bandwidths of about 10%. A surface wave device module which implements the triple product convolution equation (38) has been built by Reeder [16] at United Technologies for the Naval Undersea Center. A schematic is shown in Figure 22. Each diode acts as a mixer (multiplier) whose gain is controlled by the bias current. Present research is addressing the problem of linearizing the relationship between bias current and tap gain.

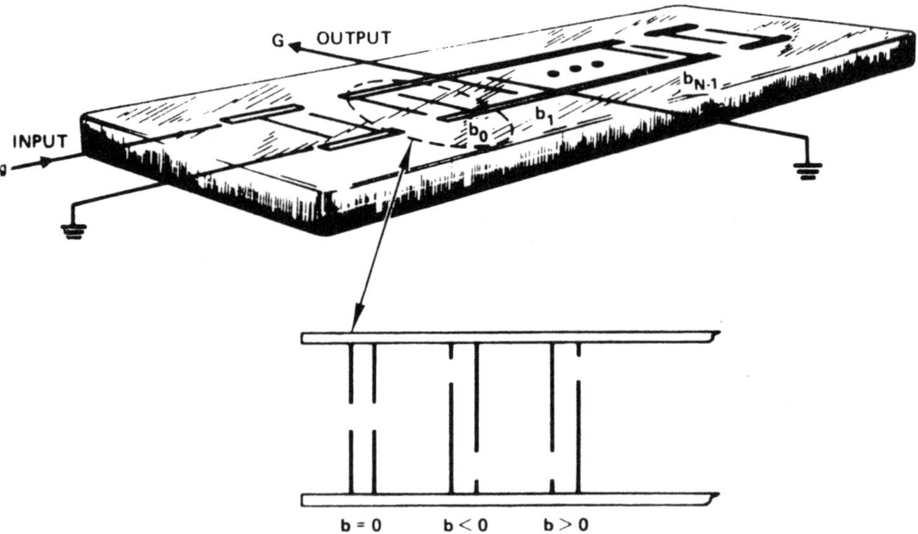

Figure 20. Surface Wave Transversal Filter

Figure 21. 32 Complex Tap SAW CZT Filter

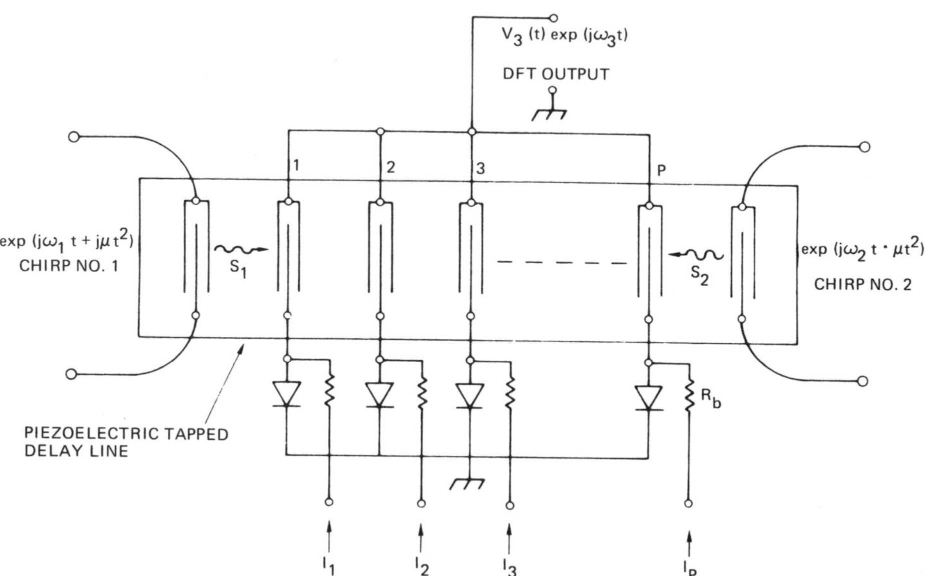

Figure 22. Diode-Correlator DFT Module

Charge Coupled Devices

CCDs are sampled data analog circuits which can be fabricated by Metal Oxide Semiconductor (MOS) technology as LSI components [17]. As such they are directly compatible with other MOS circuits. Current CCD transversal filters have operated as video devices with sample rates up to 5 MHz. CCDs operate by the manipulation of injected minority carriers in potential wells under MOS capacitors and thus behave as capacitive reactances with low power dissipation. However, since the potential wells which contain the minority carriers also attract thermally generated minority carriers, there is a maximum storage time for the analog signal which depends on the dark current associated with the temperature of the silicon. Under normal conditions at room temperature, dark currents are tens of nAmps/cm^2 and storage times of hundreds of milliseconds can be achieved.

There are many ways in which unidirectional charge transfer can be achieved. The first developed was a three-phase clocking structure which is illustrated in the transversal filter of Figure 23. The three electrode CCD structure is planar, much like the SAW devices, and the direction of charge propagation is determined by the sequence of potentials applied to the three electrodes. Unfortunately, if the minority carriers are allowed to collect at the semiconductor-oxide boundary, poor charge transfer efficiency will result due to minority carriers getting caught in trapping sites. This means that the CCD will behave nonlinearly unless there is sufficient propagating charge present to fill all of the traps. By biasing the operating condition of the CCD so that about 10% of the dynamic range is used for the injection of a "fat zero," the traps are kept continuously filled and the device has over a 60 dB dynamic range. In practice, a video signal representing the signal to be processed is added to a fixed bias somewhat larger than one-half of the peak-to-peak value of of the signal. Since the effective storage time of the device is long relative to the time required to execute a convolution, CCDs can be considered to be interruptible signal processors and as such are more compatible with the executive control required for signal processing. A 64-point CCD filter with discrete cosine transform sine and cosine chirps are shown in Figure 24. This chip was developed by Texas Instruments for the Naval Undersea Center for image processing via the DCT.

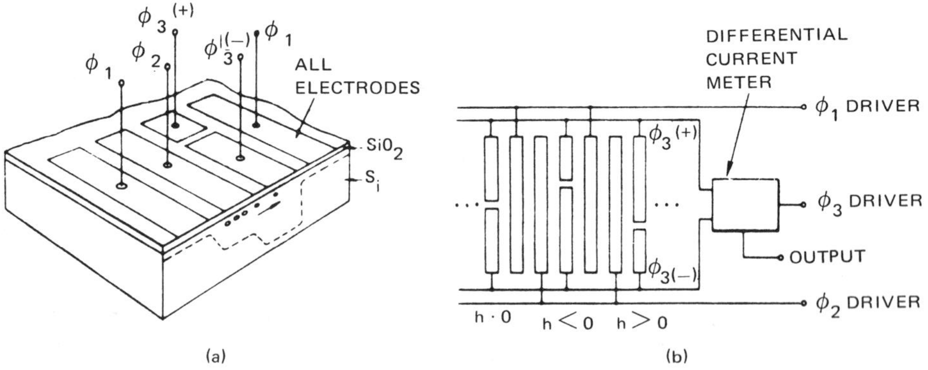

Figure 23. Schematic of the Sampling, Weighting, and Summing Operation

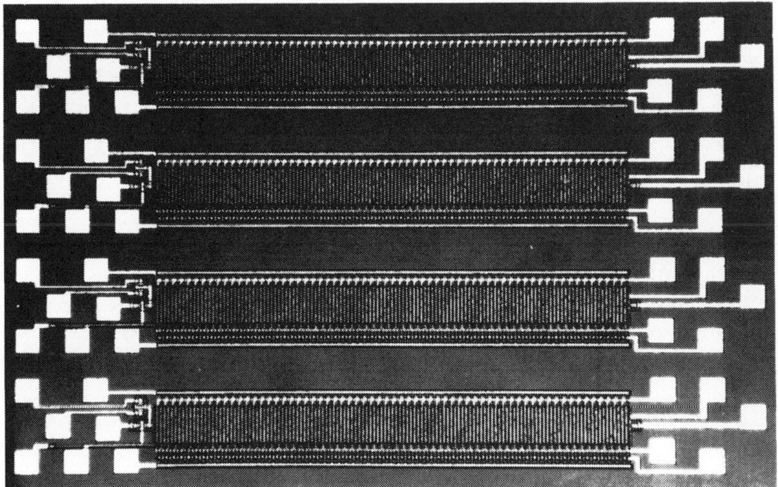

Figure 24. 64-Point CCD Filters

A hybrid binary-analog CCD correlator module with 32 analog taps [18] has been made for the Naval Undersea Center by General Electric and is shown diagrammatically in Figure 25. The maximum clock rate of the module is 4 MHz. These modules may be cascaded to increase the correlation length to provide an analog versus multilevel cross correlator with analog output. The correlator module

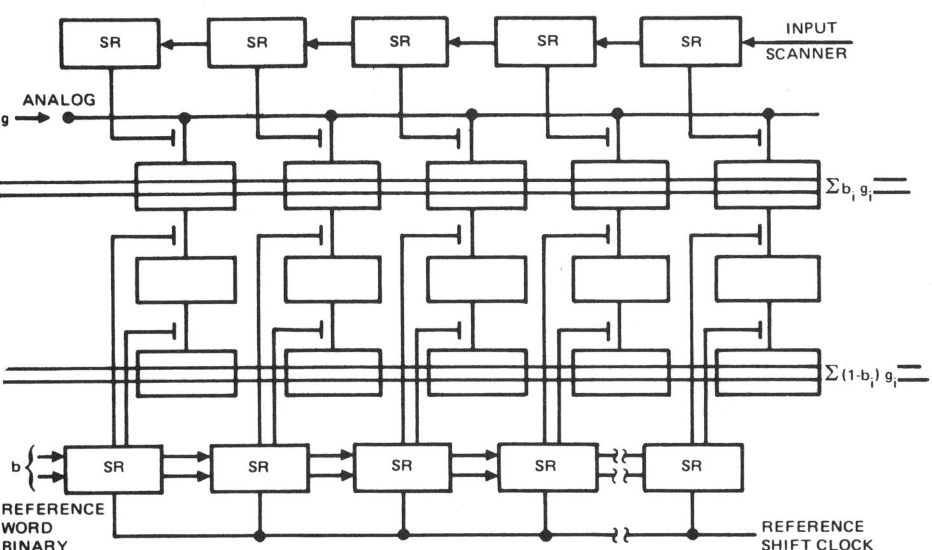

Figure 25. Schematic Diagram of a Binary-Analog CCD Correlator Module

uses charge propagation through only 3 stages since the CCD is arranged in 32 stages of 3 samples each. By transverse shifting of the charge instead of longitudinal shifting, charge transfer inefficiency degradation is avoided and the modules can be cascaded to very large sizes. However, with this configuration, dark current nonuniformity must be controlled to prevent time variable pattern noise from contributing to the output. Systematic errors in the tap weights can be measured and stored and the analog signal corrected for the measured variation in uniformity before it is stored in the CCD registers.

Devices using charge storage on MOS capacitors together with field effect transistor switches can be configured as serial access or random access memories. One such analog permuter memory, similar to Fig. 4, was developed for NUC by the Reticon Corp., and has been used together with a CCD transversal filter to implement the DCT [19].

ADVANCED METHODS

Digital correlation using multiplication by binary convolution

It has long been realized that the coefficients of the product of two polynomials is the convolution of their respective coefficients. This is in fact the basis for the convolution using the Z-transform or generating function.

$$A(z) = \sum_n a_n z^n$$

$$B(z) = \sum_k b_k z^k$$

$$A(z)\,B(z) = \sum_n \sum_k a_n b_k z^{n+k} = \sum_s \left(\sum_{n+k=s} a_n b_k \right) z^s = \sum_s (A * B)_s\, z^s$$

This equation may be applied to binary multiplication by letting $z = 2$ in the above identities.

If

$$A = \sum_{n=0}^{N_1-1} a_n 2^n$$

and

$$B = \sum_{k=0}^{N_2-1} b_k 2^k$$

then

$$AB = \sum_{k=0}^{N_1+N_2-1} [\sum_n a_n b_{s-n}] \, 2^s = \sum_s c_s 2^s \tag{52}$$

If the digits a_n, b_k take on the values of 0 or 1, this permits the multiplication of the two integers A, B where A may be $0, 1, \ldots 2^{N_1}-1$ and B may be $0, 1, \ldots$ $2^{N_2}-1$. In order to perform signed arithmetic, we may use symmetric offset binary representation where the digits a_n and b_k take the values ± 1. Then the range of values possible for A and B is $\pm 1, \pm 3, \ldots \pm(2^n-1)$ where n is the number of bits in the corresponding number. The convolution required in Eq. (52) may be performed in analog form by a correlator and the subsequent summation performed digitally after analog to digital conversion as shown in Figure 26. The number of bits of resolution required for the A/D conversion is only $\log_2 (1 + \min (N_1, N_2))$.

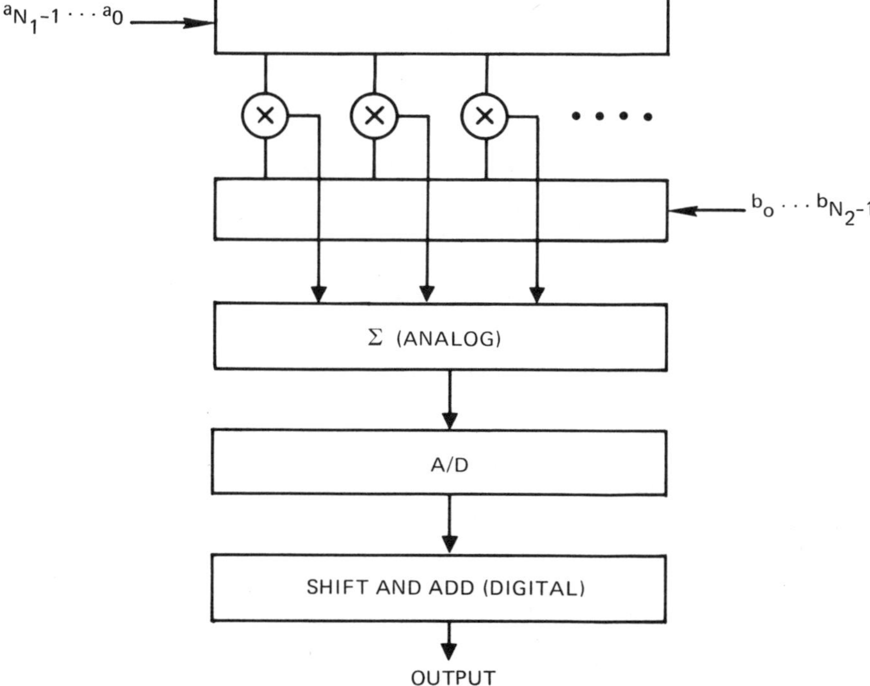

Figure 26. Analog Convolver Used as a Digital Multiplier

Real-time adaptive filters

An extremely powerful class of spectral estimators and noise cancellers may
be based on adaptive filters using the Widrow LMS (Least Mean Squares) algorithm.
The computations required by the algorithm are given in equations 53 and 54,
where W_j is the weight vector at time j, X_j is the input vector at time j, and d_j is
the training signal at time j. The difference e_j is a feedback "error" signal. Different applications of the adaptive algorithm differ mainly in the way in which the
training signal is provided. An adaptive linear combiner and configurations for
noise cancelling and adaptive line enhancing are shown in Figures 27, 28 and 29. [20].

$$e_j = d_j - (W_j, X_j) \qquad \text{(generation of feedback signal)} \qquad (53)$$

$$W_{j+1} = W_j + 2\mu \, e_j X_j \qquad \text{(tap update)} \qquad (54)$$

In adaptive filtering applications, $X_j = \text{col} (x_j, j_{j-1}, \ldots x_{j-N+1})$.

Note that at each time, N multiplications are needed to generate the feedback
signal, and an additional N multiplications are needed to update the tap weights.
These computations are not well suited to implementation via an FFT. Even the

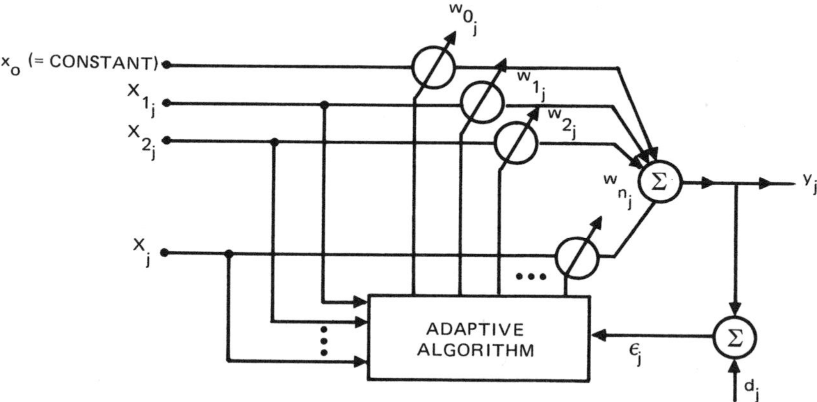

Figure 27. The Adaptive Linear Combiner

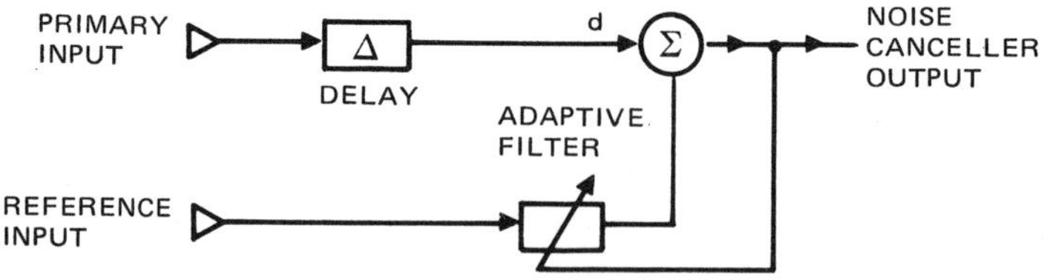

Figure 28. Adaptive Noise Canceller with Delay in Primary Input Path

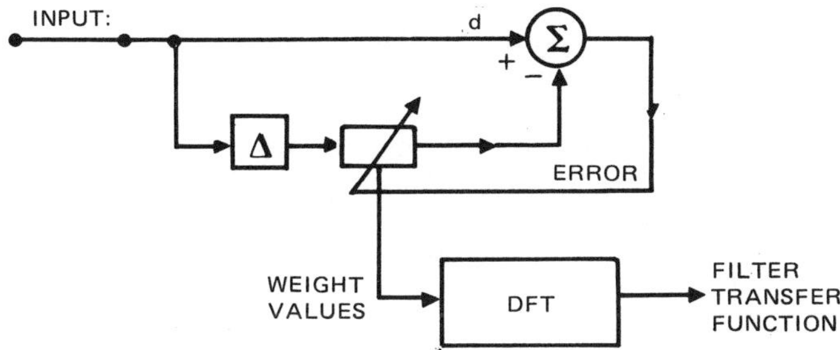

Figure 29. The Adaptive Link Enhancer

generation of the feedback signal is not a convolution because the weights change
from one time sample to the next. Present research is directed toward developing
parallel architectures to provide the large number of multiplications required for
real time adaptive filtering applications [21].

FUTURE TRENDS

It is expected that future signal processing architecture development will provide efficient processing with a greater degree of flexibility/adaptivity. The two
most immediate needs are for effective data compression for data with unknown
statistics, and beamforming for a randomly time-varying array.

REFERENCES

[1]   Bellman, R. Introduction to Matrix Analysis, McGraw-Hill Book Company,
       Inc., New York, 1960 (pp. 197-198).
[2]   Ray, W. D. and Driver, R. M. Further Decomposition of the Karhunen-
       Loeve series Representation of a Stationary Random Process, IEEE Trans-
       actions on Information Theory, Vol. IT-16, pp. 663-668, Nov. 1970.
[3]   Squire, W. D., H. J. Whitehouse, and J. M. Alsup, Linear Signal Processing
       and Ultrasonic Transversal Filters, IEEE Transactions on Microwave Theory
       and Techniques, Vol. MTT-17, pp. 1020-1040, Nov. 1969.
[4]   Bluestein, L. T., A Linear Filtering Approach to the Computation of the
       Discrete Fourier Transform, IEEE Transactions on Audio and Electro-
       acoustics, Vol. AU-17, pp. 86-92, 1969.
[5]   Rader, C. Discrete Fourier Transforms When the Number of Data Samples
       is Prime, Proc. IEEE, Vol. 56, pp. 1107-1108, 1968.
[6]   Gauss, Carl Friedrich, Disquisitiones Arithmeticae, Translated by Arthur A.
       Clark, Yale University Press, New Haven, 1965 (pp. 29-37).
[7]   Ahmed, N. and K. R. Rao, On Image Processing and a Discrete Cosine
       Transform, IEEE Transactions on Computers, Vol. C-23, pp. 90-93, 1974.

[8A] Jain, A.K., Image Restoration By An Operator Factorization Of The Point Spread Function And The Image Covariance Function, Tech. Rept. 75-001, Oct. 1975, Dept. of Electrical Engineering, State University of New York at Buffalo (also submitted to IEEE Transactions on Computers).

[8B] Jain, A.K., The Sinusoidal Family of Orthogonal Transforms, to appear

[9]   Speiser, J. M., High Speed Serial Access Implementations for Discrete Cosine Transforms, Naval Undersea Center TN 1265, 8 Jan. 1974.

[10] Cooley, J. W., P. A. W. Lewis, and P. D. Welch, Historical Notes on the Fast Fourier Transform, IEEE Transactions on Audio and Electroacoustics, Vol. AU-15, pp. 76-79, 1967.

[11] Gold, B. and T. Bially, Parallelism in Fast Fourier Transform Hardware, IEEE Transactions on Audio and Electroacoustics, Vol. AU-21, no. 1, pp. 5-16, Feb. 1973

[12] Williams, J. R., Fast Beam-Forming Algorithm, Journal of the Acoustical Society of America, Vol. 44, pp. 1454-1455, 1968.

[13] Alsup, J. M., D/A Correlator Structures, Proceedings of the Ocean '73 Symposium, IEEE Publication 73-CHO 774 OCC, pp. 112-117.

[14] Hays, R. M. and C. S. Hartmann, Surface Acoustic Wave Devices for Communications, Proceedings of the IEEE, Vol. 64, No. 5, May 1976, pp. 652-671.

[15] Alsup, J. M., R. W. Means, and H. J. Whitehouse, Real-Time Discrete Fourier Transforms Using Surface Acoustic Wave Devices, Proc. IEE International Specialist Seminar on Component Performance and Systems Applications of Surface Acoustic Wave Devices, Aviemore, Scotland, Sept. 24-28, 1973.

[16] Reeder, T. M., J. M. Speiser, and H. J. Whitehouse, Real-Time Discrete Fourier Transforms Using a Programmable Diode-Convolver Module, Proceedings of the IEEE Microwave Symposium, pp. 365-367 (May 1975).

[17] Buss, D. D., D. R. Collins, W. H. Bailey, and C. R. Reeves, Transversal Filtering Using Charge Transfer Devices, IEEE Journal of Solid State Circuits, Vol. SC-8, pp138-146, 1973.

[18] Tiemann, J. J. et al, Intracell Transfer Structures for Signal Processing, IEEE Transactions Electron Devices, Vol. ED-21, pp. 300-308, 1974.

[19] Kerber, G. Analog Implementation of the Prime Cosine Transform, to appear in GOMAC '76 digest.

[20] B. Widrow, J. R. Glover, Jr., J. M. McCool, J. Kaunitz, C. S. Williams, R. H. Hearn, J. R. Zeidler, E. Dong, Jr., and R. C. Goodwin, Adaptive Noise Cancelling: Principles and Applications, Proceeding of the IEEE Vol 63, No. 12, December 1975.

[21] Whitehouse, H. J. and J. M. Speiser, Signal Processing with Combined CCD and Digital Techniques, to appear in Proceedings of the IEEE/IEE Conference on the Impact of New Technologies on Signal Processing (Aviemore, Scotland, Sept. 20-24, 1976).

SURFACE-ACOUSTIC-WAVE DEVICES FOR SIGNAL PROCESSING

G. Manes, L. Masotti

I.R.O.E., Consiglio Nazionale delle Ricerche
Facoltà di Ingegneria, Università di Firenze
Italia.

ABSTRACT. A brief review of the fundamental principles
of the design of Surface-Acoustic-Wave(SAW)transversal
filters is reported and SAW devices, which are relevant to signal processing,are presented.

1. INTRODUCTION

Since the development of the first transducers capable
of launching or detecting acoustic waves propagated
along the surface of a piezoelectric substrate,scien-
tists working in the field of signal processing envi-
sioned the potential applications of a delay line offe-
ring a new fundamental degree of freedom: the possibility of a continuous access to the traveling signal in
any point of the propagation path. Signal processing devices had previously been implemented by tapping bulk
acoustic waves propagated in suitable three-dimensio-
nal structures [1]; however, the technique of tapping
surface acoustic waves by transducers consisting of
metal strips directly deposited on the propagation surface [2] led one to conceive the capability of achieving the same filtering functions with planar structures, whose fabrication required techniques analogous to
those employed for integrated circuits, and thus appeared easy and highly reproducible.

The interest was first attracted by the possibility of implementing delay lines having dispersive characteristics; by arranging on the propagation surface a planar array of suitably spaced selective transducers, one could be able to obtain a delay variable with frequency depending only on the transducer geometry. This new technique of achieving a dispersive characteristic was first devised by Rowen [3]. Successively a great amount of work was devoted to the implementation of surface-acoustic-wave (SAW) dispersive signal processors, in particular to meet the signal-processing needs of matched filters for radar systems [4].

The further basic degree of freedom offered by SAW tapping was later pointed out in the possibility of weighting the amplitude of the detected signal simply by varying the aperture of the transducer parallel to the acoustic wavefront. Based on this property, the possibility was realized of weighting the frequency response of SAW devices by implementing arrays of transducers with suitably graded apertures [5]. Planar transducers having two-dimensional structures were so designed, consisting of arrays of parallel electrodes (fingers) whose spacing along the propagation direction determined the time delay of the generated/detected signal outputs and whose aperture normal to the propagation direction determined the corresponding weights [6,7,8,9]. It is largely the many geometrical degree of freedom available for designing arrays of transducers that make possible the variety of signal processing functions that can be achieved with SAW devices.

The design procedure is based on the principle that discrete tapping transducers perform a sampling of the travelling surface acoustic wave, so that the desired filter transfer-function can be synthesized by spatial transducer geometries corresponding to a suitably time-sampled version of the filter impulse response. This synthesis procedure, referred to as " transversal filtering" is analogue to that used in digital filters design. Of course digital techniques based on integrated circuits and,more recently, analog techniques based on charge coupled devices offer cur-

rent and potential competition. However, with bandwidths of several hundred megahertz and dynamic ranges
extending towards 100 dB passive SAW devices form strongly competitive signal processing in the IF frequency
band.

2. INTERDIGITAL TRANSDUCERS

An acoustic surface-wave, often referred to as Rayleigh
wave [10], is a mechanical disturbance localized on the
free surface of an elastic material. Rayleigh waves have a certain resemblance to waves on fluid surfaces: in
both cases the particle motion occurs in elliptical or-
bits in a plane perpendicular to the surface and parallel to the direction of propagation. In piezoelectric
crystals the elastic wave is associated with an electric field which extends into the medium as well as
outside the medium with a decay rate of an acoustic
wavelength [11]. Surface-waves can be generated by
applying an alternate potential difference between two
metal plates deposited on opposite sides of a piezoelectric substrate (Fig.1) [12]. Greater efficiency is

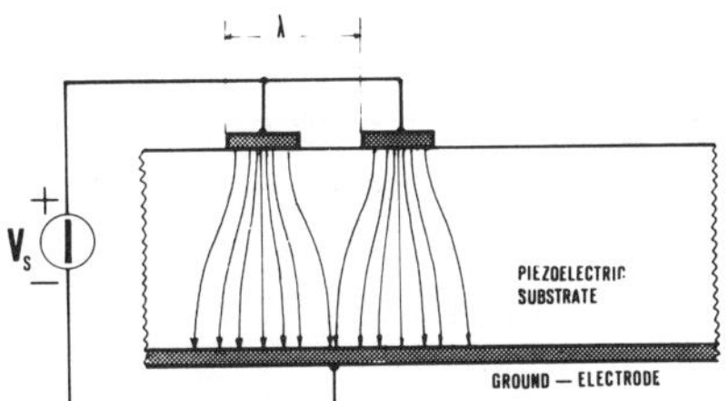

Fig.1. Periodic electric field produced by grating
       transducers with a ground electrode below the
       substrate.

achieved, however, by applying an alternate voltage,
parallel to the desired propagation direction, between
two metal strips, formed normally photolithographi-
cally in an aluminum film less than 0.5 μm thick for
frequency between 100 MHz and 1 GHz. (Fig.2). In fact
in this case the electric field is mainly localized

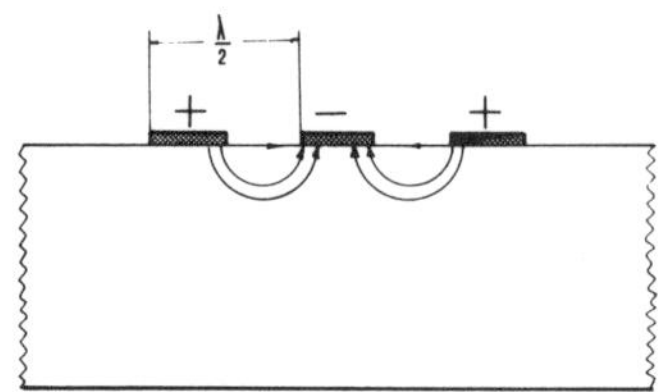

Fig.2. Periodic electric field produced by an interdi-
       gital transducer.

in the surface region interested by the elastic wave.
This type of transducer is commonly referred to as
" interdigital" (IDT) [11] and is the most popular
means of generating (or, conversely, detecting) an elastic surface-wave.
        For effective excitation at a given frequen-
cy f_o the spatial period of the transducer has to be
chosen equal to the corresponding Rayleigh wavelength
$\lambda = v/f_o$, where v is the velocity of the SAW, typically, 3.10^6 mm/s, [13,14].
        The complete theoretical solution of acoustic
-wave propagation on arbitrary anisotropic piezoelectric
media is accomplished by solving the continuum equations of motion together with Maxwell's equations under
the " quasi-static"assumption, the strain-mechanical
displacement relations, the piezoelectric constitutive
relations, and the appropriate boundary conditions
[19,20]. Bulk-wave modes are generally associated with
Rayleigh wave [15]. The main bulk propagation mode is
longitudinal, and has propagation velocity about twice
greater than that of the surface-wave. The presence of
bulk waves is generally ignored in the design of interdigital transducers and taken in account only as spurious signals to be minimized in SAW devices. Further
propagation modes can be predicted, whose effect, however, can be made negligible by a proper choice of the
crystallographic orientation of the substrate.
        The electric field associated with a surface-

-wave has two components, one normal and one parallel
to the propagation direction. The decay of these components as a function of the distance from the surface,
expressed in wavelengths, is shown in Fig.3. Outside
the solid the electric fields normal and parallel to
the surface decay exponentially with increasing distance from the surface.

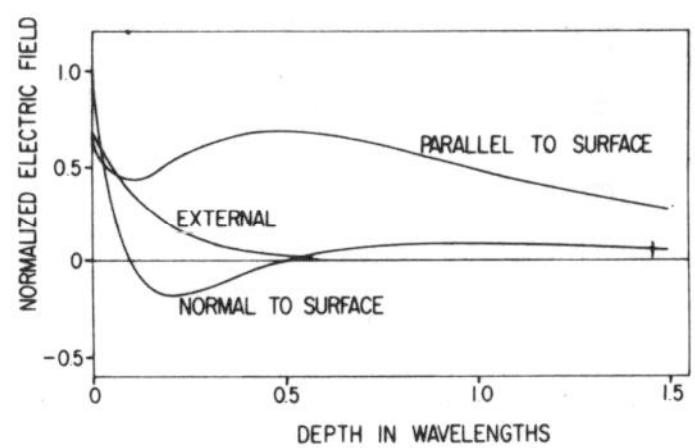

Fig.3. Decay of the electric field associated with
       surface-wave vs. distance from the surface ex-
       pressed in wavelengths (after White [15])

        Because of the two-dimensionality of the elec-
tric field and of the complexity of the behaviour of a
piezoelectric anisotropic substrate, the use of the
piezoelectric equations is extremely laborious and, from
an engineering viewpoint, unrewarding.
        The problem is very simplified by introducing
first-order approximations. The system can be reduced
to a one-dimensional configuration by assuming that
the electric field is merely parallel to the propagation
vector (" in-line" model) or normal (" crossed-field"
model) [16,17]. This second model is most popular,since
it leads to predictions in good agreement with experimental measurements for materials having a strong
electro-mechanical coupling. In the following we will
refer essentially to the crossed-field approximation.
        The enormous simplification introduced by the
one-dimensional model lies in the fact that the acoustic
behaviour of the medium can be described electrically
in terms of an equivalent circuit, originally introduced by Mason for bulk-resonant transducers [8]. Consider

a delay-line, consisting of a pair of identical transducer, each composed of N periodic sections of period L (Fig.4). For each periodic section, the interaction between the electric and acoustic portions is described

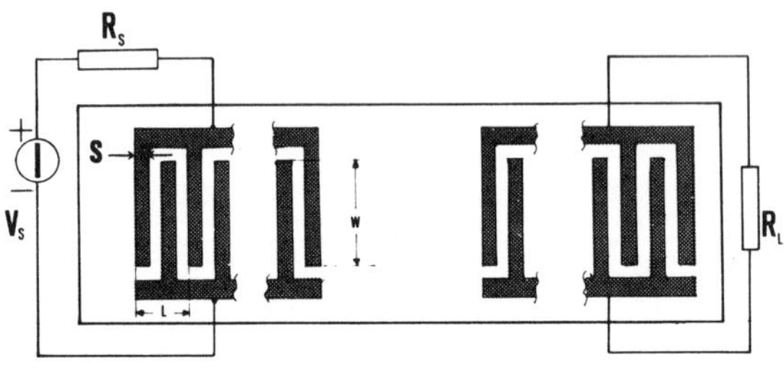

Fig.4. Configuration of two interdigital transducers
        arranged in a transmitting-receiving pair.

by the circuit shown in Fig.5. The circuit equivalent to the acoustic medium is coupled to the electrical

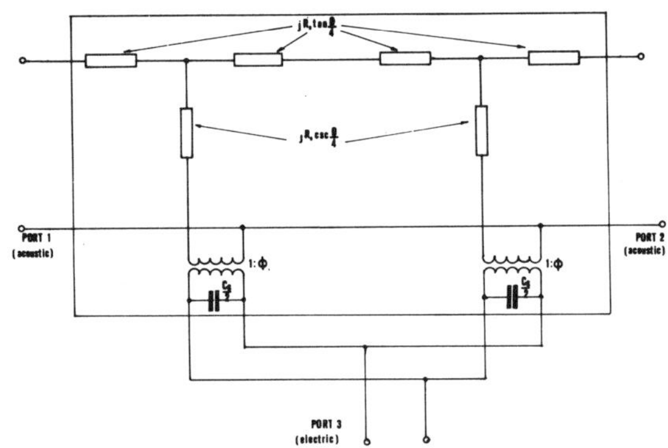

Fig.5. Equivalent circuit of one section of a periodic
        interdigital transducer.

circuit by a transformer, whose ratio ϕ is proportional to the acoustic aperture w of the finger and depends on the piezoelectric coupling constant, k^2, characteristic of the material, which for surface waves is taken $k^2 = 2\Delta v/v$, where $\Delta v/v$ is the fractional velocity change resulting from placing an infinitesimally thin,

massless perfect conductor on the substrate surface
[19,20]. $C_S/2$ is the capacitance between two adjacent
fingers, which is proportional to the finger aperture,
w and depends on the metallization ratio S/L [18].
R_O is the electric equivalent of the characteristic
impedance of the acoustic medium; $f_O = v/L$ is the "synchronism frequency" given by the ratio of the surface-
-wave velocity and the length of a periodic section.
        An interdigital transducer is therefore re-
presented by a system having three ports, one electric
and two acoustic. When an alternate potential difference is applied at the electric port, an acoustic
wave is launched, for symmetry, in opposite direction
away from the transducer.
        More periodic interdigital sections are re-
presented by the combination of more identical circuits
such as that in Fig.5, connected acoustically in cascade (to represent generation from fingers) and electrically in parallel. The admittance and scattering matrices of a transducer of N sections can be so calculated. In Fig.6 is shown the equivalent circuit of the

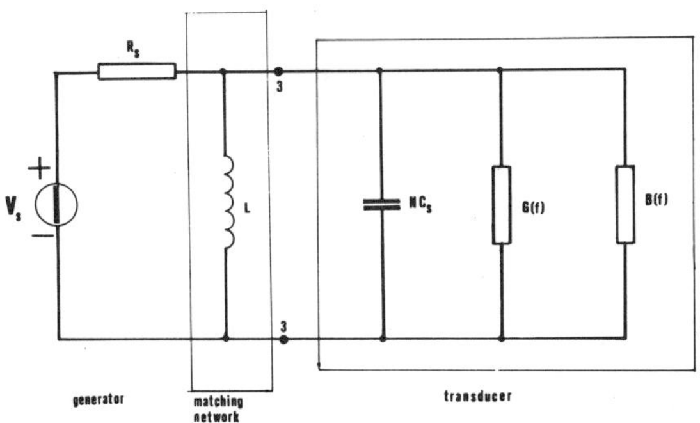

Fig.6. Equivalent circuit of a matched transducer at
       the electric port.

transducer at the electric port. $C_T = NC_S$ is the total transducer capacitance; G(f) is the "radiation
conductance" , which represents the flow of acoustic
power launched from the transducer; the susceptance
B(f) accounts for the elastic energy storage. At the

synchronism frequency [16]

$$G(f_o) = 8\, k^2 f_o C_S N^2 \tag{1}$$

$$B(f_o) = 0 \tag{2}$$

The radiation conductance results therefore proportional to the finger aperture w and to N^2.

For frequencies near the acoustic synchronism

$$G(f) = G(f_o) \left(\frac{\sin x}{x}\right)^2 \tag{3}$$

$$B(f) = G(f_o)\, \frac{\sin 2x - 2x}{2x^2} \tag{4}$$

$$x = N\pi\, \frac{f - f_o}{f_o}$$

It is seen from (3) that $G(f)$ is maximum at $f = f_o$ and its fractional bandwidth is inversely proportional to the number N of periodic interdigital sections. Since, as said above, $G(f)$ represents the flow of generated acoustic power, its frequency dependence may be regarded as the transducer bandshape. This is just the fundamental frequency response that can be expected if, as discussed further, the " fingers" of the transducer are considered as spatial samples at intervals $1/2f_o$ of a pulse of duration N/f_o.

From the equivalent circuit in Fig.5 it can be recognized that matching between the electrical and acoustic networks for efficient transducer operation is simply achieved with a parallel inductor that resonates the transducer capacitance at the synchronism frequency f_o. At this frequency the tuned transducer appears as a resistor of resistance $1/G(f_o)$ that must be made equal to the generator internal resistance R_S in order to minimize the conversion loss. By inspection of (1), this condition can be achieved by suitably choosing the transducer aperture w, that determines the capacitance

C_S between the fingers. A radiation resistance of 50 Ω can be achieved at f_o by choice of aperture in the range 50 λ to 100 λ depending on the piezoelectric material.

The overall frequency response of the matched transducer results the product of the acoustic band-shape measured by (3), and the bandshape of the resonant matching circuit. For small values of N, and hence low values of C_T, the electrical bandshape is predominant, so that the transducer fractional bandwidth is determined by the electrical Q_e of the matching circuit, i.e.

$$\frac{\Delta f}{f_o} \approx \frac{1}{Q_e} = \frac{G(f_o)}{2\pi f_o N C_S} = \frac{4}{\pi} k^2 N \tag{5}$$

Conversely, for large values of N the transducer bandwidth follows the frequency variation of $G(f)$, and the acoustic Q_a of the periodic structure predominates:

$$\frac{\Delta f}{f_o} \approx \frac{1}{Q_a} = \frac{1}{N} \tag{6}$$

The maximum bandwidth is achieved when the electric and the acoustic bandwidths are equal. By equating (5) and (6), this condition is met for

$$N = N_{opt} = \frac{\sqrt{\pi}}{2k} \tag{7}$$

This result shows that the value N_{opt} providing maximum transducer bandwidth, given roughly by N_{opt}^{-1}, is a characteristic of the piezoelectric material. For lithium niobate and quartz N_{opt} is about 4 and 20 respectively. The transducer also reflects surface waves, and the corresponding " triple transit" signal cannot be minimized by matching because the IDT is a three-port device.

The Mason model has been found to work well in predicting the properties of interdigital structures with constant finger-spacing and constant finger-length. As already mentioned, interdigital structures having different finger-lenghts must be sometimes considered.

To analyze this structure having different finger-lengths
the array is broken into imaginary strips parallel to
the propagation direction [8]. Each strip is indipendently considered as a constant-length-finger transducer. The sections where the fingers do not overlap are
treated as an acoustic delay. Those sections in the
strip which have finger overlap have the equivalent
circuit of Fig.5. The acoustic terminals of the individual circuits are cascaded in series to represent each
strip. The electrical terminals of each section are all
connected in parallel to the driving source. This approach
can give good results, but, in some cases, a considerably more complicated analysis would have to be used because of diffraction effects of the wavefront [21,44]
and other second order effects [39,40] arising because
of both electrical and mechanical discontinuities at
the edges of the electrodes.

The description of the transducer equivalent-
-circuit using the Mason model works well also for finger spacing with a smooth variation and constant width-
-to-gap ratio over the whole transducer. A generalized
circuit model combining the advantage of the three-port
network and spatial harmonic analysis was recently
introduced [18].

3. SAW DEVICES

3.1. Design by synchronous-frequency sampling

The idea that is on the basis of SAW device design is
that the transfer function of an interdigital array can
be determined by identifying the array configuration
with the impulse response of the device. In fact, when
the array is excited by an impulse, the output consists
of a sequence of pulses that is the time-image of the
spatial configuration of fingers. Conversely, when an
impulse, propagating along the line, passes an array
of interdigital electrodes, the resulting response is a
set of samples reproducing in time the spatial configuration of the array. The discrete form of this impulse-
-response corresponds to a sampled version of a continuous impulse-response. On the basis of this model,

procedures were developed [8,9] that, for a given impulse-response, provided the criteria of designing the interdigital structure synthesizing this impulse-response in a sampled form.

In order to synthesize samples of different magnitude tap weights were implemented directly by means of the interdigital structure by grading the overlap length of each finger with respect to the adjacent ones [8,18,22], to the first order of approximation the signal generated or detected from this finger results proportionally weighted. These structures are briefly referred to as " apodized" . Amplitude and/or frequency modulated impulse-responses were so synthesized merely by designing planar configurations of tapping fingers, whose separation along the surface-wave propagation direction controls the tap delays and whose aperture normal to the propagation direction controls the tap weights. Obviously, only impulse-responses of finite duration can be synthesized by SAW devices.

Consider first the case that the device characteristics are assigned in the time-domain. If the filter impulse-response h(t) has constant carrier-frequency f_0 the design procedure consists in sampling h(t) at intervals multiple of $\tau = 1/2f_0$ and determining the samples amplitudes

$$A_n = h(\frac{n}{2f_0}) \qquad n = 1,2,\ldots,\underline{N} \qquad (8)$$

To the first order of approximation the impulse-response of an array of fingers of interdigital spacing $v/2f_0$ and overlap-lengths proportional to the amplitudes A_n is given by

$$h_S(t) = \sum_{1}^{\underline{N}} A_n \delta(t - \frac{n}{2f_0}) \qquad (9)$$

The corresponding transfer function is

$$H_S(f) = \sum_{1}^{\underline{N}} A_n e^{-j\pi fn/f_0} \qquad (10)$$

This function is periodic with respect to frequency with
period $2f_o$, and consists of the periodic repetion of
the Fourier transform $H(f)$ of $h(t)$. Since the fractional
bandwidths of interest are generally small, the distorsion of the desired frequency response caused by overlapping of adjacent responses is negligible.

When $h(t)$ is a frequency-modulated waveform,
the sampling rate, as mentioned above, varies with the
instantaneous frequency. The investigation of the resulting transfer function shows that also in this case
spectral aliasing is avoided for a large range of fractional bandwidths [9].

When conversely the filter characteristics
are specified in the frequency domain, even in numerical
or graphical form, the synthesis problem consists in
determining a frequency function of the form

$$F(f) = \sum_{1}^{N} \alpha_n e^{-j\pi fn/f_o} \tag{11}$$

that approximates the filter transfer function in the
assigned band. The degree of approximation depends
on the number $\underline{N}$ of components. The impulse-response
has finite duration, and contains $\underline{N}$ samples of amplitude α_n at times $n/2f_o$.

SAW devices, however, require two separate
transducers, one for lauching the elastic wave along
the substrate surface, the other for tapping this
elastic wave. In the time domain, the overall response
of the device results from the convolution of the impulse responses of the transmitting and the receiving
transducers; in the frequency domain, the response is
the product of the respective transfer functions. If
one of the transducers is designed so as to provide
the desired transfer function, the transfer function
of the other must be nondistorting in amplitude and
phase over the operation bandwidth B. This result is
usually achieved using an interdigital array having
uniform-length fingers, corresponding to the samples
of a pulse of frequency f_o and duration suitably less
than $1/B$.

3.2. Signal processors

Considerable interest has been aroused in the potential
signal processing functions offered by SAW devices.
The improved key signal processing functions have included simple delay; store of information; frequency filtering; waveform generation; correlation and matched
filter reception and spectrum analysis, with associated Fourier domain processing [26].

3.2.1. Delay lines

The use of acoustic waves with their low velocity (relative to electromagnetic waves) has been a feature
of delay line developments. Delay-line for fixed delay
was presented in section II (Fig.4), variable delay
may be achieved discretely or continuously.Discrete
variable delay is obtained by switching between taps
of a typical delay line. Continuous variable delay
can be achieved using correlators (Section 3.2.5.),
disersive delay lines with a frequency variable local
oscillator [25], analogue chirp-transformation [26].

3.2.2. Oscillators

A SAW delay-line oscillator consists of a SAW delay
line with an amplifier providing positive feedback
from the delay-line output transducer to the delay-
-line input transducer (Fig.7) [27,28,29]. After

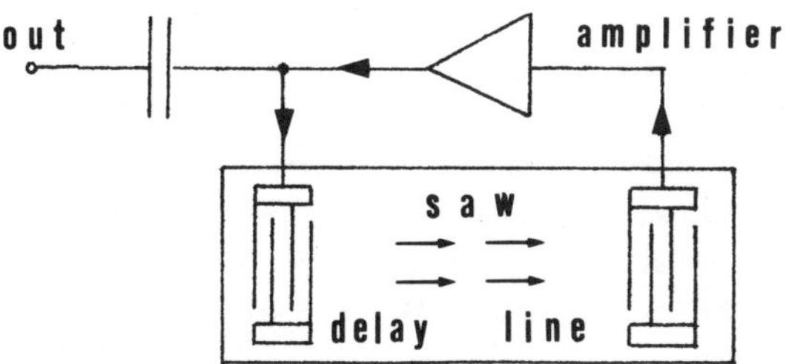

Fig.7. Scheme of a SAW delay-line oscillator.

oscillator switch-on the amplitude of the oscillation
builds up until it is limited by the non-linearity of
the amplifier.

Advantages attributed to the SAW oscillator include: operation up to low gigahertz frequencies without multipliers, freedom from satellite modes, rugged, modulable, the usual advantage of SAW devices.

The allowed oscillation frequencies of a delay line oscillator are those having excess gain and a net phase shift around the loop of $2\pi n$, where n is an integer. The phase condition is

$$\varphi_{el} + 2\pi f_n \frac{L_o}{v} = 2\pi n \qquad (12)$$

where L_o is the acoustic path length, f_n the frequency of the n-th mode and φ_{el} the phase shift through the feedback loop and that due to the reactive part of the transducers. The allowed modes therefore form a comb of frequencies from which all but one may be suppressed by suitable choice of transducer geometry. The flexibility of SAW delay lines allows such single-mode oscillators to be designed with varying sensitivity to the electrical part of the circuit by suitable choice of the relative size of $2\pi f_n L_o/v$ and φ_{el}. For high stabilities the acoustic phase shift should dominate while oscillators requiring modulation capability will demand a lower phase shift ratio (the oscillator can be tunable using, for example, a varactor phase shift network).

In contrast to conventional quartz oscillators, the frequency is not determined by the dimensions of the crystal but by the transducer geometry, so that devices can be made as rugged as desired, even at the highest frequency. Oscillators have been produced operating in the frequency range 20 MHz to 1.5 GHz with transducers operating at their fundamental.

The temperature stability of an SAW oscillator using ST-X quartz is adequate for many applications (variation $\sim$ 3 parts in 10^6 per °C). The short-term stability of these devices was 10^{-7} measured over 1 s. Single sideband FM noise of -110 dB per Hz (relative to carrier), 10 KHz away from carrier and frequency deviation up to 2% are typical features. Aging rates measured at the different establishment vary upward from approximately 10 ppm per year. The stability of

SAW oscillators lies between that of the LC oscillator and the bulk-wave resonator-stabilized oscillator, although the SAW resonator-stabilized oscillators,recently introduced [30,31], are now approaching the stability of bulk-wave resonator-stabilized oscillators [32].

3.2.3. Frequency filters

The synthesis of a SAW band-pass filter is an example of the design procedure employed when the filter characteristics are assigned in the frequency domain [5,8,33]. The degree of approximation to the desired band-pass characteristic is controlled by the number $\underline{N}$ of components in (11). Since this number is finite, the well--known-Gibbs'phenomenon arises, that can be controlled by a suitable choice of the coefficients α_n. A band--pass " apodized" finger structure is shown in Fig.8 together with the frequency response.
        Further complication to the design of any SAW filter arise for effective suppression of second order effects [18,34,39,44]. However, excellent results have been achieved with out of band response 60 db down, minimum insertion loss 6 dB, controlled to a fraction of dB,and phase ripple of $\pm$ 1.5° [35]. Finite size of the SAW device and attenuation of the wave at highest frequencies make it impossible to achieve very narrow--band with a passive filter. Multiple transit schemes, like"resonators" [36,37], provide compact periodic structure with Q in excess of 10^4.

3.2.3.1. Programmable filters

An attractive feature of SAW filters is the possibility of externally programming the filtering function. This can be achieved in the typical transversal filter scheme by controlling the phase and amplitude of each tap [44,45]. More recently the possibility of implementing an adaptive processor by using a SAW Fourier transformer has been shown [26,46,47,48].

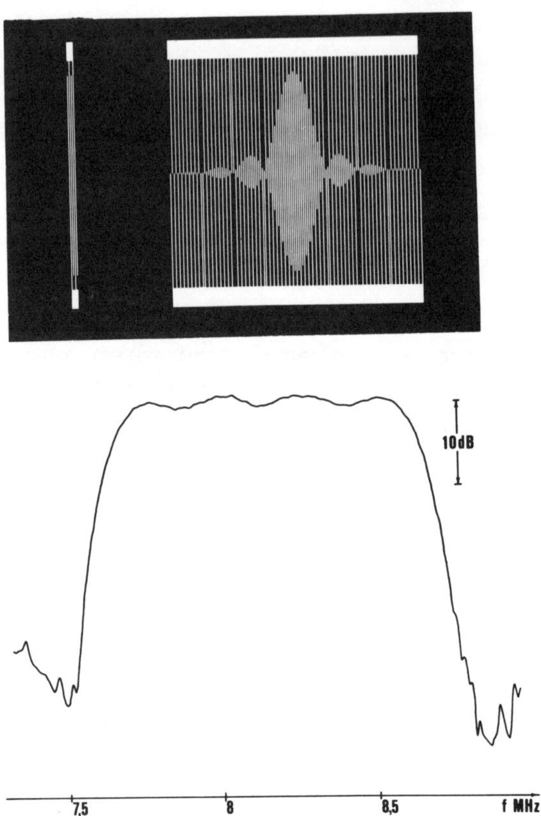

Fig.8. Apodized transducer for the synthesis of a band-
       -pass characteristic (top) and experimental fre-
       quency response (bottom).

3.2.4. Dispersive filters

The impact of SAW devices technology on system design
is mainly due to the possibility of synthesizing dispersive phase characteristics in a relatively simple
way. These devices can be used for both generation and
reception of the coded waveform in pulse compression
radar. Linear frequency modulation is the commonest
form of encoding in this application, and its characteristic parameters are pulse duration T and bandwidth B

[23,24,26,49]. The filter characteristics are given
in the time-domain in the form of an impulse-response
of finite duration, constant amplitude and linear frequency-modulation law. The fingers of the interdigital
transducer synthesize samples taken in correspondence
of the impulse-response maxima and minima (Fig.9). The
same array can be used to introduce the phase distortion,
by stretching an impulse over a pulse of long duration,
and, conversely, to equalize these phases, compressing
this pulse into the original impulse.

3.2.4.1. Reflecting array compressors (RAC)

Time-bandwidth products which exceed a few hundred
can be achieved by using a reflecting array structure
(Table 1) [51,52]. It consists of a periodic set of
reflectors normal to the incident beam or inclined
at some other angle. Typical reflectors are metal strips,
dielectric strips, or grooves. The intensity reflected
from each reflector is low (-40 dB), however constructive interference from a large number ($\sim$ 100) leads
to a sufficient reflectivity ($\sim$ 98%) for a particular
frequency, and over a narrow bandwidth centered on it,
where the reflector periodicity corresponds to the
wavelength. Weighting can be achieved by varying the
length of the reflectors.

TABLE I

B	Centre freq.	BT	Dispersive delay
100 MHz	300 MHz	1000	10 ns
512 MHz	1 GHz	5120	10 ns[*]
500 MHz	1.3 GHz	250	0.5 ns[†]
17 MHz	70 MHz	1548	88 ns[*]

[*] RAC (reflecting array compressor)
[†] IDT (interdigital transducer)

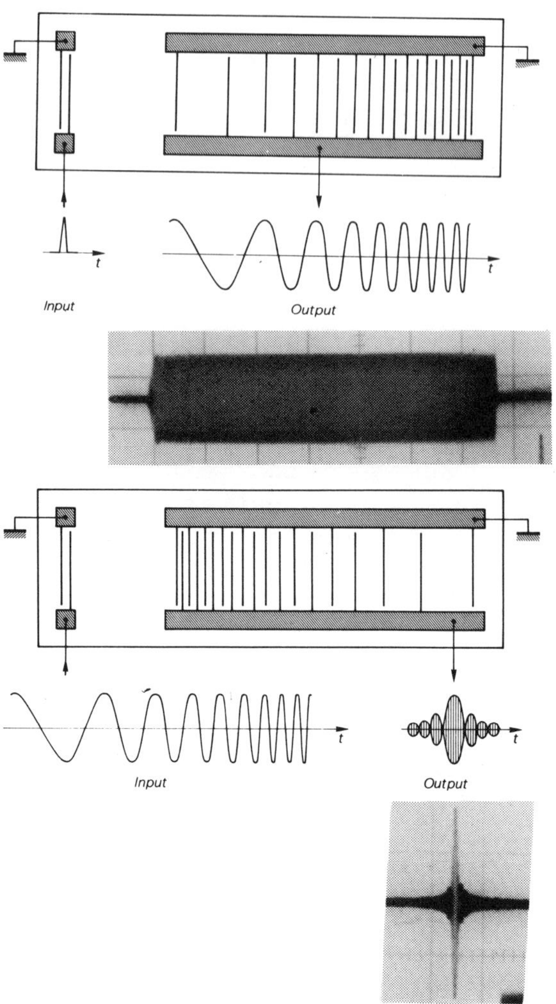

Fig.9. Schematic representation of a SAW dispersive
      filter used for phase dispersion (top) or
      phase equalization (bottom).

        These structures are important because
together with multistrip components [25,53] and re-
sonators belong to a new philosophy in which the response of the device is not dictate by the transducer
itself but by suitable SAW multipath arrangements.

An important area of signal processing in which large
time-bandwidth product dispersive delay lines are reque-
sted is that of Fourier transformers [47,48,54].

Waveforms other than linear FM can be generated or correlated by SAW dispersive devices [55].
In Fig. 10 is schematically shown the array for the generation of a phase-reversal 13-element Barker code.

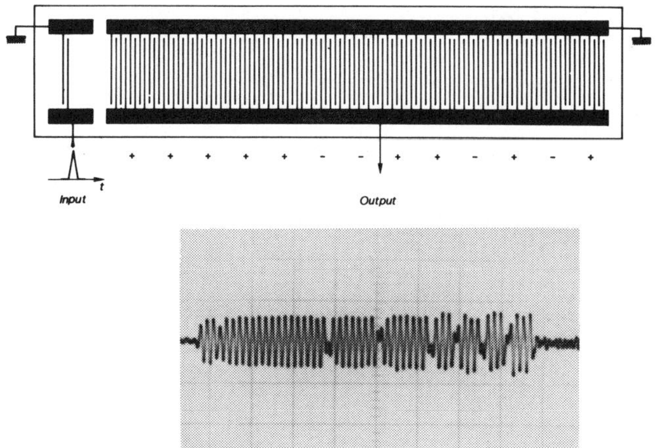

Fig. 10. Schematic representation of a SAW filter for
        generation of 13-element Barker code [22].

3.2.5. Convolution using parametric interaction of SAW

The basic process involved is the parametric interaction between two SAW beam propagating in opposite directions, and obtaining a product term by means of a non
linearity present in the device. An output can be obtained proportional to a time integral of the product
of the two inputs, i.e. the convolution of the two inputs [56]. If the two signals are identical, but reversed in time the device acts as a correlator. The significance of the device for signal processing applications lies in the fact that any signal can be correlated
with its stored replica, reversed in time. The device
becomes equivalent to a variable, real time matched filter. In acoustic convolvers parametric interaction is
due to the non linearity in the propagation medium, and

the efficency is low (90 dB insertion loss for 0 dB reference level). Better efficiency (20 dB insertion loss
for 0 dB reference level) can be achieved by using the
non linear interaction in external diodes connected to
a series of taps. It is also possible to store a signal
by using the traps of a semiconductor coupled to the
SAW propagating surface, and then, at a later time, either read it out or correlate with a signal. Such a device offers both the advantages of storage of the reference signal and of time reversal.

4. SAW signal processors for time compressed waveform

The application of SAW devices to radar systems has be-
en so far the most extensive, mainly because SAW device
parameters are suitable to radar. The attractive characteristic of SAW devices of processing large bandwidth
(> 100 MHz) signals makes it possible, when operated
in conjunction with CCDS, to substantially remove the
intrinsec limitation due to the relatively low (100 μs
<) propagation delay, compatible with physical dimens-
ion of available substrates and to enlarge the field
of applications to communication and sonar systems.
In a recently reported experiment [57] the combined
use of CCD shift registers and a SAW real-time spectrum analyzer permits a variable spectral resolution up
to 20 Hz to be achieved.
Moreover, it has been shown the possibility of correlating pseudo noise code and linear FM waveforms using
SAW devices buffered by a 1000:1 CCD shift-register
 [58].

REFERENCES

1.   J.H.Eveleth, Proc. IEEE, 53, 1406, 1965.
2.   R.M.White and F.W.Voltmer, Appl. Phys. Lett., 7,
      314, 1965.
3.   J.H.Rowen, U.S. Patent 3 289 114, 1966.
4.   C.E.Cook and M.Bernfeld, Radar Signals, Acade-
      mic Press, New York, 1967.
5.   P.Hartemann and E. Dieulesaint, Electron. Lett.,
      5, 219, 1969.

6.   P.Hartemann and E.Dieulesaint, Electron. Lett.,
      5, 657, 1969.
7.   W.D.Squire, H.J.Whitehouse and J.M.Alsup, IEEE
      Trans. Microwave Theory Tech. (special Issue on
      Microwave Acoustics), MTT-17, 1020, 1969.
8.   R.H.Tancrell and M.G.Holland, Proc. IEEE, 59,
      393, 1971.
9.   C.Atzeni and L.Masotti, IEEE Trans. Aerosp.
      Electron. Syst., AES-7, 662, 1971.
10.   Lord Rayleigh, Proc. London Math. Soc., 17,
      4, 1885.
11.   R.M.White and F.M.Voltmer, Appl. Phys. Lett.,7,
      314, 1965.
12   R.M.Artz,E.Saltzmann and K.Dransfeld, Appl. Phys.
      Lett., 10, 165, 1967.
13   G.A.Coquin and H.F.Tiersten, Jor. Acoust. Soc.
      Am., 41, 921, 1967.
14.   C.C.Tseng, IEEE Trans. Electron Devices, ED-15,
      586, 1968.
15.   R.M.White, Proc. IEEE, 58, 1238, 1970.
16.   W.R.Smith et alii, IEEE Trans. Microwave Theory
      Tech. (special Issue on Microwave Acoustics),
      MTT-17, 859, 1969.
17.   W.R.Smith et alii, IEEE Trans. Microwave Theory
      Tech. (special Issue on Microwave Acoustics),
      MTT-17, 865, 1969.
18.   W.R.Smith, Wave Electronics (special Issue on
      Workshop on CAD of SAW devices, Bologna, Italy),
      2, 25, 1976.
19.   J.J.Campbell and W.R.Jones, IEEE Trans. Sonics
      Ultrasonics, SU-15, 209, 1968.
20.   G.W.Farnell, Wave Electronics (special Issue on
      Workshop on CAD of SAW devices, Bologna, Italy),
      2, 1, 1976.
21.   A.J.Slobodnik jr., Proc. IEEE (special Issue on
      surface acoustic wave devices and applications),
      64, 581, 1976.
22.   C.Atzeni and L.Masotti, IEEE Trans. Microwave
      Theory Tech., MTT-21, 505, 1973.
23.   D.P.Morgan, Ultrasonics, 11, 121, 1973.
24.   C.Atzeni, L.Masotti and E.Teodori, Alta Frequen-
      za, XL, 506, 1971.

25.   J.D.Maines and E.G.S.Paige, IEE Rev., 120, 1078, 1973.
26.   C.Atzeni, G.Manes and L.Masotti, Proc. 1975 IEEE Ultrasonics Symp., 371, 1975,(n.75 CHO 994-4SU).
27.   L.W.Davies and M.W.Laurence, Proc. IREE Aust., 32,61, 1971.
28.   J.Crabb, M.F.Lewis and J.D.Maines, Electron. Lett., 9, 195, 1973.
29.   M.F.Lewis, Proc. 1973 IEEE Ultrasonics Symp., 344, 1973.
30.   R.C.M.Li, J.A.Alusow and R.C.Williamson, Proc. 1975 IEEE Ultrasonics Symp., 279, 1975.
31.   F.G.Marshall, Proc. 1975 IEEE Ultrasonics Symp., 290, 1975.
32.   M.W.Laurence, Wave Electronics (special Issue on Workshop on CAD of SAW devices, Bologna, Italy), 2, 199, 1976.
33.   C.Atzeni and L.Masotti, Alta Frequenza, XLII, 24E, 1973.
34.   R.E.Mitchell, Wave Electronics (special Issue on Workshop on CAD of SAW devices, Bologna, Italy), 111, 1976.
35.   H.Engan, Wave Electronics (special Issue on Workshop on CAD of SAW devices, Bologna, Italy), 2, 133, 1976.
36.   V.H.Haydl and R.S.Smith, Proc. IEEE (special Issue on surface acoustic wave devices and applications), 64, 682, 1976.
37.   Y.Koyamada, F.Ishibara and S.Yoshikawa, Proc. IEEE (special Issue on surface acoustic wave devices and applications) , 64, 685, 1976.
38.   J.D.Maines and E.G.S.Paige, Proc. IEEE (special Issue on surface acoustic wave devices and applications), 64, 639, 1976.
39.   W.S.Jones, C.S.Hartmann and T.D.Sturdivant, IEEE Trans. Sonics Ultrasonics, SU-19, 368, 1972.
40.   H.M.Gerard, 1971 IEEE Ultrasonics Symp., paper J-1.
41.   R.C.Williamson, MIT LINCOLN LAB. Cambridge, Mass., Rep. 1971.
42.   R.H.Tancrell and R.C.Williamson, Appl. Phys. Lett., 19, 456, 1971.

43.   C.Atzeni, G.Manes and L.Masotti, Proc. 1973 IEEE
      Ultrasonics Symp., 414, 1973.
44.   C.Atzeni, G.Manes and L.Masotti, Alta Frequenza,
      XLIII, 601 E, 1974.
45.   P.J.Hagon and L.R.Adkins, Proc. 1974 IEEE Ultraso-
      nics Symp., 177, 1974.
46.   J.D.Maines, G.L.Moule and E.G.S.Paige, Proc. 1975
      IEEE Ultrasonics Symp., 355, 1975.
47.   C.Atzeni, G.Manes and L.Masotti, this issue.
48.   H.J.Whitehouse, this issue.
49.   G.A.Armstrong, Wave Electronics (special Issue
      on Workshop on CAD of SAW devices, Bologna, Italy),
      2, 155, 1976.
50.   B.J.Darby, Wave Electronics(special Issue on
      Workshop on CAD of SAW devices, Bologna, Italy),
      2, 266, 1976.
51.   R.C.Williamson, Proc. IEEE (special Issue on sur-
      face acoustic wave devices and applications), 64,
      702, 1976.
52.   J.Melngailis et alii, Wave Electronics (special
      Issue on Workshop on CAD of SAW devices, Bologna,
      Italy), 2, 177, 1976.
53.   C.Maerfeld, Wave Electronics (special Issue on
      Workshop on CAD of SAW devices, Bologna, Italy),
      2, 88, 1976.
54.   C.Atzeni, Wave Electronics (special Issue on
      Workshop on CAD of SAW devices, Bologna, Italy),
      2, 238, 1976.
55.   C.Atzeni and L.Masotti, Electronics Lett., 7, 693,
      1971.
56.   G.S. Kino, Proc. IEEE (special Issue on surface
      acoustic wave devices and applications), 64, 724,
      1976.
57.   J.B.G.Roberts, Electronics Lett., 11, 525, 1975.
58.   P.M.Grant, Proc 1976 CCD Int. Conf.(to be publi-
      shed).
Suggested books
      B.A.Auld, Acoustic Fields and Waves in Solids, I
      and II, J.Wiley and Sons, New York, 1973.
      E.Dieulesaint and D.Royer, Ondes elastiques dans
      le solides, Masson et C., Editeurs, Paris, 1974.

ONE/TWO DIMENSIONAL DIGITAL SIGNAL PROCESSORS[*]

V. Cappellini

Istituto di Elettronica – Ingegneria, Università di
Firenze and IROE – C.N.R., Firenze, Italy

ABSTRACT. Digital methods and techniques for the processing of
one/two dimensional signals are considered. Spectral estimation,
digital filtering and data compression operations are in particular described. It is shown how digital signal processing operations
can be practically performed by means of standard computers (software definition) or by using special type digital processors (hardware definition). The use of recently developed microprocessors
and charge-coupled devices is also considered. Some digital processors performing digital filtering and spectral estimation are
described in detail, as typical examples. Applications to processing of signals (audio , ECG...) and images (in biomedicine, from
earth resource satellites...) are presented.

1. INTRODUCTION

In the last decade the processing of one–dimensional (1-D)
and two–dimensional (2-D) signals (images) presented great development and expansion in all branches of science and practical
activity.

A peculiar aspect of this impressive progress is represented
by the increasing importance, both from a theoretical and practi-

---

[*] This work has been supported in part by italian C.N.R.

cal viewpoint, of digital methods and techniques for processing
of 1-D/2-D signals. Digital signal processors are used ever more
in many different areas as telecommunications, sonar-radar, geophysics, bioengineering, meteorology, space activity... This fact
is due to several reasons: high efficiency which is obtainable by
using digital methods and techniques; great expansion of fast digital computers and minicomputers; decreasing cost of hardware
implementation for the development of high integration digital circuits and more recently of microprocessors and charge-coupled devices (CCD).

Among the different methods and techniques of 1-D/2-D signal
digital processing, spectral estimation, digital filtering, data
compression and coding are becoming of increasing interest [1].
After the description of the more important properties of these
digital signal operations, general criteria are here given for
designing digital signal processors performing in particular these operations. Standard computers and minicomputers (software definition) and special type digital processors are considered. Attention is also given to recently developed microprocessors and
CCD. Particular processors performing digital filtering and spectral estimation are described in detail. Examples of application
to signal and image processing are presented.

2. DIGITAL SIGNAL PROCESSING OPERATIONS

To perform digital processing the signals must be in digital
form. If analog signals and images are to be processed in a digital way, they can be converted through the following transformations:

(a) "sampling", according to the sampling theorem: for 1-D
signal $f(t)$ having f_M maximum spectral frequency, a sampling frequency $f_s \geq 2f_M$ is required, obtaining the discrete time signal samples $f(nT)$ with $T=1/f_s$; for a 2-D signal (image) $f(x,y)$ having the maximum space frequencies
f_{M1} for x and f_{M2} for y, sampling frequencies $f_{s1} \geq 2f_{M1}$
for x and $f_{s2} \geq 2f_{M2}$ for y are required, obtaining the discrete space image samples $f(n_1 X, n_2 Y)$ with $X=1/f_{s1}$ and
$Y=1/f_{s2}$ (in general a unique vale f_s equal to the greater of f_{s1} and f_{s2} and sampling interval $X=1/f_s$ are used);

(b) "quantization", according to the required amplitude accu-

racy for the discrete representation of 1-D/2-D signals:
if q is the quantization step, the signal amplitude range is represented through Nq discrete amplitude levels,
being N a suitable integer;
(c) "coding", according to the selected digital representation: for the binary form, in general used, the quantized
amplitude values are converted to words of n bits, being
$N=2^n$ [1][2] .

Once the 1-D/2-D signal is set, through the analog-digital
(A-D) conversion, in digital form, one can perform on it different
digital processing operations. Fig. 1 shows a simplified structure
of a digital processor (receiving data from the input A-D converter)
having three main parts: an arithmetic unit performing digital operations on input data, a memory maintaining input-output data for
the required time intervals, a general programmer controlling the
correct time sequence of operations (according to given instructions)
and assuring the synchronization of different parts. The output data from the digital processor can be converted again in analog form,
if required, through a digital-analog (D-A) converter performing
the inverse transformations (a) (b) (c) above considered.

Some important digital operations which can be performed on
1-D/2-D signals are represented by: spectral estimation, digital
filtering, data compression and coding [1][2] .

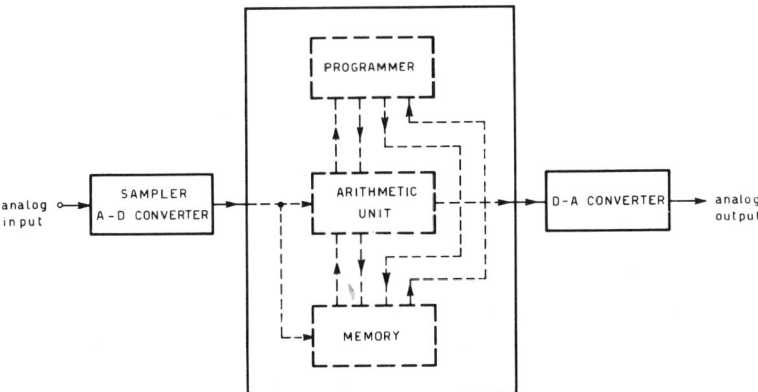

Fig. 1. Block diagram of a simplified structure of a digital processor with input-output units (A-D and D-A conversion).

2.1 Spectral estimation

There are several methods to perform spectral estimation on
1-D/2-D signals in a digital way. Three important methods correspond to the analog ones: Fourier transform, autocorrelation, band-
-pass analysis [1][2] .

For the digital evaluation of the Fourier transform, the more used method as in the analog case, the Discrete Fourier Transform (DFT) is defined. By considering the 1-D signal f(t) represented by the sequence of N samples $f(n)=f(nT)$, $0 \leq n \leq (N-1)$, the DFT
of this sample sequence is defined as another sequence [2][3]

$$F(k) = \sum_{n=0}^{N-1} f(n)e^{-jkn(2\pi/N)} \tag{1}$$

where $F(k)=F(k\Omega)$ represent samples of the continuous Fourier spectrum $F(\omega)$, being $\Omega=(2\pi)/(NT)$ the angular frequency increment between samples in the frequency domain. There are only N distinct
values F(k) computable by (1), those for $0 \leq k \leq (N-1)$. There exists
also an inverse DFT (IDFT), a transformation that maps a DFT back
into the sequence from which it was obtained: it is given by

$$f(n) = \frac{1}{N} \sum_{k=0}^{N-1} F(k)e^{jnk(2\pi/N)} \tag{2}$$

where n and k can assume only N distinct values, as above outlined.

By considering the 2-D signal f(x,y) represented by the ma-
trix of $N_1 \cdot N_2$ samples $f(n_1,n_2)=f(n_1 X, n_2 Y)$, $0 \leq n_1 \leq (N_1-1)$ and $0 \leq n_2 \leq (N_2-1)$
the DFT of this sample matrix is defined as [3][4]

$$F(k_1,k_2) = \sum_{n_1=0}^{N_1-1} \sum_{n_2=0}^{N_2-1} f(n_1,n_2)e^{-jk_1 n_1 (2\pi/N_1)} e^{-jk_2 n_2 (2\pi/N_2)} \tag{3}$$

where $F(k_1,k_2)= F(k_1\Omega_1, k_2\Omega_2)$ represent samples of the continuous Fourier spectrum $F(\omega_1,\omega_2)$, being $\Omega_1=(2\pi)/(N_1 X)$ and $\Omega_2=(2\pi)/(N_2 Y)$ the
angular frequency increments between samples in the frequency domain.
There are $N_1 \cdot N_2$ distinct values $F(k_1,k_2)$ computable by (3), those
for $0 \leq k_1 \leq (N_1-1)$ and $0 \leq k_2 \leq (N_2-1)$. The inverse DFT (IDFT) is now given by
the following relation (extension of the 1-D case)

$$f(n_1,n_2) = \frac{1}{N_1 N_2} \sum_{k_1=0}^{N_1-1} \sum_{k_2=0}^{N_2-1} F(k_1,k_2) e^{jn_1 k_1 (2\pi/N_1)} e^{jn_2 k_2 (2\pi/N_2)} \tag{4}$$

where n_1 and k_1 can assume N_1 distinct values, while n_2 and k_2 can assume N_2 distinct values.

From a practical viewpoint it is interesting to observe that DFT and IDFT can be evaluated in a fast way (FFT) through suitable decimation procedures [2-4] : for 1-D FFT , being N a power of 2, a number of computations proportional to $N\log_2 N$ is required, while for 2-D FFT, being N_1 and N_2 power of 2, a number of computations proportional to $2(N_1 N_2)\log_2(N_1 N_2)$ is required.

A useful method of spectral estimation for many applications is also represented by the band-pass analysis performed through digital filtering: by evaluating the r.m.s. values of each output band filtering, a power spectral estimation is easily obtained.

2.2 Digital filtering

Digital filtering represents in general linear filtering operation performed on sampled data. 1-D digital filter can be defined by a relation as [1-4]

$$g(n) = \sum_{k=0}^{N} a(k)f(n-k) - \sum_{k=1}^{M} b(k)g(n-k) \tag{5}$$

where: f(n)=f(nT) are the input data, g(n)=g(nT) are the output data; a(k) and b(k) are the coefficients defining the digital filter; N and M are integers. If a(k) and b(k) are different from zero, the digital filter is said "recursive", while if all b(k) are zero it is called "non-recursive" (or transversal).

An important classification of digital filters is connected to the time duration of the filter impulse-response; digital filters can be divided in two main types: filters with impulse response of finite duration (FIR) and filters with impulse response of infinite duration (IIR) [1-4] .

By using the z-transform, the transfer function H(z) of the digital filter results expressed by

$$H(z) = \frac{G(z)}{F(z)} = \frac{\sum_{k=0}^{N} a(k)z^{-k}}{1 + \sum_{k=1}^{M} b(k)z^{-k}} \qquad (6)$$

where $F(z)$ and $G(z)$ represent the z-transforms, respectively, of the input and output sequences of data and z^{-1} is the unity delay operator. The frequency response $H(\omega)$ results setting in (6) $z=e^{j\omega T}$.

2-D digital filters can be defined by a relation as [1-4]

$$g(n_1,n_2) = \sum_{k_1=0}^{N_1} \sum_{k_2=0}^{N_2} a(k_1,k_2)f(n_1-k_1,n_2-k_2) -$$

$$- \sum_{\substack{k_1=0 \\ k_1+k_2\neq 0}}^{M_1} \sum_{k_2=0}^{M_2} b(k_1,k_2)g(n_1-k_1,n_2-k_2) \qquad (7)$$

where: $f(n_1,n_2)=f(n_1 X,n_2 X)$ are the input data, $g(n_1,n_2)=g(n_1 X,n_2 X)$ are the output data; $a(k_1,k_2)$ and $b(k_1,k_2)$ are the coefficients defining the digital filter. As in the 1-D case, if $a(k_1,k_2)$ and $b(k_1,k_2)$ are different from zero, the digital filter is said "recursive", while if all $b(k_1,k_2)$ are zero it is called "non-recursive" (or transversal). Further considering the impulse response we have FIR filters (with finite extension impulse response) and IIR filters (with infinite extension impulse response).

The transfer function $H(z_1,z_2)$ of the 2-D digital filter results now

$$H(z_1,z_2) = \frac{\sum_{k_1=0}^{N_1} \sum_{k_2=0}^{N_2} a(k_1,k_2)z_1^{-k_1}z_2^{-k_2}}{1 + \sum_{\substack{k_1=0 \\ k_1+k_2\neq 0}}^{M_1} \sum_{k_2=0}^{M_2} b(k_1,k_2)z_1^{-k_1}z_2^{-k_2}} \qquad (8)$$

The frequency response $H(\omega_1,\omega_2)$ can be obtained by the relation (8) setting $z_1=e^{j\omega_1 X}$ $z_2=e^{j\omega_2 X}$.

Many methods and algorithms are available for designing 1-D and 2-D digital filters. These methods are in general different for FIR digital filters (usually of non-recusrive type) and for IIR digital filters (of recursive type).

FIR digital filters can be designed with an exact linear phase and with non-recursive implementation are ever stable; the errors due to the finite word-length of coefficients and data are not so important as in recursive IIR filters. Among the several methods for designing FIR digital filters, there are [1-5] :
  (a) "window method", which consists in modifying the coeffi-
      cients (multiplying them by suitable weighting factors)
      to reduce the Gibbs oscillations in the frequency domain;
  (b) "frequency sampling method", in which samples of the fre-
      quency response are used as DFT of the filter impulse re-
      sponse (a recursive structure is obtained);
  (c) "equiripple filter design method", to obtain in particu-
      lar near optimum Chebyshev approximation to imposed fre-
      quency response (an efficient design method was defined
      by Parks and McClellan [6]).
The above methods can be used for designing 1-D FIR digital filters (for which other methods are also available) and (with some modifications or extensions) for designing 2-D FIR digital filters. For this last purpose a useful method was proposed by McClellan [7]: it permits to transform 1-D filter to a 2-D filter through the relation

$$\cos\omega = A\cos\omega_1 + B\cos\omega_2 + C\cos\omega_1\cos\omega_2 + D \tag{9}$$

with A,B,C,D suitable coefficients.

For IIR digital filter design a clear separation must be done between 1-D and 2-D filters, because the fundamental theorem of algebra does not extend to 2-D case. Among the different methods for designing 1-D IIR digital filters, there are [1-5]:
  (a) "design through direct z-transform", in which a transfor-
      mation from Laplace s-plane to z-plane is used being inva-
      riant impulsive, that is giving a digital filter with impul-
      se response equal to the sampled impulse response of the
      starting analog filter;
  (b) "design through bilinear z-transform", in which a bilinear
      z-transform is used, giving a digital filter with a frequen-
      cy response equal to that of the analog filter, but with a

non linear contraction of the frequency scale.
For designing 2-D IIR digital filters, there are some approaches
giving stabilized recursive filters: particular mapping from continuous domain is used [8], differential correction procedures [9] ,
suitable 2-D spectral factorization [10] , McClellan transformation
(9) with four stable one-quadrant filters [11] .

2.3 Data compression

An important problem in the digital processing of 1-D/2-D signals is represented by the big amount of data to be considered and
used in the actual processing operations and in the store or archi-
ving of data. A data transformation that can be performed to solve
(at least in part) this problem is represented by the data compression: a transformation of this type reduces the amount of not useful or redundant data. If the data are to be transmitted, a reduction
of the bandwidth required for the transmission is also obtained [12].

Many types of data compression methods and techniques have been
introduced. As a large classification, we can divide them in two
main groups:
    (a) "data compression methods with entropy conservation", that
        can be reversible, permitting to recover all the original
        data (1-D/2-D samples) after the decompression [1];
    (b) "data compression methods with entropy reduction", that
        are unreversible, loosing in any case some original data
        after decompression.
Some important reversible methods are: adaptive sampling, prediction, interpolation, encoding (delta modulation, differential pulse
code modulation...), multiple filtering covering the complete spectrum, Fourier analysis, use of transformations (Hadamard, Walsh,
Haar, Slant, Karhunen-Loeve...). Some unreversible methods are: parameter extraction, thresholding, power spectrum [13] .

Prediction-interpolation algorithms are of particular interest
due to their relative simplicity of implementation. In the process

---

[1] As it is usually defined, we call reversible a method giving, as
limit case, zero error reconstruction, but that, in general, can
be used with a given error reconstruction (depending on the values of the method parameters).

of prediction the knowledge of preceding samples is used, while in
interpolation the knowledge of preceding and subsequent samples is
utilized. In both types of operation the predicted or interpolated
sample or data is compared with the actual sample: if the resulting
difference is within the allowable error tolerance, then the prediction or interpolation is considered correct and the actual sample is eliminated; otherwise the actual sample remains in the compressed data [12] [13] . Two simple cases of prediction are in
particular zero order prediction (ZOP), in which only the preceding sample is used, and first order prediction (FOP), in which
two preceding samples are used. A more complex algorithm of prediction type is the linear adaptive prediction, in which the predicted sample is evaluated as a linear weighting of a given set of
preceding samples, that is

$$g(n) = \sum_{k=1}^{N} a(k)f(n-k) \tag{10}$$

where: $f(n)$ are the preceding samples or data, $g(n)$ are the predicted samples , $a(k)$ are the weighting coefficients (in general
defined to minimize the mean square prediction error), N is an integer. It is clear how the relation (10) has the same structure as
1-D non-recursive digital filter.

The transformations (Hadamard, Walsh, Haar, Slant, Karhunen-
-Loeve...) can be used, as above observed,to obtain 1-D or 2-D data
compression. In general some data reduction is performed in the
transformed domain, with thresholding, variable length coding of
groups of transformed data (for images, corresponding to sub-areas
of the processed image)... Also the simple thresholding and annihi-
lation (elimination of transformed values under a given value) give often appreciable compression ratios due to the fact that the
most part of transformed values are concentrated in small intervals
or regions.

3. IMPLEMENTATION OF DIGITAL SIGNAL PROCESSING

Any digital signal processing operation, as in particular above considered, can be implemented following two main solutions:
    (a) use of standard computers or minicomputers (software defi-
        nition);
    (b) use of special type or special-purpose digital processors

(hardware definition).

The use of the recently introduced microprocessors and micro-computers can be considered an intermediate or hybrid solution, because the software and hardware approaches are strictly connected and a good knowledge of both (software and hardware) is required to design and build satisfactory digital signal processors. For this reason three solutions are in the following separately considered as the main important to implement digital signal processing: use of standard computers or minicomputers, use of special type processors, use of microprocessors and minicomputers.

3.1 Use of standard computers or minicomputers

A digital signal processing operation can be implemented by using a standard computer or minicomputer, through a suitable software package (set of programming instructions) defining completely the considered processing operation [2].

An advantage of using standard computers for digital signal processing is represented by the fact that by modifying the program the same machine or computing system can be applied to perform any digital signal processing. Further in the actual program some preliminary parts can be inserted not for processing, but for defining the structure of the desired processing operation; as example, if we have to perform digital filtering, a part of the program can compute at first the more suitable digital filter for the particular considered application: Fig. 2 shows the flow-chart of a program defining recursive digital filters of Chebyshev type for 1-D signal processing.

However other type problems and limitations can occur by using standard computers for digital signal processing. Apart the size and cost of this solution, problems can appear due to the limited speed of the computing systems (for the internal organisation), such that continuous signals or sampled data (binary words) cannot be processed in "real time" with the general structure of Fig. 1, but the digital processing is done, after the acquisition of a block of signal data, in a subsequent delayed time interval. The-

---

[2] The general structure of the computer is like that shown in Fig. 1: now the three main parts (arithmetic unit, memory, programmer) are divided in many sub-units to obtain great processing flexibility.

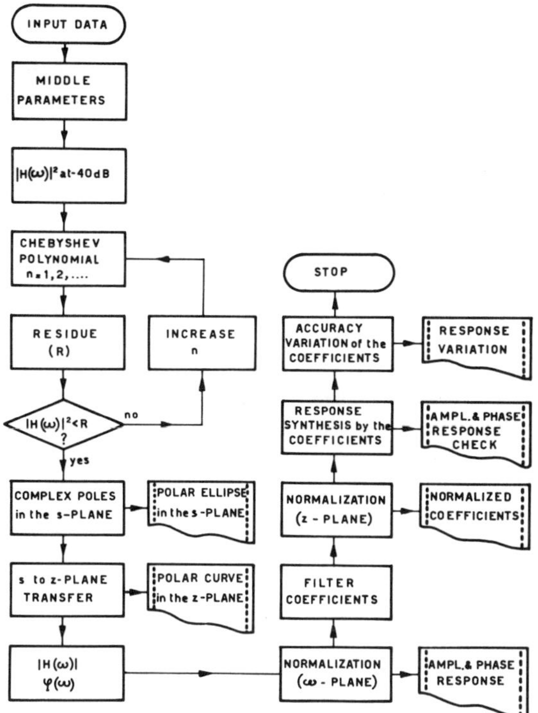

Fig. 2. Flow-chart of a program defining 1-D recursive digital filters of Chebyshev type through several steps (sub-routines).

se problems are partially solved by last generation computers and in particular by some last types of minicomputers, which are characterized by smaller size (one standard rack unit or less) and higher processing speed (cycle time of 0.1-1μs).

Another aspect of using digital computer for signal processing is represented by the fact that different computers and processors can be interconnected for a better efficiency: one or more minicomputers can be connected to a larger computer, to which the data are sent after a suitable pre-processing; a special fast processor can perform pre-processing on data, sending them after to a minicomputer.

3.2 Use of special type or special-purpose digital processors

A special type or special-purpose digital processor has a
structure designed to solve in the best way a particular or some
specific problems of signal processing. Due to this more specific
approach, this solution is complementary to the preceding one with
standard computers.

Now we have limited flexibility of processing and very complex
operations are in general not performed. However lower size and cost
with fast processing can be obtained.

The general structure of the special type processor is also
now like that shown in Fig. 1. The specific architecture of the
processor depends on the operation to be implemented and on the
used hardware components. According to the form in which elementary operations are performed inside the processor we have two
different types:
      (a) "serial processors", in which the elementary operations
           are performed in serial, in subsequent time intervals;
      (b) "parallel processor", in which the elementary operations
           are performed in parallel, in the same time interval.
It is clear that (a) processors are more economic than (b) processors, while these last assure faster processing. An example of
special type digital processor, performing 1-D non-recursive digital filtering or adaptive linear prediction (in this last case
the output sample memory is in general not used), with a parallel
organisation of the multiplier (multiplication of the corresponding bits of incoming words in parallel) is shown in Fig. 3.

The serial-parallel processing organisation can be extended
from elementary inside operations to the general structure of
the processor. Fig. 4 shows an example of special purpose processor performing spectral estimation through digital filtering, as
described in 2.1: a single digital filter is used giving in time
sequence the output samples for each band-pass filter (the first
set of m values, the second set of m values...). The processor
is designed in such a way that the r.m.s. values representing the
spectral estimation can be transmitted through a multiplexer in a
PCM format to a receiving terminal. If the information contained
in the short-time spectral estimation is sufficient for a specific
signal identification, the processor of Fig. 4 can be very useful
to obtain large compression ratios or reduction of the bandwidth

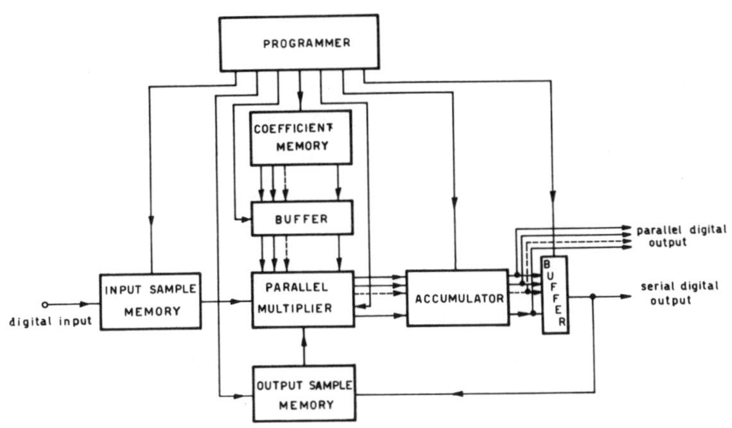

Fig. 3. Block-diagram of a special type digital processor, performing 1-D non-recursive digital filtering or adaptive linear
prediction, with a parallel organisation of the multiplier.

required for the transmission [12][13].

An important aspect of digital processor development is represented, as already pointed out, by the high speed of processing
which can be obtained. By extending the idea of parallel processing, above considered, "array processors" have been developed
and are currently designed and built, especially for 2-D signal
or image processing. Array processors use hardware parallelism
in all the structure: we have elementary special-purpose processors acting in parallel on input data. In the image processing,
the purpose is to construct a large array of elementary processors

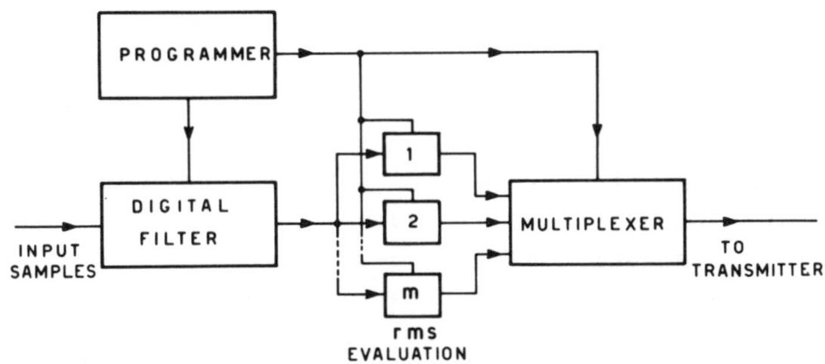

Fig. 4. A digital processor for spectral estimation, data compression.

so that each picture element has assigned to it one processor or
"cell". The first practical proposal for an array of programmable
cells for image processing was made by Unger [14] in 1958; subsequently many other approaches have been done, and some systems
(CLIP arrays) were in particular designed and built by Duff [15].
These array processors are really very fast and seem well oriented to solve several problems in the digital signal processing
area (as pattern recognition).

The development of digital signal processors of special type
is greatly connected (in a sense, more than standard computers) to
the technology progress and advance in electronics components. In
last years an impressive progress of semiconductor technology and
in particular of digital integrated circuits occurred: from standard single-function integrated circuits mean scale integration
(MSI) circuits and large scale integration circuits (LSI) were
introduced, arriving to thousand of elementary circuits putted
together on a single chip performing also complex functions (microcomputers and microprocessors, as in the following considered).
An interesting aspect of new high integration components is that
in several cases an increase of processing speed is assured.

In parallel to the progress in integration of circuits, new
components or special components have been developed, which are
very useful for signal and image processing. Three important components and devices on this line are: electro-optical components,
surface-acoustic-wave (SAW) devices and charge-coupled devices (CCD).

Electro-optical components were developed permitting easy
conversion of electrical signals to optical signals: light emitting diodes (LED), photodiodes, avalanche photo-detectors (APD),
semiconductor lasers, thin-film acoustooptic devices, optical fi-
bers... [1] [16]. In this way processing structures can be defined in which a very fast pre-processing is done with optical systems and accurate processing is after done by digital processors.

SAW devices provide high frequency (VHF-UHF) signal processing functions, often not available by other means. Fast acousto-
electric convolvers (replacing FFT processors in high frequency),
planar crystalline SAW filters (substituting analog filters in
the VHF-UHF range) can be built with modified microcircuit techniques in a form compatible with integrated circuits. [16] .

CCD devices are presently appearing as interesting discrete
time components permitting to build discrete time processors having low cost,low size and low power absorption [16].
They are essentially analog delay lines of sampled data. A fundamental building block with CCD is a CCD transversal filter, whose
operation is described by the relation (5) in which all b(k) are
zero and f(n) g(n) are now analog samples of the continuous analog
signals f(t) g(t). They represent therefore alternative solution
for performing transversal filtering with respect to digital filtering described in 2.2. Further CCD can be easily combined with
integrated circuits (MOS) to obtain complex signal processing functions. By adding CCD transversal filters to MOS multipliers the
chirp z-transform (CZT) algorithm for DFT can be implemented, resulting in low-cost spectral estimators (as described in 2.1). Many applications of CCD to image analysis and processing are also
currently developed.

As a general trend in hardware signal processing, in particular with discrete methods and techniques, we see that the frequency
range of applications can be divided in three main regions:
a lower   frequency range in which digital processing can be performed by using high integration digital circuits and discrete time
processing can be done with CCD; medium frequency range in which
both types of processing above considered can be developed, but
with increasing convenience of CCD solution; a high frequency range
in which SAW devices represent the more convenient approach. As
indication the medium frequency range can be 1-20MHz.

3.3 Use of microprocessors and microcomputers

As already above observed, the advent of microprocessors and
microcomputers (computers on a chip) has recently opened a new type approach to digital signal processing, for the fact that both
software and hardware aspects are strictly connected and their
knowledge is required to implement efficient digital signal processors.

Basically a microprocessor consists, according to the general
structure of Fig. 1, of a set of registers and arithmetic operators
interconnected by appropriate data paths controlled by a decoder
and a sequencer. This basic structure is shown in Fig. 5, where
with a simplified microprocessor also a memory for storing the

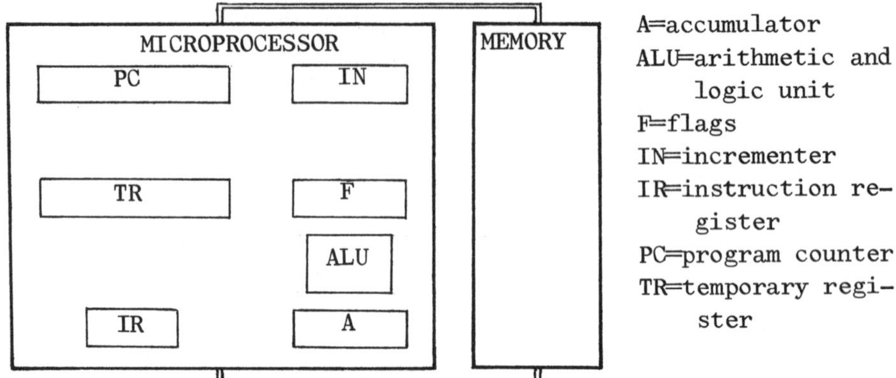

Fig. 5. Basic structure of a microprocessor with a memory for storing the program.

program is reported [17]. The program instructions are stored
in the memory and executed in time sequence under the control of
the program counter (PC). In a typical working step the content
of the memory location pointed by the PC is transferred to the
instruction register IRand the content of PC is simultaneously in-
cremented (PC is acting as program register). A control unit, syn-
chronized by an external clock, triggers all the transfers and gives
the necessary signals for the memory. The sequence of operations
is in general represented on a timing diagram.

With the development of the microprocessors a microprogram-
ming terminology has been introduced: microprogram (a type of program that directly controls the operation of each functional element of the microprocessor), microinstruction (a bit pattern that
is stored in a microprogram memory), microinstruction sequence (the
series of microinstructions that the microprogram control unit (MCU)
selects from the microprogram to execute a single macroinstruction
or control command), macroinstruction (either a conventional computer instruction or device controller command).

Microprocessors are available with 4 bits, 8 bits, 16 bits
as word-length, working over 10 MHz and with capabilities of programming over 100 instructions. They can be easily combined with
standard LSI memories (dynamic, static) to build high performance
digital processors with a minimum of added logic circuitry. Several microprocessors can be arrayed in cascade or in parallel to form

multi-microprocessor systems solving more complex problems.

Microprocessors can solve at the present time, in a single
or multi configuration, many problems of digital signal processing (spectral estimation, digital filtering, data compression).
Further they can be put in analog processing systems to perform
special functions, as adaptivity: interesting hybrid processors
can result for specific problems (adaptive filtering, adaptive
equalization, adaptive data compression...).

4. SPECIAL TYPE PROCESSOR PERFORMING DIGITAL FILTERING AND SPECTRAL ESTIMATION

As an example of special type digital processor implementation,
we describe a processor, called ANPAD, designed and built at IROE-
-C.N.R. of Florence. The digital processor can perform band-pass
analysis, spectral estimation and signal recognizing. The processor is based on a particular algorithm of frequency shift of the
sampled signal spectrum and sample decimation and uses a single digital filter (with an approach near to that shown in Fig. 4).

The considered algorithm requires an angular sampling frequen-
cy $\omega_s = 2\omega_M$, ω_M being the maximum angular frequency limiting the
input signal spectrum. The property is used that a multiplication
of the input samples $f(n)$ by $(-1)^n$ corresponds to a frequency shift
of the sampled signal spectrum equal to ω_M [5] . Thus, by means
of a single low-pass digital filter with cutoff angular frequency
$\omega_c = \omega_M/2$, it is possible to obtain in a first operation on the original input samples $f(n)$ the low-frequency part (0 to $\omega_M/2$) of the
spectrum, and in a second operation on the $f(n)$ samples multiplied
by $(-1)^n$ the high frequency part ($\omega_M/2$ to ω_M) of the spectrum shifted into the low frequency region. The same procedure can be repeated on the two sets of obtained output samples by considering,
in each set, only one of two successive samples (decimation procedure). In this way, reducing the sampling rate by two each time, different band-pass filterings can be performed by means of a
single low-pass filter. In particular we can separate, with a tree
structure, the signal spectrum in many bands: if r subsequent operations are performed, 2^r bands with constant bandwidth are obtained
(if only one band output is taken at each subsequent operation, a
proportional bandwidth band-pass analysis with r bands is performed).

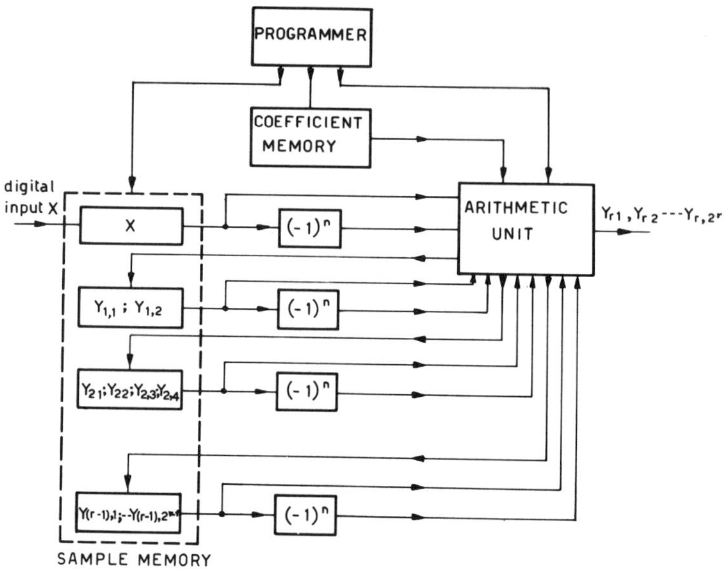

Fig. 6. Block diagram of the digital band-pass analyzer.

According to the above described algorithm a digital band-pass
analyzer (Fig. 6) was designed and built, performing an analysis
up to 16 bands [5] . The analyzer has an input A-D converter giving sampled data as words of 8 bits (7 bits for the amplitude
value and 1 bit for the sign) and 16 D-A output converters (a special type digital linear interpolator, inserting a staircase with
many steps between two subsequent samples can also be used in output [5]). With the used MSI and LSI integrated circuits input
signals with a maximum frequency of 4 KHz can be processed (real time).

An interesting aspect of the above analyzer is that a spectral
estimation can be easily obtained, by evaluating the r.m.s. value
of each band output (through an hardware constituted by a multiplier,
an accumulator, a divider by N - if N is the number of samples on
which the r.m.s. is evaluated - and a circuit performing the square root). Further a signal recognizing can be obtained if we set
in a memory the 16 r.m.s. values characterizing the short-time
spectrum of a known signal and we compare these values with that
obtained in real time for the incoming signal. Two other digital
processors were indeed designed and built to be inserted in cascade after the band-pass analyzer, obtaining a unique system with

three digital processors.

5. APPLICATIONS OF 1-D/2-D DIGITAL SIGNAL PROCESSORS

Digital processors can be applied, as observed in the Introduction, in many areas of science and practical activity. Efficient
processing of 1-D signals (audio, ECG, EEG, radar-sonar...) and of
2-D signals or images (in biomedicine, from earth resource satellites, in meteorology...) can be obtained.

An example of 1-D signal digital processing (in hardware) is
shown in Fig. 7, where a low-pass filtering (cutoff frequency of
30 Hz) performed by the above described ANPAD system on ECG signal
is reported. Another example of 2-D signal digital processing (in
software) is shown in Fig. 8, where the result of data compression
is reported (b) by applying 2-D Fast Walsh Transform on an image
received from earth resource satellite (given by TELESPAZIO): according to the consideration presented in 2.3, a method of variable
length encoding of the transformed values in 4x4 squares is used,
giving a mean word-length of 1.30 bits/sample in the compressed image with respect to the original word-length of 5 bits/sample (the
images are presented with 10 grey levels).

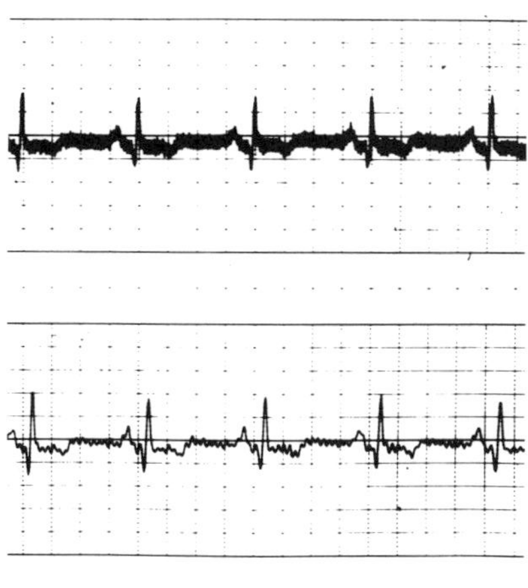

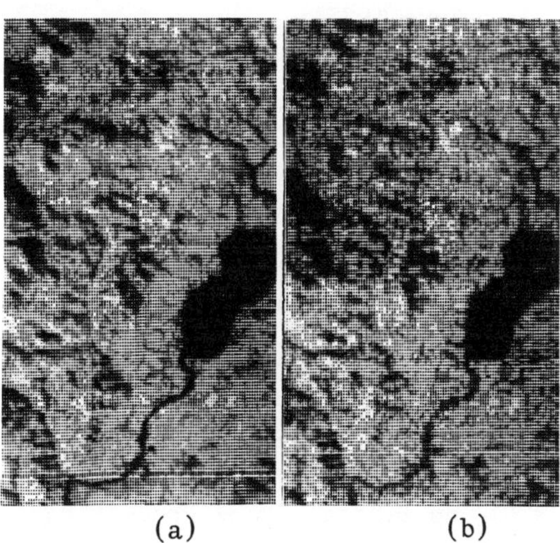

Fig. 7. Example of low-pass   Fig. 8. Example of data compression
        digital filtering of   applied to an image received
        ECG signal by ANPAD.   from earth resource satellite.

REFERENCES

[1] Special Issue on Digital Signal Processing, Proceedings of the IEEE, vol. 63, n. 4, April 1975.

[2] B. Gold, C. M. Rader, "Digital Processing of Signals", Mc-Graw Hill, New York, 1969.

[3] A. Oppenheim, R. Schafer, "Digital Signal Processing", Prentice-Hall, Englewood Cliffs, N. J., 1975.

[4] "Picture Processing and Digital Filtering", T. S. Huang Ed., Springer Verlag, Berlin-Heidelberg-New York, 1975.

[5] V. Cappellini, P. L. Emiliani, "Design of Some Digital Filters with Application to Signal and Image Processing", Proceed. of Summer School on Circuit Theory, vol. 1,p. 161-176, Praha, 1974.

[6] T. W. Parks, J. H. McClellan, "Chebyshev Approximation for Non-recursive Digital Filters with Linear Phase", IEEE Trans. Circuit Theory, vol. CT-19, p. 188-194, November 1972.

[7] J. H. McClellan, "The Design of Two-Dimensional Digital Filters by Transformations", Proc. 7th Ann. Princeton Conference on Information Sciences and Systems, p. 247-251, 1973.

[8] M. Ahmadi, A. G. Constantinides, R. A. King, "Design Technique for a Class of Stable Two-Dimensional Recursive Digital Filters", Proc. of Acoustics, Speech and Signal Proces. Conf.,1976.

[9] D. E. Dudgeon, "Two-Dimensional Recursive Filter Design Using Differential Correction", IEEE Trans. ASSP-23, n. 3, p. 264-267, June 1975.

[10] M. P. Ekstrom, J. W. Woods, "Two-Dimensional Spectral Factorization with Applications in Recursive Digital Filtering", IEEE Intern. Symposium on Circuits and Systems, Munich, 1976.

[11] F. Bernabo', V. Cappellini, P. L. Emiliani, "A Method for Designing 2-D Recursive Digital Filters Having a Circular Symmetry", Proc. Florence Conference on Digital Signal Processing,1975.

[12] Special Issue on Redundany Reduction, Proc. IEEE,vol. 55,n.3,1967.

[13] V. Cappellini, F. Lotti, F. Pieralli, "Application of Some Data Compression Methods to ESRO Satellite Telemetry Data", NATO ASI N. 12, Darlington, August 1974, Proc. p. 219-225.

[14] S. H. Unger, "A Computer Oriented Towards Spatial Problems", Proc. IRE, vol. 46, n. 10, 1958.

[15] M. J. B. Duff, "Geometrical Analysis of Image Parts", NATO ASI on Digital Image Processing and Analysis, Bonas-France,June 1976.

[16] Special Issue on Surface Acoustic Wave Devices and Applications", Proc. IEEE, vol. 64, n. 5, May 1976.

[17] J. D. Nicoud, "Basic Microprocessor Structure I", Microscope, vol. 1, n. 1, January 1976, p. 5-8.

SURFACE–ACOUSTIC–WAVE FOURIER–TRANSFORM PROCESSORS

C. Atzeni, G. Manes and L. Masotti

I.R.O.E., Consiglio Nazionale delle Ricerche
Facoltà di Ingegneria, Università di Firenze,
Italia.

ABSTRACT. Chirp-transformation is a new processing technique capable of providing the Fourier transform of a signal in real time by mapping the frequency domain into the time domain. The implementation of this technique has been made possible by the availability of high-quality large time-bandwidth product surface-acoustic-wave (SAW) chirp filters. SAW chirp-transform processors provide a powerful tool for performing a large variety of signal processing functions: spectral analysis, network analysis, signal generation and filtering.

1. FOURIER TRANSFORM PROCESSORS

Fourier transform of a signal can be obtained using two basic chirp-transform configurations.

The most known configuration, shown in Fig. 1.a, will be referred to as MCM (multiplication-convolution-multiplication) transform-unit. The input signal is premultiplied by a waveform whose frequency is varying linearly with time (commonly referred to as a " chirp"), and convoluted in a chirp filter of opposite frequency-modulation slope.

The output from the chirp filter can be written as

$$g(t) = \left[s(t)\, e^{-j\mu t^2} \right] * e^{j\mu t^2} =$$

$$= e^{j\mu t^2} \int s(\tau)\, e^{-j2\mu t\tau}\, d\tau \tag{1}$$

where * indicates convolution. It can be recognized

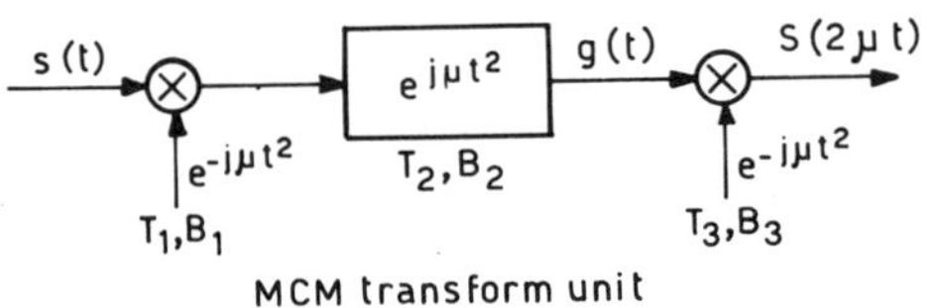

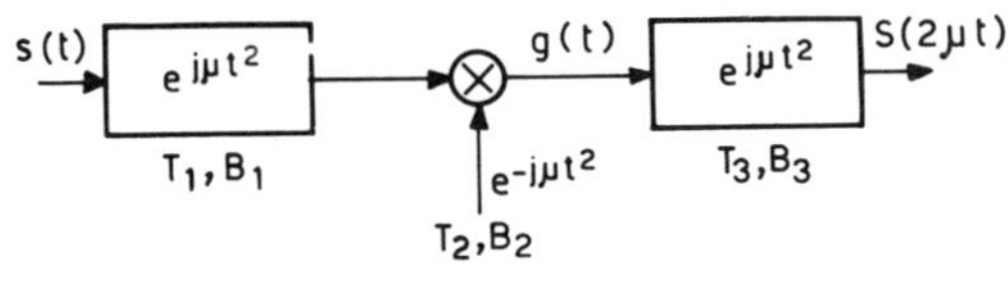

Fig. 1. Fourier transform processors

that the integral appearing in the right member of (1) has the form of the Fourier-transform $S(\omega)$ of the input signal $s(t)$, the frequency variable ω being mapped into the time variable t according to the relation

$$\omega = 2\mu t \tag{2}$$

The frequency-to-time conversion factor, 2μ, is equal to the chirp frequency-to-time slope.

The role of the final multiplication is to remove the undesired quadratic phase distorsion appearing in (1). It can be shown that the system bandwidth B_s and the frequency resolution $\Delta\omega$ are related to the durations of the chirps involved in the transformation (Fig. 1) by

$$B_s = 2\mu\left[T_2 - T_1\right] \tag{3}$$

$$\Delta\omega = \frac{2\pi}{T_1} \tag{4}$$

The second configuration, shown in Fig.1.b, interchanges the roles of multiplication and convolution, and thus will be referred to as CMC (convolution-multiplication-convolution) transform-unit. Its operation can be easily investigated by dualising to frequency-domain the time-domain operation of the MCM processor, making use of duality of convolution and multiplication and of the approximate Fourier-transform relationship

$$FT\left[e^{\pm j\mu t^2}\right] = e^{\mp j\omega^2/4\mu} \tag{5}$$

The system bandwidth and frequency resolution result in this case

$$B_s = B_1 \tag{6}$$

$$\Delta\omega = \frac{2\mu}{B_2 - B_1} \tag{7}$$

Although the principle of chirp-transformation has been known for a number of years [1;2;3;4], the application of this technique to analog signal processing has received only recently a renewed interest, owing to the development of new devices for the practical implementation of analog chirp-filters, which represent the key element of the transformation. Surface-acoustic-wave (SAW) devices are today available capable of synthesising chirp filters operating in the frequency range from a few MHz up to 1 GHz, with typical fractional bandwidths of 30%, dispersive delay of tens of microseconds, and reasonable insertion loss [5]. Charge-transfer devices, and in particular charge-coupled devices (CCD), have provided a powerful means for the implementation of chirp filters operating from baseband up to a maximum frequency of about 10 MHz [6].
Both the above system involve complex multi-

plications and convolutions. When processing is performed at baseband, each complex operation must be decomposed into the corresponding pairs of real operations. In phase and quadrature channels are required, yielding four mixers for each multiplication and two couples of chirp filters for each convolution [7;8;9]. Considerable simplification is obtained when chirp-transform is implemented using narrow-band r.f. chirps. In this case in fact spectral amplitude is given by the envelope, and spectral phase by the modulation of the r.f. carrier.

SAW devices are the most powerful means of implementing the r.f. chirp filters needed in the above transformations. They are commonly used as well for passive generation of the chirp signals, as in this manner coherent operation and exact matching of dispersive slopes are ensured. The main limitation of SAW filters is the relatively poor dispersive delay attainable with the presently available size of piezoelectric substrates. While dispersive delays as long as possible are needed for maximum frequency resolution, large bandwidths require properly high sweep-rates μ. These requirements demand for sophisticated high time-bandwidth product SAW devices. The time-bandwidth product $B_s T_s$ of both transform configurations results

$$B_s T_s = 2\mu(T_2 - T_1)\, T_1 \qquad (8)$$

The maximum $B_s T_s$ is thus achieved when

$$T_2 = 2T_1 \qquad (9)$$

2. SPECTRUM ANALYSER

Real-time spectral analysis, which has great interest in communication systems and radar, is the most direct application of chirp-transform. Its more simple implementation employs the MCM configuration, where final multiplication is omitted when only spectral amplitude is desired. Spectral analysis of analog signals has been demonstrated using conventional SAW filters [10;11;12], SAW convolvers [13], and SAW reflective arrays [14].

At the present state of the art, bandwidths of some hundreds of MHz might be processed.

3. NETWORK ANALYSER

Real-time network analysis by SAW Fourier-transform was first demonstrated by Grant et al. [15] using a system based on the MCM processor. The principle of operation is chirp-transformation of the network impulse response. An alternative implementation, based on the CMC processor, has been recently introduced by the authors [16]. Its most attractive feature is that the first step of the processing, i.e. convolution of the network impulse response in the first chirp filter, can be performed by directly impulsing the network with a chirp, so ensuring maximum energy density over the explored bandwidth. The same waveform used for network excitation can be used for successive multiplication as well, yielding the simple configuration shown in Fig.2. The product is processed in the chirp filter, of opposite dispersive slope, which selects the difference frequency component and equalises the quadratic phase distorsion of the signal from the mixer. The amplitude of the network transfer function is directly obtained by envelope detection. Phase information is obtained by amplitude limiting and removing the carrier by mixing with the local oscillator.

In this system coherent operation is ensured even if active techniques are employed for generating the reference chirp. In fact, phase incoherence and frequency drift of the active- generated chirp are cancelled in the difference frequency component out from the mixer. The advantage of active generation is that chirps of long duration can be obtained, thus allowing large bandwidths to be analysed using relatively moderate sweep rates. The explored bandwidth can therefore be larger than the SAW filter bandwidth.

The experimental result shown in Fig. 3 was obtained using a 180 μsec active-generated chirp and a 2.7 MHz bandwidth, 30 μsec dispersion SAW chirp filter.

This network analyser offers the additional possibility of compensating network delay for detailed

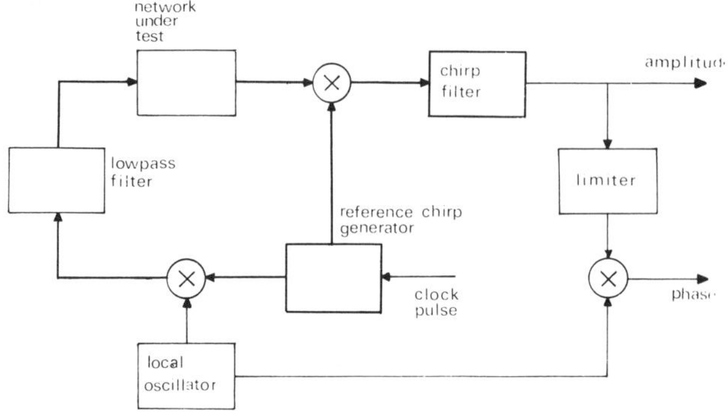

Fig. 2. Block diagram of network analyser employing CMC processing

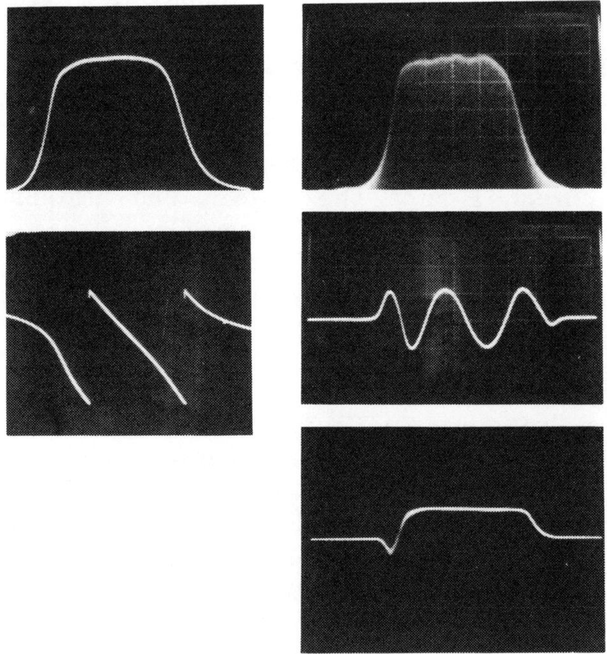

Fig. 3. Comparison of a 4 MHz bandwidth bandpass-filter transfer-function characteristic on (left) a conventional network analyser and (right) SAW CMC network analyser

investigation of phase linearity with frequency (lower
trace in Fig. 3.b).

4. SIGNAL FILTERING

Chirp-transformation, making accessible Fourier-transform, offers a unique means of processing a signal in
real-time by directly modifying its spectrum. Linear
filtering can be implemented by multiplying the signal
Fourier-transform by the desired transfer function
$H(\omega) = H(2\mu t)$, and inverse-transforming the product
using a cascaded transform unit. The MCM unit is more
convenient for this purpose, as the signal from the
chirp filter, given by

$$g(t) = S(2\mu t)\ e^{j\mu t^2} \tag{10}$$

contains Fourier-transform already multiplied by the
chirp needed for successive inverse-transform. The output (10) can therefore be directly processed without
need of usual post-multiplication, and inverse-trans-
formed without need of further premultiplication, thus
reducing the number of components (Fig. 4.a).
        The transfer function $H(2\mu t)$ can be directly
synthesised, or obtained by Fourier-transforming the
impulse-response $h(t)$ using a second MCM processor
(Fig.4.b). In both cases the most attractive feature
is the possibility of externally programming the filtering function, which offers a unique capability of
implementing an adaptive processor.
        A great variety of processing functions can
be performed by processing spectral amplitude and/or
phase. Bandpass/bandstop filtering [11;12;17], variable delay [12], crosscorrelation, matched filtering
[11;14], CW interference rejection [11] have been demonstrated.

5. FREQUENCY SYNTHESIS

The architecture in Fig.5 [12;18] can be considered
as derived from the MCM processor omitting premultiplication or from the CMC processor omitting final convo-

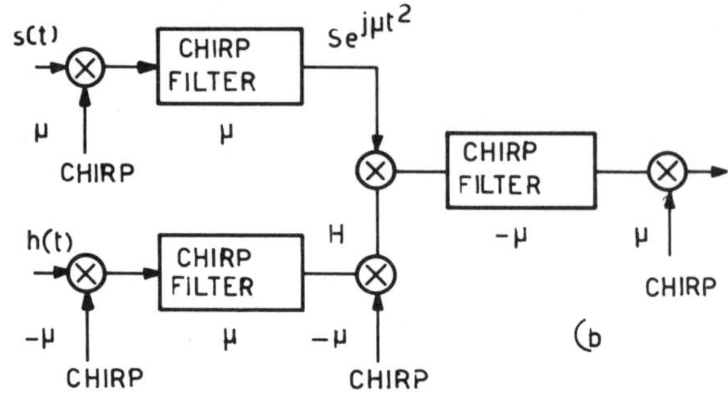

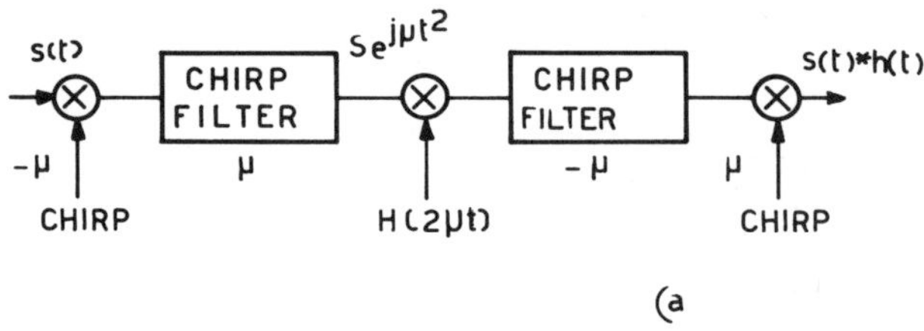

Fig. 4. Signal processing architectures

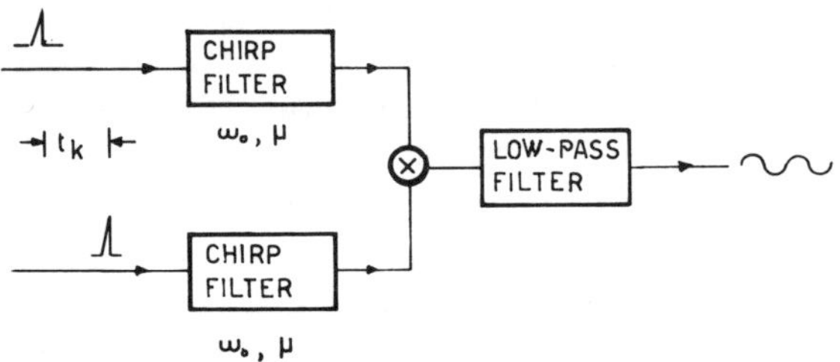

Fig. 5. Chirp-interferometer frequency-synthesiser

lution. The output from the mixer is the product of the chirps passive-generated by the two input impulses, spaced by t_k. By filtering the low-frequency component, the output over the time-interval where the chirps over-lap is

$$u(t) = \cos(2\mu t_k t + \omega_0 t_k) \tag{11}$$

The frequency $2\mu t_k$ of this pulse is linearly related to the delay of the driving impulses, and can therefore be varied by controlling this delay.

The above configuration can be simplified by applying both the impulses to the same chirp filter and using a non-linear detector to obtain mixing of the generated chirps.

An attractive application is programmable generation of frequency-hopped (FH) waveforms, which have interest in spread-spectrum communications. FH waveforms consist of a train of contiguous equal-length pulses, whose frequency can be varied rapidly from pulse to pulse. Frequency hops are generated by controlling the spacing t_k in a sequence of impulse pairs. If T_1 is the maximum value of t_k and T_2 the dispersive delay of the chirp-filter, a gate of width $T_2 - T_1$ must be used to restrict all the output pulses to a common length. Since the spacing between the centres of adjacent impulse pairs cannot be less than T_2 to avoid unwanted products within the gate, the output pulses result separated by a dead time-interval of duration T_1. If $T_1 = T_2/2$, contiguous FH pulses can be generated using two parallel channels. The number of orthogonal frequencies is equal to one-quarter of the chirp filter time-bandwidth product.

A parallel-channel system has been employed by Grant et al. [19] to implement a digitally-controlled coherent FH generator. One single 25 MHz bandwidth chirp-filter was used for each channel, and mixing was obtained in a soft limiting amplifier. Frequency hops up to 12.4 MHz were generated (Fig.6). The coherence between pulses was demonstrated by correlating the FH waveform in a SAW convolver.

A limit of the above technique is the short duration of the frequency-hopped waveforms. Manes and Grant [20] have overcome this limitation by detecting two contradirected active-generated chirp signals at one tap in a SAW tapped delay-line. Multiplication was obtained in a diode mixer. Output waveform duration is thus controlled by the input chirp duration, while out-

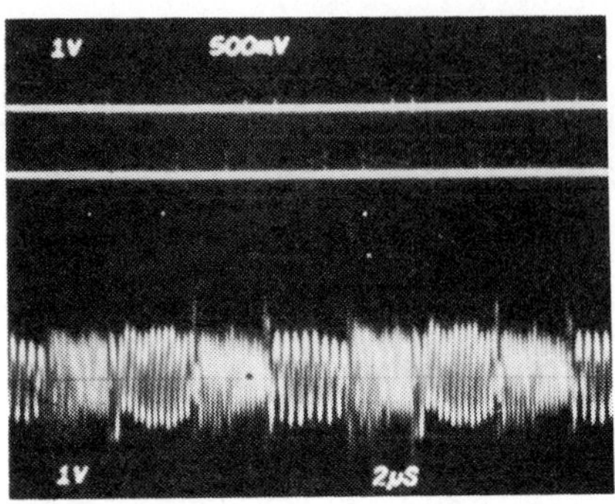

Fig. 6. Frequency-hopped waveform generated by a digi-
        tally-controlled SAW chirp frequency-synthesiser
        (courtesy of Grant et al., Ref. 19)

put frequency is determined by the propagation delay
between the inputs and the tap. An M-tap SAW delay-line
is capable of synthesising simultaneously M different
r.f. signals. The minimum frequency hop depends on the
propagation delay between adjacent taps. An additional
advantage is that the output frequencies can be chan-
ged by varying the sweep-rate μ of the input chirps.
By controlling the external chirp inputs, waveforms
programmable in frequency from a few hertz to tens of
megahertz with durations extending from many seconds
to a few microseconds were generated.

REFERENCES

1.   S. Darlington, Bell Syst. Tech. J., 43, 339, 1964
2.   L.Mertz, Transformations in Optics, Wiley, New York,
      82.
3.   A.Papoulis, Systems and Transforms with Applica-
      tions in Optics, McGraw-Hill, New York, 1968.
4.   L.R.Rabiner et al., Bell Syst.Tech.J., 48,1249,1969.
5.   H.M.Gerard et al., IEEE Trans., SU-20, 94, 1973.
6.   D.D.Buss et al., IEEE J., SC-8, 138, 1973.
7.   J.M.Alsup et al., IEE Conf. Publ. 109, 278, 1973
8.   H.J.Whitehouse et al. Proc. on all Applications

    Digital Computer Symposium, Orlando, Florida, 84, 1973.

9. R.W.Means et al., Proc. CCD Applications Conference, San Diego, 95, 1973.

10. G.R.Nudd and O.W.Otto, IEEE Trans. MTT-24, 54,1975.

11. R.M.Hays et al., Proc. IEEE Ultrasonics Symp.,1975, (n.75 CHO 994-4SU), 363

12. C.Atzeni, G.Manes and L.Masotti, ibidem, 371

13. O.W.Otto, Electron.Lett., 8, 623, 1972.

14. G.R.Nudd and O.W.Otto, Proc.IEEE Ultrasonics Symp. 1975 (n.75 CHO 994-4SU), 350.

15. P.M.Grant, M.A.Jack and J.H.Collins, Electron Lett. 11, 460,1975.

16. C.Atzeni, G.Manes and L.Masotti, Electron.Lett. 12, 1976.

17. J.D.Maines et al., Proc. IEEE Ultrasonics Symp. 1975 (n.75 CHO 994-4SU), 355.

18. J.M.Alsup and H.J.Whitehouse, Proc. IEEE, to be published in May 1976.

19. P.M.Grant, D.P.Morgan and T.H.Collins, Proc.IEEE, to be published

20. G.Manes and P.Grant, Electron.Lett., 12, 39, 1976.

<u>DISCUSSION</u>

Comment : R.G. TAYLOR

Regarding the chirp-Z transform, could you say how weighting functions for sidelobe reduction would be implemented; and also what would be a reasonable dynamic range to expect from a practical device?

Reply   : C. ATZENI

Sidelobe reduction can be achieved at the expense of resolution using weighted chirp filters. There is a number of efficient weighting functions (Taylor,Hamming "Cosine on Pedestal ...") reported in the literature. Weighting can be directly implemented by properly dimensioning the length of fingers, the synthesize, the impulse response of the SAW device. The problem of implementing SAW weighted chirp filters has been extensively investigated for application to pulse-compression radar and excellent results have been achieved. A reasonable dynamic range to expect is 40 dB.

HIGH RESOLUTION SONAR SIMULATION TECHNIQUES

D. L. Folds and N. H. Anderson

Naval Coastal Systems Laboratory,
Panama City, Florida   32401
USA

ABSTRACT   A number of factors influence the performance of coherent sonar systems designed to achieve a high degree of spatial
resolution for location of small objects on the ocean bottom.
First order factors include sonar beamwidth, sidelobe levels,
temporal bandwidth, and display dynamic range. Factors over
which the sonar designer has no control are target roughness,
level of backscatter from the bottom, and refractive effects of
the medium. For the computer simulations described in this
paper, first order sonar parameters are varied in order to
establish their relative importance in the detection of targets.
The simulation is accomplished by assigning complex reflectivity
values to a two-dimensional array of points representing the bottom and target. The points are weighted according to the directional response of a specified projector, and the complex
distribution at the receiving aperture is computed by discrete
summations. Receiver array characteristics such as number of
elements, length, and shading coefficients and bandwidth are
imposed upon the computed aperture distribution and a Fourier
Transform operation is performed on the resultant. The magnitude of the transformed distribution represents the sonar image
of the bottom and target degraded from the originally assigned
values of reflectivity by the characteristics of the sonar. In
this paper, a description of the simulation model will be given.

1.   INTRODUCTION

High resolution sonar systems are used for bottom mapping, small
object location, and other underwater tasks requiring high quality images. The term *high resolution sonar* is used here to

describe systems operating in the frequency range of 100 kHz and
greater, with beamwidths of less than 2 degrees and range resolution of better than 1 metre. Unfortunately, most theoretical,
high resolution sonar design and performance evaluations are
based on solution of the "sonar equation," and many important
phenomena are not considered when only energy relationships are
evaluated. Examples are: the relative importance of sidelobe
levels and sonar beamwidth, display dynamic range, and range
resolution.

The work presented here is the result of a preliminary investigation of advanced simulation of high resolution sonar systems
and the operating environment. The purpose of this basic model
is to demonstrate the feasibility and versatility of such a computer simulation and to provide a foundation for a more detailed
model.

The major elements of the model are the target area, the
projector-hydrophone arrays, the beamformer, and the display. A
schematic of the simulation concept is shown in Figure 1. The

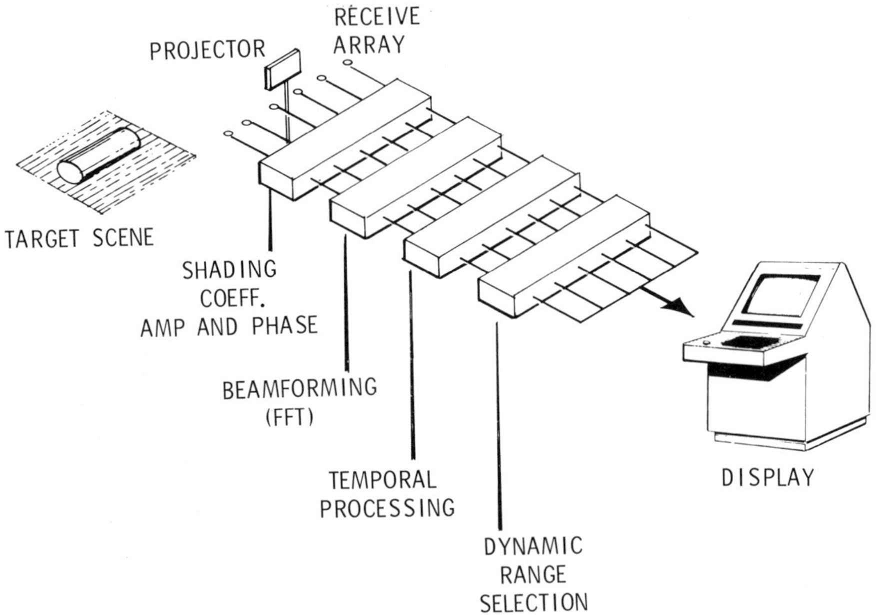

Fig. 1. Basic elements of sonar simulation.

target area is modeled by defining a reflectivity amplitude and
phase for incremental elements within the target region. The
projector array is modeled by describing a line array length in
coordinates of the array center. The hydrophone array is modeled
by a number of discrete elements of finite length and a 3-D

coordinate for each element. The beamformer is modeled as either
an FFT algorithm using outputs of the discrete hydrophone elements
as inputs, or as a convolution of the hydrophone array spatial impulse response with the target reflectivity distribution. The display is driven by a mini-computer which can be programmed to scale,
expand, or compress image data as desired.

2. TARGET ECHO SIMULATION

The geometry used is shown in Figure 2. The target area is
modeled by defining a reflectivity amplitude and phase for incremental elements within the target region. These arrays are designated $T_A(m,n)$ and $T_p(m,n)$ for the amplitude and phase components

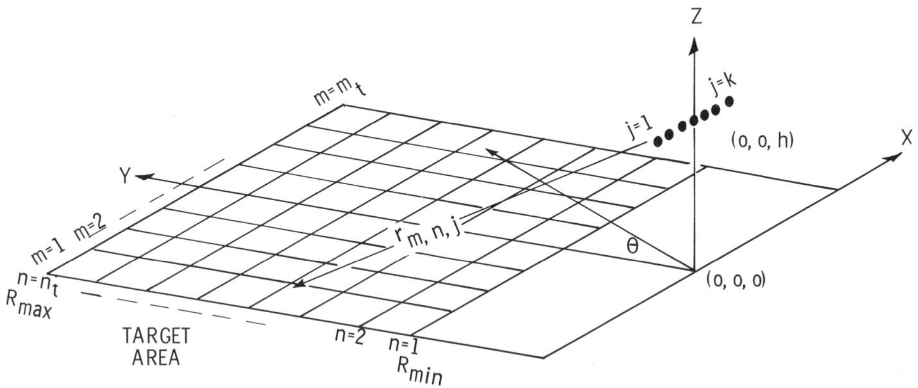

Fig. 2. Geometry of sonar target and array simulation.

respectively, where m, n represent coordinate points in the target area. Empirical data were used to establish echo levels from
various bottom types as a function of frequency, grazing angle,
and bottom type; i.e., sand, mud, etc.; however, the statistics
of this reverberation are not well defined. Therefore, a random
number procedure with only upper and lower bounds specified was
used to fill the target arrays with the background amplitude;
with numbers in the range 0.05 to 0.2 and the target by numbers
in the range 0.5 to 1.0. Phases were taken as constant or random
from 0 to π.

The projector is modeled by assuming a line array of length L_p
positioned at a point $(0,0,h)$. The directional response of the
projector array is computed by

$$P_A(\theta) = \sin(\pi(L_p/\lambda)\sin\theta)/(\pi(L_p/\lambda)\sin\theta)$$

$$P_p(\theta) = \text{sign}(P_A(\theta)) \tag{2.1}$$

where θ is measured from the origin with respect to the y axis, and λ is the wavelength. By using Equation (2.1), the assumption is made that the target area is in the projector far field; i.e., $R_{max} > L_p^2/\lambda$; where range R is measured from the origin and is given by $R = (x^2 + y^2)^{\frac{1}{2}}$. The complex amplitude, A_t, reflected from a target point as measured in the plane of the target, is given by

$$A_t(m,n) = \frac{T_A[m,n]\,|P_A(\theta)|}{r_p}\, 10^{-\frac{r_p \cdot \alpha\, S_p}{20}}\, e^{-i\left(\frac{2\pi}{\lambda}r_p + P_p(\theta)\right)} \tag{2.2}$$

where r_p is the slant range from projector center to target point, α is absorption coefficient in dB/unit distance, and S_p is the projector source level in dB.

3. TARGET ECHO PROCESSING

Echo processing simulation includes sonar beamforming and effects of a finite pulse length. Included are the parameters of the hydrophone, aperture apodization, FFT characteristics, pulse shape, and display characteristics.

The receive hydrophone is modeled by assuming k number of elements each of receiving length ℓ_h, and center-to-center separation, s_h. The total length L_h of the hydrophone is given by $L_h = k \cdot s_h + \ell_h$.

Discrete finite length elements are assumed so the reflected wavefront may be properly sampled and beamforming may be simulated using a Fourier Transform algorithm. Focusing may be accomplished by placement of element coordinates to achieve array curvature.

The complex amplitude, $A_h(n,j)$, output of a discrete element j resulting from returns from reflectors in target row n, may be computed as

$$A_h\,n,j = \sum_{m=1}^{m_t} A_t[m,n]\, 10^{-\left(\frac{r_{m,n,j}\cdot\alpha}{20}\right)}\, e^{-i\left(\frac{2\pi}{\lambda}r_{m,n,j}\right)} \tag{3.1}$$

where $r_{m,n,j}$ is the slant range from target element m,n, to hydrophone element j.

In many systems, aperture apodization or shading is used to reduce sidelobe (first diffraction order) levels[1]. Shading is accomplished normally by amplitude shading, but any complex operation on the outputs of discrete elements may be simulated by multiplying $A_h(j)$ by the appropriate complex function $S_h(j)$.

Beamforming is accomplished by the Cooley-Tukey FFT algorithm[2]. The FFT, of single dimension N, is performed using as input data one range increment n at a time. Using the values calculated for discrete transducer element outputs A_h as the input array, the operation performed by the discrete FFT operation is

$$Q[n,q] = \sum_{j=1}^{k} A_h[n,j] \, S_h[j] \, e^{-i\left(\frac{2\pi kq}{N}\right)} \tag{3.2}$$

where Q is a complex number representative of the beamformer output. In the computer model the complex value of Q is contained in the real and imaginary arrays Q_R and Q_I, each having dimension n_t x N.

Each element of the array represents a sonar beam output in the direction given by the expression

$$\theta = \arcsin(q \cdot \lambda/N \cdot S_h) \tag{3.3}$$

where q is the qth array element in the FFT output and takes on values $-N/2$ to $+N/2$.

The beamforming or FFT operation is carried out separately for each target row, n. Therefore, the arrays Q_R and Q_I are two-dimensional where the parameter n is representative of range and q is representative of azimuth. To simulate a pulse echo system having a pulse length exceeding the length of incremental range elements, it is necessary to compute a running sum or convolution on Q, with a rectangular function representing the length of the pulse. If a pulse of length v incremental range units is to be simulated, the values contained in the Q arrays are modified as follows

$$Q'[n,q] = \sum_{n}^{n+v} Q[n,q] \tag{3.4}$$

where Q represents the results of the running sum over the pulse length v.

The image of the target region can now be written as

$$I[n,q] = (Q_R^2[n,q] + Q_I^2[n,q])^{\frac{1}{2}} \qquad (3.5)$$

This image represents a degraded version of the target region
reflectivity which has been degraded first by the impulse response of the array, and secondly by the finite pulse length.
An image of the target area may be obtained by converting the
numerical values in array I to density variations on photographic
film or by intensity modulation on a CRT. The original and
undegraded target data can be compared by translating the numerical values in array T to object amplitudes $O[m,n]$ by

$$O[m,n] = \left| T_A\, e^{-i\left(\frac{2\pi}{\lambda}\, T_p\,[m,n]\right)} \right| \qquad (3.6)$$

and using these values to intensity modulate a CRT.

4.   CONVOLUTION MODEL

The model previously described is suitable for forward-looking,
multibeam sonar systems. In sidelooking systems, a single beam
is formed by the sonar array and is caused to scan the sea
bottom by forward motion of the sonar platform. For this geometry, a model based on convolution is more suitable.

In this model, a target array is defined in Equation (2.1). A
directional response of the projector is calculated and the complex reflectivity distribution, A_t, is determined from Equation
(2.2). A receiving array directional response, $P_h(\theta)$, is required and is calculated independently, or experimentally
measured patterns are used if they are available. The beamforming process is described by the convolution of $P_h[m,n]$ and
$A_t[m,n]$ where $P_h[m,n]$ is the hydrophone array response transformed to target coordinates with the response curve maximum
centered at m_o. The convolution is accomplished through solution of

$$Q[m_o,n] = \sum_m A_t[m,n]\, P_h[m-m_o,n] \qquad (4.1)$$

where $[m_o,m]$ represents the image of the point m_o in the target
area.

This summation must be carried out for each of the points (m,n)
in the target array. The results of this calculation are stored

in arrays Q_R and Q_I, and pulse length effects can be simulated
by Equation (3.4). The image is then described by Equation (3.5).

5. APPLICATION OF SONAR MODELS

A simulation was made of a sonar target configuration to demonstrate the results of applying the FFT model.

$$m_t = 60, \ n_t = 40, \ N = 128$$

$$h = 10 \text{ metres}, \ R_{min} = 20 \text{ metres}, \ R_{max} = 60 \text{ metres}$$

$$L_p = 20\lambda$$

$$s_h = 0.5\lambda, \ \ell_h = 0.1\lambda, \ k,v = \text{variable}$$

$$0.05 < T_A < 0.2, \quad 0 < T_p < \pi; \ \text{background}$$

$$T_A = 1, \ T_p = 0; \ \text{target} \ .$$

For the simulation results shown in Figure 3, a target of 8 x 8
metres was simulated by reflectivity value of unity with a constant phase angle. The background was assigned a random amplitude and phase value where the amplitude was taken in the range
0.05 to 0.2 and the range of phases taken was 0 to π. Figure
3(a) illustrates the image resulting from simulating a receiving
array with a beamwidth of 1.3 degrees at −3 dB and a 1-metre
pulse length. Figures 3(b) and (c) show the same image assuming
increases by a factor of 4 and 8, respectively, in both beamwidth and pulse length. The target area shown represents a sec-
tor coverage of 60 degrees. The target is stretched in azimuth
due to the reduced effective aperture and broadened beamwidth
for off-axis beams; i.e., BW = $\lambda/(L_h \sin \theta)$. Note that interference effects in this coherent sonar system can be seen in
Figure 3(a) even where the target is approximately eight
"beamwidths" wide. Recognition of the target as a square or
rectangle is not possible in Figures 3(b) and 3(c) because of
the loss of sharp edge information, although target detection
is obviously possible.

The receiving and projector directional responses shown in
Figure 4 were used in simulating the convolution model. Target
information is given below.

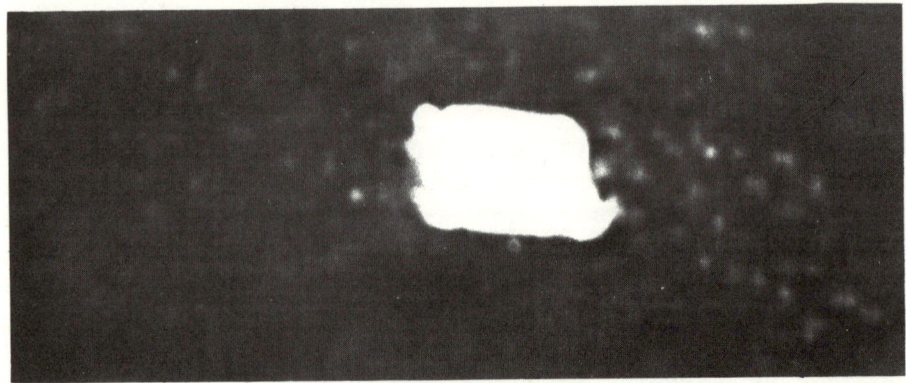

a.   k = 86, v = 1.

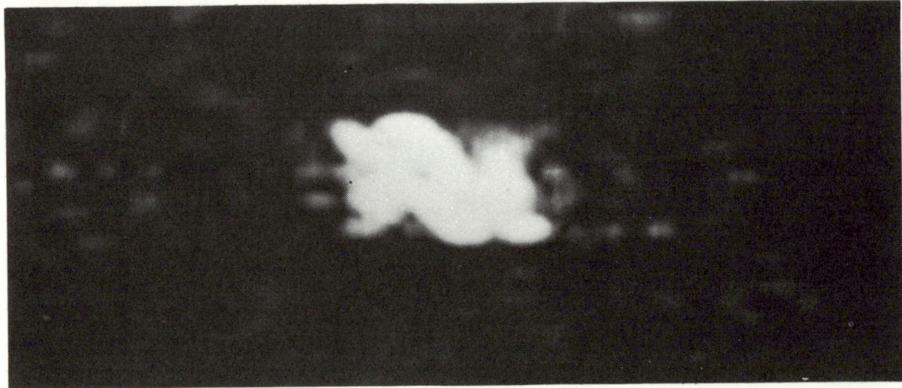

b.   k = 43, v = 2.

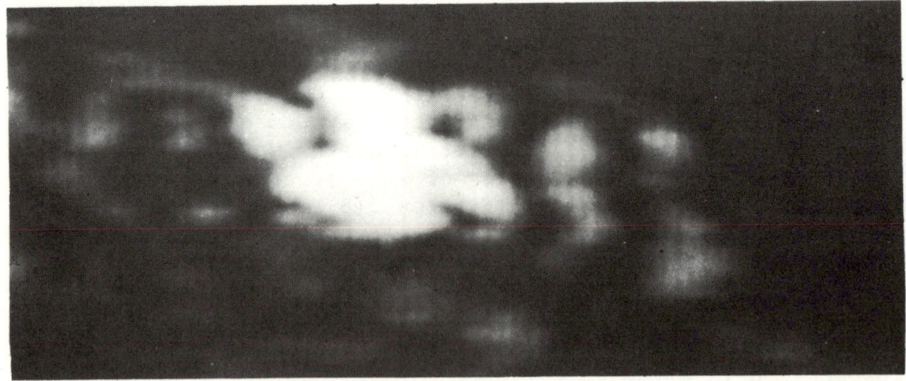

c.   k = 21, v = 4.

Fig. 3.   FFT model simulation.

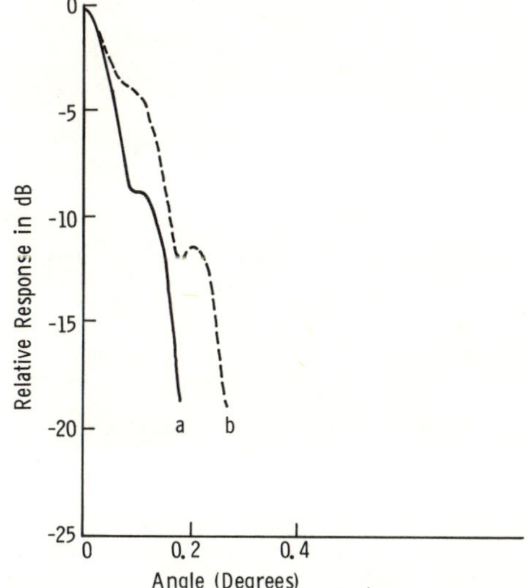

Fig. 4.   Directional responses used in convolution model simulation.

$$n_t = 64, \quad m_t = 256$$

$$h = 10 \text{ m}, \quad R_{min} = 106.5 \text{ m}, \quad R_{max} = 110 \text{ m}$$

$$0 < T_A < 1.0, \qquad T_p = 0; \quad \text{background}$$

$$0.7 < T_A < 1.0, \qquad T_p = 0; \quad \text{target} \quad .$$

The dimensions of each of the three targets were 30 by 96 cm.
The -3dB beamwidth of the receive directional response was
approximately 20 cm.

Figure 5 illustrates the effect of using the directional response
(curve b) of Figure 4 with range resolution of 6, 12, and 24 cm
respectively.

Figure 6 illustrates the effect of using the directional response
(curve a) of Figure 4 with range resolution of 6 and 12 cm
respectively.

6.   CONCLUSIONS

The effects of beamshape and pulse length can be observed in the
target images.   Subtle degrading effects, difficult to express
mathematically, are observed as range resolution is degraded and

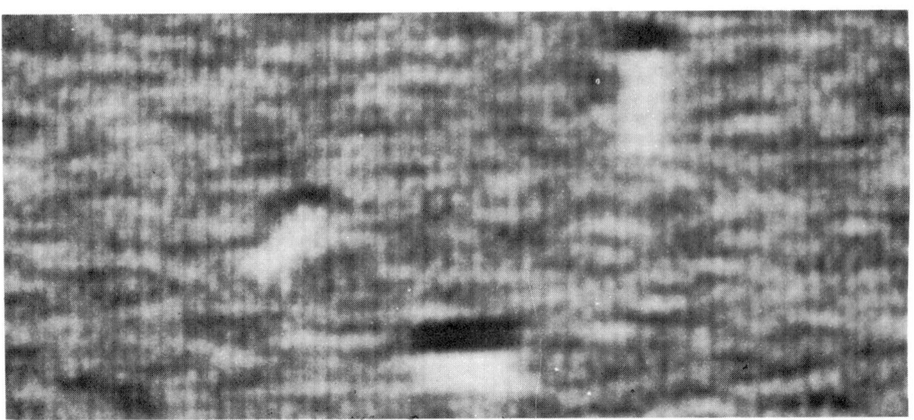

a.   Range resolution of 6 cm.

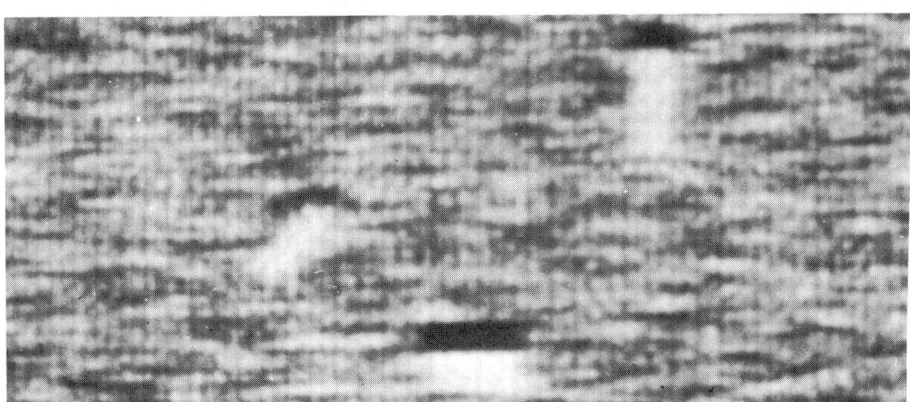

b.   Range resolution of 12 cm.

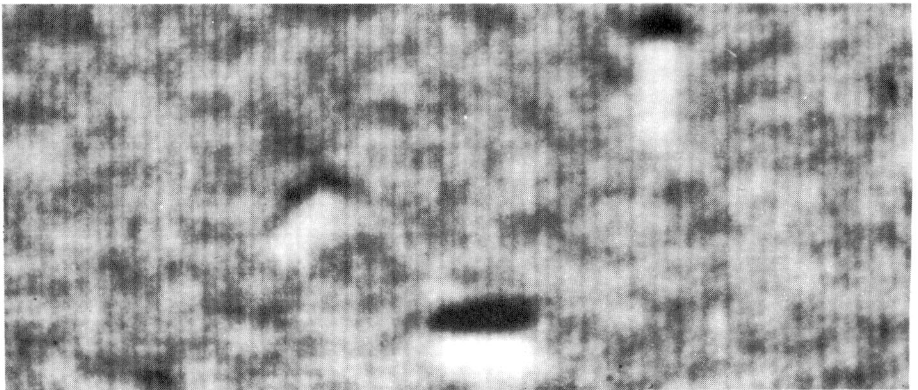

c.   Range resolution of 24 cm.

Fig. 5.   Convolution model simulation . Directional response
          of fig. 4(a).

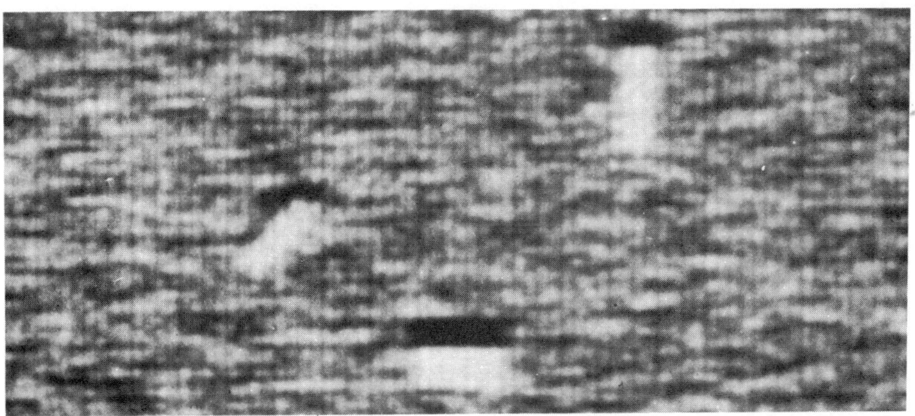

a. Range resolution 6 cm.

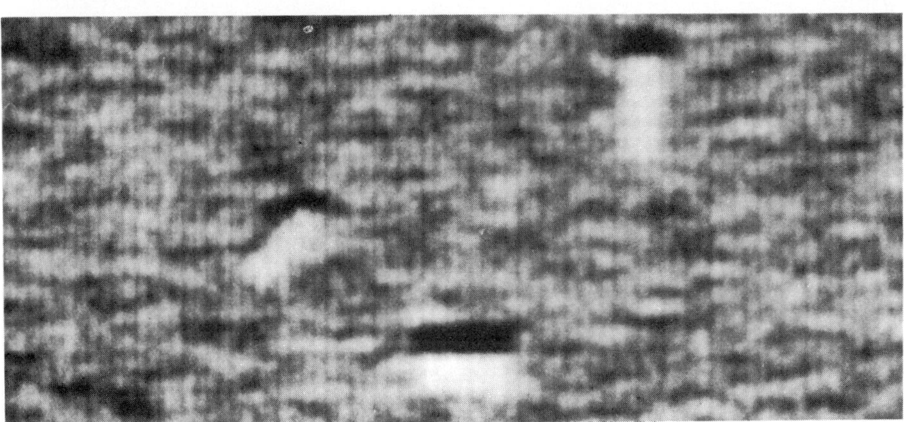

b. Range resolution of 12 cm.

Fig. 6. Convolution model simulation directional response of
         Fig. 4(a).

as sidelobe levels are increased. It is conceded that the image
is "fuzzier" or defocused as these degrading effects are applied.
This obviously is due to the absence of sharp gradients in the
degraded image. Though these effects can be observed sub-
jectively, it remains to establish criteria for "adequacy" of
resolution for detection of specific object shapes in the
presence of the reverberation,

REFERENCES

1. R. C. Hansen, *Microwave Scanning Apertures,* Vol. 1,
    Academic Press, New York (1964).

2. J. W. Cooley and J. W. Tukey, *Mathematics of Computation,*
    Vol. 19, p. 297 (1965).

SUMMARIES OF WORKSHOPS

SUMMARY OF WORKSHOP ON PROPERTIES OF TIME/SPACE VARIANT CHANNEL

chaired by
A. Wasiljeff, University of Bremen

The workshop was mainly concerned with the basic assumptions for
the theory and measurement of scattering functions.

Wide-sense, stationary, uncorrelated scattering (WSSUS) is an
idealization which might not always be valid in reality. In
order to draw conclusions from experiments it is necessary to
check whether the condition of WSSUS has been fulfilled approximately within the range of the variables in question.

It is not easy to give simple explanations of a four- or even
five-dimensional scattering function that satisfy the physical
intuition of applied engineering. Suggestions were made to use
new processing methods that had proved to be very powerful in
modern astronomy, like the so-called speckle technique.
However, the problem of detection and classification of more or
less random targets in a random time and space variant medium
might be much more complicated than the disturbance of astronomi-
cal observations by a turbulent atmosphere.

Previous measurements of spatial cross-correlation have been
mainly performed in order to determine the decorrelation of
signals caused by the transmission medium. The concept of multi-
dimensional scattering functions, as developed by Laval, Hovem
and others, gives much more insight into the complicated
structure of time, frequency and space correlation. It is
practically impossible to measure the required correlation
functions as ensemble averages. Therefore, ergodicity in all
variables has to be assumed. This hypothesis has not been proved
so far for the underwater channel. Therefore the averages taken
during experiments might be based on over-simplified assumptions.

But, even with the assumption of ergodicity, it is still hardly
feasibile to measure the required five-dimensional correlation
functions. Therefore projects of the correlation functions on one
plane or another have to be considered.

A theoretical and an intuitive approach is required at the same
time to construct a model of the time and space variant channel.
Multipath structures in the near field have to be taken into
account as well as normal mode propagation and scattering by
surface and bottom roughness.

A complete solution to the problem is not yet at hand and further
work is required in order to obtain reliable results that will
enable the prediction of the performance of space-time processing
systems in a space-and time-variant, random medium.

SUMMARY OF WORKSHOP ON AUTOREGRESSIVE SPECTRAL ANALYSIS

chaired by
O.L. Frost, III; ARGO Systems, Inc.
and
L.J. Griffiths; University of Colorado

The purpose of this workshop was twofold: First, to provide
basic background material and information relating to the application of AR methods to the analysis of practical data and second,
to discuss current, state of the art theoretical and practical
questions relating to a detailed understanding of the AR technique.
Among the questions raised in the latter category were the
following:

1) In many applications of AR analysis, one is faced with
    the problem of non-continuous input data. This problem
    may arise either through dropouts (missing data) or by
    the presence of excessively large noise bursts. What is
    the correct method of handling these cases? Suggestions
    included filling with zeros or using extrapolation
    techniques.
2) The influence of sampling rate on AR performance is not
    well understood at this time, particularly for the case
    of broadband random data.
3) What is the ultimate resolution of AR spectra? It has
    been pointed out that although the spectral estimates
    have a much more spiked appearance than do conventionally-
    generated spectra, resolution is not determined by the
    width of this spike. Rather, one must consider the
    effect of signal-to-noise ratio and, for AR, the peak-
    shift dependence on SNR.

In addition, topics related to extensions of AR processing were
also discussed. A recent paper by A. Papoulis entitled "A new
algorithm in spectral analysis and band-limited extrapolation,"

which appeared in the September, 1975 issue of the IEEE Transactions
on Circuits and Systems was metnioned. This article presents an
alternative approach for extrapolating autocorrelation functions.
Professor Van Schooneveld mentioned that he and his colleagues
have used AR method to extrapolate $r(1)$, $r(2)$, $r(N)$ to values
at lags greater than N and the approach appears promising. One
advantage of this extrapolation is that the AR model is not required to match the white-noise component in $r(0)$.

Overall, the workshop can certainly be characterized as successful. It was well attended, the discussions were lively, and the
participants indicated their approval of the manner in which it
was conducted.

SUMMARY OF WORKSHOP ON ADAPTIVE ARRAY PROCESSING

Chaired by Dr N.L. Owsley

Naval Underwater Systems Center
New London, Conn., U.S.A.

Three major topics were considered at this workshop: constrained
adaptive arrays; time domain versus frequency domain processor
implementations; and adaptive processing for coherent signal
fields. The following presents a summary of each discussion and
attempts to point out topics for future work.

1. CONSTRAINED ADAPTIVE ARRAYS

It is well known that mismatch between an assumed signal model
and an actual signal can result in the suppression of such a
signal by an adaptive beamformer using the model. To preserve
signal techniques which desensitize an adaptive beamformer to
mismatch incorporate constraints on either the array filter
vector or directly on the sensor output data. Preprocessing of
the array sensor data prior to adaptation appears to offer some
simplification in the adaptive filter control algorithm.
Specifically, proper and often simple preprocessing of the sensor
data involving only additions, subtractions and scale operations
can allow the use of unconstrained filter vector control
algorithms. Analogous preprocessing occurs in the frequency
domain application of window functions in discrete Fourier spectrum analysis. Specifically, higher order windows decrease
resolution in both spectrum analysis and beamforming. This
effect in adaptive beamforming is tantamount to signal model
robustness. The impact of bad sensors on constraint effectiveness was also discussed. The fact that the output from a single
sensor appears in an increasing number of constraint preprocessor
outputs for higher-order constraints suggested that higher-order
constraints might be more sensitive to the "bad" sensor.
Constrained sub-arrays which operate as "out-of-the beam" noise

estimate-and-cancel processors were considered. Such adaptive
noise estimation beamformers work to cancel the coherent component at the output of a conventional beamformer. Interest in
this technique is based on the flexibility of such a realization
in terms of the control of the number of adaptive degrees of
freedom. Such processors have been referred to as auxiliary
array beamformers having the number of adaptive channels significantly less than the actual number of sensors in the array.

2. TIME DOMAIN AND FREQUENCY DOMAIN IMPLEMENTATIONS

Time domain adaptive system implementations are characterized by
an adaptive finite impulse response filter component for each
adaptive channel. Frequency domain realizations employ pre-
beamformer discrete Fourier transform processing with subsequent
adaptive filtering performed independently from frequency cell
to frequency cell. It was suggested that a time domain adaptive
component must simultaneously perform temporal and spatial "pre-
whitening" whereas a frequency domain element must perform only
the spatial function. The possibility of extreme temporal
coherence capturing an inordinate proportion of the available
time domain degrees of freedom at the expense of spatial filter
performance was discussed. Adaptive effects such as steady
state misadjustment and adaptation noise should be considered in
such a comparison. It was pointed out that frequency domain
adaptation allows the distribution of a prescribed number of
degrees of freedom as a function of frequency. In contrast, the
time domain adaptive filter is controlled by a total bandwidth
energy minimization criteria which, for a dynamic (adaptive)
system, might not be the optimum spatial filter for logical
spectral regions.

3. COHERENT SIGNAL FIELDS

A coherent signal field was defined as that portion of a sensor
array environment consisting of wavefronts for which the associated modulating wavefronts exhibit some degree of coherence.
A multipath environment is a specific important example of a
coherent signal field wherein the different arrivals may or may
not be spatially resolvable. A fundamental problem exists in
the application of optimum-adaptive processors to coherent signal
fields. Namely, signal arrivals from one direction and or frequency can be correlated with signal arrivals from another direction and or frequency. In this situation, the idea of independent signal and noise (interference) which is implicit in an
optimum beamformer is invalid and it can be shown that signal
suppression can result. The interest in signal fields stems
from the possibility of coherent recombination of related arrivals

with obvious resultant gains against incoherent noise. Without
a priori knowledge of the multipath structure it is necessary,
however, to implement an adaptive processor which can measure
arrival correlation in real time and provide the proper recombination filter. One approach suggested was the real-time
measurement of the array acoustic environment eigenvectors.
It was shown that these orthogonal vectors are a linear combination of all wavefront vectors with coherent wavefront modulation.
Furthermore, the associated eigenvalues contain the total energy
in the coherent signal field. Work should continue on the application of adaptive orthogonal decomposition techniques to both
spatial and spectral processing applications.